Caves

Caves: Processes, Development, and Management

David Shaw Gillieson

Second Edition

WILEY Blackwell

Registered Offices
John Wiley & Sons, Inc., 111 River Street, Hoboken, NJ 07030, USA
John Wiley & Sons Ltd, The Atrium, Southern Gate, Chichester, West Sussex, PO19 8SQ, UK

Editorial Office
9600 Garsington Road, Oxford, OX4 2DQ, UK

For details of our global editorial offices, customer services, and more information about Wiley products visit us at www.wiley.com.

Wiley also publishes its books in a variety of electronic formats and by print-on-demand. Some content that appears in standard print versions of this book may not be available in other formats.

Library of Congress Cataloging-in-Publication Data

Names: Gillieson, David S., author.
Title: Caves : processes, development, and management / David Shaw
    Gillieson.
Description: Second edition. | Hoboken, NJ : Wiley-Blackwell, 2021. |
    Includes bibliographical references and index.
Identifiers: LCCN 2020028479 (print) | LCCN 2020028480 (ebook) | ISBN
    9781119455578 (paperback) | ISBN 9781119455592 (adobe pdf) | ISBN
    9781119455622 (epub)
Subjects: LCSH: Caves.
Classification: LCC GB601 .G5 2021 (print) | LCC GB601 (ebook) | DDC
    551.44/7–dc23
LC record available at https://lccn.loc.gov/2020028479
LC ebook record available at https://lccn.loc.gov/2020028480

Cover design: Wiley
Cover image: © Steven Bourne, used with permission

Set in 9.5/12.5pt STIXTwoText by Straive, Chennai, India

SKYD941CE4A-67EA-4092-BD9F-C8C77D156DA2_052421

This book is dedicated to Gabriel Crowley, in thanks for her unfailing support.

Contents

Preface and Acknowledgements

We carry caves inside us:
The heart's dark chambers,
Water-washed cavern of the womb,
Limestone pockets of the brain.

Adrienne Eberhard
Earth, Air, Water, Fire: A Love Poem in Four Elements

This is a substantial revision of my earlier book *Caves: Processes, Development and Management*, which was first published in 1996. It has been updated, significantly expanded, and largely rewritten. The first edition attempted to take a contextual and holistic approach to both the geomorphology and biology of caves, as part of wider karst landscapes, and as ecosystems. It also provided an overview of management issues – both above and below ground – that affect caves.

Over the last 20 years, there has been a dramatic increase in karst and cave research globally, with significant advances in our understanding of fundamental processes, in our ability to extract proxy climatic and environmental data from cave deposits, and in our understanding of the complexity of cave management. Today, scientists from many institutions across the planet are studying caves and the literature is vast in the English language alone. Where possible, I have tried to provide a very broad international perspective in the cited literature and the examples used. I have deliberately used many examples from the tropical world and the southern continents to counteract the bias to Europe and North America in the speleological bibliography.

I am deeply grateful to Jeremy Garnett for his considerable editorial skills, which greatly eased the production of this book, especially in the later stages. His attention to detail and good organization of a large volume of text and illustrative material made my life much easier. I am also greatly in debt to my wife Gabriel Crowley for her love and support, her perceptive editorial comments, and her sound advice when my computing skills proved inadequate. Finally, I wish to thank Rosie Hayden, Athira Menon and Mathangi Balasubramanian for their editorial skills and tolerance throughout the writing process.

I have been fortunate enough to visit many caves over the last 50 years; among the goodly band of international speleologists I must acknowledge my particular debt to the late Neil

Anderson, Mike Bourke, Steve Bourne, John Brush, Brian and Sue Clarke, Marj Coggan, Gareth Davies, Derek Ford, Dave Gill, the late Ken Grimes, John Gunn, Steve Harris, Ernst Holland, Julia James, the late Joe Jennings, Andrej Krancj, Kevin Kiernan, the late Jill Landsberg, Lana Little, Andrej Mihevc, Armstrong Osborne, the late Jim Quinlan, Henry Shannon, Geary Schindel, Andy Spate, Mia Thurgate, Alan Warild, John Webb, Nick and Susan White, Paul Williams, Yuan Daoxian, and Nadja Zupan Hajna.

The following people kindly provided photographs to supplement my own: Andrew Baker, John Brush, Steve Bourne, Paul Caffyn, Neil Collinson, Gabriel Crowley, Gareth Davies, Stefan Eberhard, John Gunn, Chris Howes, Leonardo Piccini, Peter Serov, Lino Schmid, Moira Prati, Andy Spate, Tony White, and Paul Williams. Their individual contributions are acknowledged in the figure captions.

1

Introduction

People have been interested in caves for a very long time. Our distant ancestors used them for shelter, as sources of water, and as places in which to conduct essential rituals. They adorned their walls with quite sophisticated artwork, at whose meaning we can only guess. Caves are featured in our mythology. They are used as places of worship in many cultures, and as places in which to store prized foodstuffs and wine throughout the world. Over the last 200 years, they have attracted scientists, artists, photographers, and recreational cavers. This book aims to provide a real understanding of how caves form, how they can inform our knowledge of past environments and climates, and the values – both environmental and cultural – that they provide to humanity.

1.1 Some Basic Propositions

This book is based around several propositions:

Firstly, that caves are a measure of the intensity and persistence of the karst (rock solution) process, and its interruption by other geomorphic processes. In limestone, as well as the continued efficiency of the solution process through time, cave development is affected by tectonic activity and sea-level change. If the solution process has operated efficiently through time, then extensive caves will be found in a limestone massif. If that process has been severely interrupted by glaciation, aridity, or sea-level change, then these processes will also be reflected in cave morphology. Thus, the extensive caves of the Nullarbor Plain in Australia formed as large phreatic tubes under increased rainfall intensity during the Pliocene. The progressive aridity of the Australian continent since that time changed the process regime, so that collapse aided by salt wedging has replaced solution as the dominant process. Deepening of the caves has been enhanced by sea-level lowering in the Pleistocene, and today there are extensive flooded tunnels under the arid plain that have been explored by divers for up to 6 km. All these processes are reflected in cave morphology, and we now have the prospect of dating these events by analysing the calcite, gypsum, and halite formations found in the caves. In regions where climatic change has been minimal, such as the ever-wet tropics, variation in cave development may reflect regional uplift patterns alone. In the karst towers of China, caves are found at various levels – right up to the summits having been abandoned by their streams as the valleys have incised into rapidly rising

Caves: Processes, Development, and Management, Second Edition. David Shaw Gillieson.
© 2021 John Wiley & Sons Ltd. Published 2021 by John Wiley & Sons Ltd.

terrain on the margins of the Tibetan plateau. A more extreme example is provided by the alpine caves of the Canadian Rockies, which hang hundreds of metres above the valley floors and have been abandoned as the cordillera has risen over geologic time. Finally, the fluctuations in sea level during the Quaternary have produced cave development well below present mean sea level, such as the Blue Holes of the Bahamas.

Secondly, that caves are a product of both surface and underground geomorphic processes. Solutes and sediments from the non-karst catchment of a cave combine with karstic solutes and sediments; these may be homogenised and lose their identity, or may be deposited in discrete units, such as flood deposits. This is somewhat akin to the way in which the products of catchment processes are integrated within a lake basin and combine with processes in the water column and on the lake bottom to produce distinctive physical, chemical, and biological properties. Quaternary science can be applied to caves as well as lakes, and the caves provide us with an array of information about landscape processes on timescales, ranging from yesterday to millions of years.

Thirdly, that once these products of surface and underground processes enter the cave system, they are likely to be preserved with minimal alteration for thousands, perhaps even millions of years. In the near-constant temperature and humidity of the cave, weathering processes are reduced in intensity compared with the surface environment. The normal deepening of valleys leads to incision within the cave, leaving deposits in abandoned higher-level passages. These passages are out of the reach of all but the largest floods. The unctuous clays common in caves have a great deal of resistance to erosion: so, once deposited, it is very difficult to erode cave sediments. Caves can be regarded as natural museums in which evidence of past climate, past geomorphic processes, past vegetation, past animals, and past people will be found by those who are persistent and know how to read the pages of earth history displayed before them. Caves have become increasingly important as sources of proxy records of the changes in atmospheric conditions over long time periods – tens to hundreds of thousands of years – and today rival ice cores in the quality of the information that can be gained. Isotopic analyses coupled with radiometric dating techniques can provide records of annual variation in temperature and other environmental parameters, while cryptotephra and pollen that reflect unique environmental conditions can be extracted from individual growth layers in a stalagmite.

Fourthly, the unusual ecosystems in caves have attracted the interest of biologists and ecologists for over a hundred years. Deprivation of light and other stimuli produce physiological adaptations to cave living, often producing somewhat bizarre-looking animals that have thus attracted attention despite their cryptic habits. Cave biota are totally dependent on periodic inputs of nutrients, usually swept in by floods. These organisms are also severely disadvantaged by quite minor disturbances. Thus, they have low resilience in the face of a change to the cave ecosystem. The spectrum of impacts on caves has serious consequences for their biology and ecology, and adequately conserving cave biota is a major challenge for protected area management.

Finally, caves are contested spaces. They are subject to the demands of various forms of recreation, both commercial and non-commercial; the exploitation of cave waters and the very rocks in which they have formed; extraction of gas and oil in karstic terrains; the alteration of the surface soils and vegetation for agriculture and other extensive land uses; and the creation of transport networks to facilitate regional economic development. Cave and karst management is dependent on the implementation of sensible public policy and legislation, with community involvement essential to maintain the values of the caves and karst.

1.2   Now the Details…

This book is organised into four sections. In the first section, contemporary processes of cave formation are examined. In Chapter 2, some definitions of caves and their host karst terrains are outlined, and consideration is given to caves as geomorphic and biological systems. Caves can be regarded as three-dimensional networks in which individual links may become abandoned or may develop through time. The nature of a cave at any time is a function of inheritance of previous states and the contemporary climatic and tectonic setting. The state of the hydrological network (Chapter 3) may be fully phreatic, fully vadose, or a combination, depending on the position of the water table in a cave. The nature of individual linkages (passages) may change, depending on the frequency of floods and their exploitation of existing or potential conduits in the rock. In caves with a temporary or permanent air space, the nature of the cave climate will determine the kind of speleothem deposits that form (calcite, ice, gypsum, and halite). There is also a feedback between climate and the karst solution process.

The roles of the kinetics of bedrock solution, and of rock architecture and cave collapse in the evolution of the position of the subterranean water table and in cavern enlargement, are outlined in Chapter 4. The effects of brine and gypsum on cave formation are regionally important. The formation of weathering caves and pseudokarst caves is another aspect of the subject. The final form of a cave owes much to both the purity of the limestone and the network of fissures that dissect the rock (Chapter 5). Following a brief treatment of limestone lithology and structural variation, examples of cave passage shapes in different limestone settings are given. Geological history plays a key role and may override other factors. Some of the largest and deepest caves are formed by combinations of these factors. The deep circulation of water in limestone leads to the development of flow networks, which are the precursors of the caves we can enter and map now. The legacy of this mode of origin is a suite of erosional forms in cave passages – bellholes, cupolas, and spongework – which can provide useful diagnostics as to the long-term history of the cave system.

The second section of the book deals with past processes and their products. The diversity of cave formations has long fascinated people. Chapter 6 considers the interior deposits of caves, principally the calcite stalactites and stalagmites familiar to most visitors and known to karst scientists as speleothems. Thus, the basic mechanisms of calcite deposition are described, including the roles of trace elements and cave biota in providing shape and color variation. Effects of tectonics and of hydrologic change on cave formations are outlined. Gypsum and halite speleothems are also briefly reviewed. In the tropics, biogenic deposits, such as guano, are dominant and have economic significance.

Caves can be seen as underground gorges and floodplains in which sedimentation proceeds in modes analogous to those of surface fluvial systems (Chapter 7). Cave sediments may be of external or internal origin; various types and their properties will be described (glacial, fluvial, aeolian, and biogenic). The deposition and alteration of entrance sediments in caves have relevance for our understanding of cave palaeontology and archaeology. Relationships exist between cave sediment structures and depositional energy, and interactions between cave sedimentation and hydrology are thus relevant to understanding past and present processes.

In the last two decades, a bewildering array of dating techniques have become available to the cave scientist, and these are briefly reviewed in Chapter 8. The most important of these is uranium series dating, with several variant forms. This technique will be outlined in detail, with examples from Europe, North America, and Australasia. Other techniques, such as palaeomagnetism and radiocarbon, will also be considered. These dating techniques have radically altered scientific thought about caves and wider landscape evolution. Calcite speleothems can be used as paleothermometers through the technique of oxygen isotope analysis. Carbon isotope analysis on speleothems has great potential for determining changes in the surface vegetation above the cave. This extraction of proxy records from cave interior deposits is the fastest developing field in karst research and is reviewed in Chapter 9.

In the third section of this book, the use of caves by various organisms, present and past, is considered. Physiological and evolutionary adaptations to cave living occur in many phyla of the animal kingdom, and also in both flowering plants and fungi. The basic characteristics of the cave ecosystem and its constituent trophic levels are described in Chapter 10. This relates back to the basic concepts of mass and energy flow in the karst system. The role of external energy sources is critical for cave life. Cave biota are dependent on periodic inputs of nutrients, usually swept in by floods. They are also severely disadvantaged by quite minor disturbances. Thus, they have low resilience in the face of a change to the cave ecosystem.

Caves have long been of importance to people – for shelter, water supply, and food, and as places of worship. Chapter 11 reviews our knowledge of cave archaeology. This exciting field has regained importance in recent years with major discoveries of hominid sites in Southern Africa and Central Asia. Much has been learnt about the processes of alteration of cave sediments and the effects on archaeological and palaeontological material. As well, re-interpretation and re-excavation of cave sites in Europe have led to new insights into the lifeways and rituals of our ancestors and our near-relatives, such as the Neanderthal people. Recent discoveries of cave art, improved dating, and new interpretations have changed our perceptions of prehistoric people and their rituals.

Recent uses of caves (Chapter 12) include the mining of cave formations and guano, for hydroelectricity, for storing gourmet foods, as refuges in times of war, and as sanatoria. Around the world, increasing cave tourism presents problems owing to the irreversible degradation of cave ecosystems and alteration of cave microclimates. The addition of energy sources (heat, lint, and dead skin cells) alters the trophic status of caverns. For appropriate use and to avoid degradation, caves should be classified in terms of their limits to acceptable change for the proposed use (recreation, resource extraction). This classification must be dynamic to allow for seasonal changes in cave function and use. Cave lighting and pathways must be designed to minimise the effects on cave microclimates and biota. Subtlety and effective interpretation are the key tools of cave managers. Only through public education can the aims of ecologically based cave management be achieved. There is now a global network of cave scientists considering these problems.

The final section of the book reviews our changing approaches to cave management (Chapter 13) and catchment management on karst terrains (Chapter 14). The rapid development of cave mapping and photography techniques have greatly enhanced our view of the world of caves. The use of 3D scanning has produced stunning visual material that has aesthetic appeal, aids cave exploration, and enhances our understanding of the development of the cave. These techniques are reviewed in Chapter 15.

2

Caves and Karst

In this chapter, caves and karst are first defined and then placed in the context of geomorphic and biological systems, allowing a brief overview of the various inputs and outputs to the cave system. The flux of materials (water, solutes, sediments) through a cave system is also considered. Finally, the current longest and deepest caves on Earth are listed and placed in their broad geomorphic and geologic context.

2.1 What Is a Cave?

A strictly scientific definition would be that "a cave is a natural cavity in a rock which acts as a conduit for water flow between input points, such as stream sinks and output points, such as springs or seeps" (White 1984). It seems that once this type of conduit has a diameter larger than 5–15 mm, the basic form and hydraulics do not change much, though the diameter can be as much as 30 m. The range of minimum diameters allows turbulent flow, optimising the solution of rock, and effective sediment transport. *Solution voids* that are not connected to inputs and outputs are types of isolated vugs; they may act as targets for developing cave systems. Small conduits less than 5 mm diameter but connected to an input or output or both are called *protocaves*. These precursors of cave systems may carry seepage water or groundwater, and at the output may allow the formation of weathering hollows. These may coalesce to form rock shelters. Caves may also be distinguished from rock shelters by having a dark zone and a different biological assemblage.

A simpler non-scientific definition would be that *caves are natural cavities in a rock that are enterable by people*. This implies a minimum size of about a 0.3 m diameter. These are the caves that we can explore, map, and directly study (Figure 2.1). Most of the caves we study are in limestone and its related carbonate rocks, but significant caves (formed by a variety of processes) are also found in sandstones, evaporites such as gypsum, basalt, and granites. However, caves can also be found in the Antarctic ice and in the partially cemented dust of the loess plateau of China. In this book, we will tend to concentrate on the caves that form in limestone, because they are the most numerous, certainly the largest, and we know most about them. The advent of cave diving has dramatically expanded the number and types of

Caves: Processes, Development, and Management, Second Edition. David Shaw Gillieson.
© 2021 John Wiley & Sons Ltd. Published 2021 by John Wiley & Sons Ltd.

Figure 2.1 Exploring a drained phreatic passage in Dan yr Ogof, Wales.

caves we can study, and information from these drowned conduits is causing some major revisions in thinking about caves and their formation. But most of the information in this book has been gained by the patient and often painful progress of cave scientists crawling, climbing, and swimming in the subterranean world over the last two centuries.

2.2   What Is Karst?

Caves are intimately associated with karst landscapes. We can define this realm of caves in terms of both a suite of landforms and associated geomorphic processes, involving the circulation of water in the rock at various depths. The term *kras* was initially applied to a region of Slovenia where stony barren ground is associated with sinkholes or dolines, underground drainage, caves, and springs (Figure 2.2). This definition of karst as a type of landform or a specific geological environment dominated the scientific literature for at least a century until the concept expanded to include solutional processes and hydrogeological systems. The concept of karst has been refined over the last three decades to include the deep circulation of water with no surface expression in the landscape (hypogene karst). Thus, karst is defined by Ford and Williams (2007) as "comprising terrain with distinctive hydrology and landforms that arise form a combination of high rock solubility and well

Figure 2.2 The Cares Gorge, Picos de Europa, Spain. A karst terrain with exposed limestone bedrock, sinkholes, springs, and many caves.

developed secondary (fracture) porosity." This has been further developed by Klimchouk (2016) to encompass a karst system in which two processes dominate:

> Speleogenesis (karstic) – the formation of voids and conduits in rocks through mainly dissolutional enlargement of initial preferential flow pathways involving self-organization due the positive feedback between flow and conduit growth.

And

> Karst (karstic) process – a geological process (an interconnected set of processes) of transformation of soluble rocks under the dominant action of coupled flow-dissolution processes and respective self-organization of the groundwater flow system.

This definition of a karst system focusses on groundwater movement at various depths and does not necessarily have a surface expression in the landscape, though one may occur at a later stage in its evolution. Although such terrain is formed principally in limestone and its close relatives, solution of rocks occurs in other lithologies. These can be in other carbonates such as dolomite, evaporites such as gypsum and halite, silicates such as sandstone and quartzites, and some basalts and granites where geochemical conditions are favourable. All these terrains are, therefore, true karst. Karstic terrain can also develop

Table 2.1 Karstic terrains and processes.

Rock type	Processes	Examples
KARST		
Carbonates, e.g. limestone, dolomite	Bicarbonate solution	Krubera, Georgia; Mammoth Cave, USA
Evaporites, e.g. gypsum, halite, anhydrite	Dissolution	Optimisticheskaja, Ukraine; Mearat Malham, Mt. Sedom, Israel
Silicates, e.g. sandstone, basalt, granite, laterites	Silicate solution	Sima Aonda and Sima Auyantepuy Noroeste, Venezuela; Mawenge Mwena, Zimbabwe
PSEUDOKARST		
Basalts	Evacuation of molten rock	Kazumura Cave, Hawaii, USA; Cueva del Viento, Islas Canarias; Leviathan Cave, Kenya
Ice	Evacuation of meltwater	Glacier caves, e.g. Paradise Ice Caves, USA; Nilgardsbreen, Norway
Soil, especially duplex profiles	Dissolution and granular disintegration	Soil pipes, e.g. Yulirenji Cave, Arnhem Land, Australia; Malabunga Caves, New Britain, Papua New Guinea
Most rocks, especially bedded and foliated rocks	Hydraulic plucking, some exsudation	Sea caves, e.g. Fingal's Cave, Scotland; Remarkable Cave, Tasmania
Most rocks	Tectonic movements	Fault fissures, e.g. Dan yr Ogof, Wales; Onesquethaw Cave, USA
Sandstones	Granular disintegration and wind transport	Rock shelters, e.g. Tassili, Algeria; Ubiri Rock, Kakadu, Australia
Many rocks, especially granular lithologies	Granular disintegration aided by seepage moisture	Tafoni, rock-shelters, and boulder caves; e.g. TSOD and Greenhorn Caves, USA

by other processes – weathering, hydraulic action, tectonic movements, meltwater, and the evacuation of molten rock (lava). Because the dominant process in these cases is not thought to be solution, we can choose to call this suite of landforms *pseudokarst*. This fundamental dichotomy is expanded in Table 2.1.

2.3 Caves as Systems

What processes operate to form these major subterranean landforms? Karst landscapes are complex three-dimensional integrated natural systems comprised of rock, water, soil, vegetation, and atmosphere elements. Nearly all of the karst solution process is moderated by factors operating on the surface of the karst and in the upper skin of the rock. Surface vegetation regulates the flow of water into the karst through interception, the control of litter and roots on soil infiltration, and the biogenic production of carbon dioxide in the root zone. The metabolic uptake of water by plants, especially trees, may regulate the quantity of water available to feed cave formations. Water is the primary mechanism for the transfer of surface actions on karst to become subsurface environmental impacts in the caves.

Pre-eminent among these karst processes is the cascade of carbon dioxide from low levels in the external atmosphere, through greatly enhanced levels in the soil atmosphere, to reduced levels in the cave passages. Elevation of soil carbon dioxide levels depends on plant root respiration, microbial activity, and a healthy soil invertebrate fauna. This cascade must be maintained for the effective operation of karst solution processes and is clearly dependent on biological processes.

2.3.1 Caves as Geomorphic Systems

A variety of materials enter caves, travels through them, or become stored temporarily or permanently, and finally leaves the cave system at springs. This range of materials is illustrated in Figure 2.3. These materials can be conveniently divided into several inputs and outputs, though in reality, the boundaries between these conceptual compartments are somewhat blurred. In addition, not all of the materials will be present in any one cave system, nor will a particular material be present for all of the cave's history. These absences

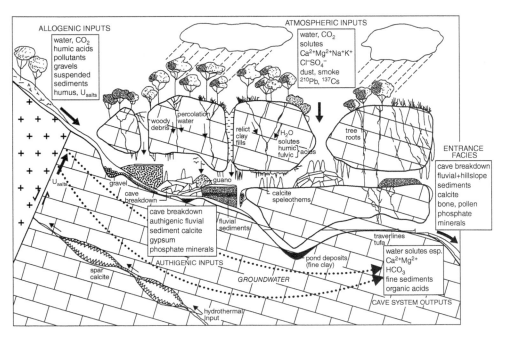

Figure 2.3 A hypothetical cave system showing the potential types of inputs and outputs from the system. Each of these may be discontinuous in space and time. Quantification of some elements of the cave system, especially those below the water table, may be difficult if not impossible (from Gillieson 1996).

raise interesting questions for geomorphologists about changes in the process regime in an area and allow inference of climatic change, tectonic activity or, latterly, human impact.

A good deal of the material to be seen in any cave is derived from its catchment, transported into the cave by fluvial, mass movement, aeolian, or glacial processes. This transport may be episodic, and the material may have been temporarily stored in small terraces or floodplains prior to entering the cave. Thus, the material may have already been altered by weathering or sorting between leaving its source area and entering the cave. The typical materials that derive from the catchment are water, of course, dissolved gases such as carbon dioxide, fluvial gravels, and fine sediments (sands, silts, and clays), humus and dissolved organic acids, and a range of solutes, which will depend on the catchment lithology. Perhaps the most important solutes for the study of karst are calcium and magnesium cations; the ephemeral bicarbonate anions; the uranium series anions; and an increasing range of pollutants including herbicides, sewage, petroleum products, and some heavy metals. Other compounds may help to characterise the karst water in tracing experiments but are of secondary importance.

A range of materials enter the cave as *atmospheric inputs:* these are transported as solutes in rainwater, as hydrometeors, or as dust. Again, water and carbon dioxide are important, along with cations and anions gained from sea spray or from cyclic salts. Radioactive fallout containing the important isotopes Lead-210 and Caesium-137 may also enter the cave directly from the atmosphere and become incorporated in cave sediments or biota. Dust and smoke particles are also important materials; at Yarrangobilly in New South Wales, Australia, bushfire smoke is often seen being sucked into caves, and thin black smoke layers can be seen in sectioned flowstones. Elsewhere in Australia, the Kimberley caves often contain red quartz sand layers, blown off the Great Sandy Desert in Western Australia during more arid climate phases than the present. Together, such materials form the *allogenic inputs* to the karst system.

Several processes operate within any cave to produce debris and chemical precipitates. These *authigenic inputs* are spatially and temporally quite variable. The most common is cave breakdown, where the angular debris may range in size from pebbles to house-sized blocks. Breakdown is heavily influenced by the mechanical strength of the limestone and the geological structure. It tends to occur after a cave passage has drained, because the roof of the passage loses the hydrostatic support of the water. The upwards collapse (stoping) of the passage produces a characteristic flat arch, with a pile of debris below it. This debris pile may later be capped by calcite flowstone or stalagmites (Figure 2.4). The subsequent erosion of breakdown by cave streams, or the slow release of clay fills from roof crevices, provides authigenic fluvial sediments. Finally, although calcite dominates the chemical precipitates in most caves, a wide range of minerals may be deposited, including gypsum, halite, and a complex array of phosphatic minerals associated with bat or bird guano.

Cave waters emerge at springs or seepages. At these points, the *cave system outputs* are water, perhaps some dissolved gases (carbon dioxide and/or methane), the products of limestone solution, some fine clay sediments, and organic acids leached from guano. The solutes may be augmented substantially from solution processes operating below the water table in the phreatic zone and carried to the outlet spring in the steady flow of groundwater.

A curious mixture of allogenic and authigenic materials becomes preserved as *entrance facies* deposits in the mouths of caves and in rock-shelters. These sites are the targets of

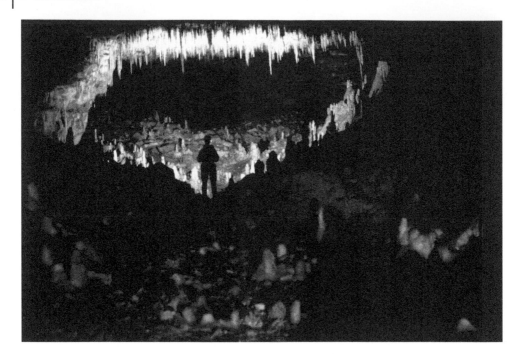

Figure 2.4 Stalagmites dated at 45000 years ago, capping mudflow deposits in Selminum Tem cave, Papua New Guinea.

intensive excavation and research by palaeontologists and archaeologists. As well as cave breakdown, whose development is enhanced by temperature and humidity fluctuations in the entrance zone, relict fluvial and hillslope sediments are well-preserved there. These sediments may become coated with calcite deposits, enhancing their preservation and providing a dateable material. The bones of animals, sometimes including people, become incorporated into the sediments, as well as plant remains such as pollen and, rarely, leaves, nuts, and seeds. The leaching of bones may provide opportunities for the formation of phosphatic minerals. Near to valley sides, tree roots may penetrate the limestone to great depths along fissures, and act to channel water and humic substances deep into the cave.

Although the flux of solutes through cave systems has been investigated worldwide, few attempts have been made to estimate the flux of all matter through a cave system. This is due partially to the extreme difficulty of quantifying some items – cave breakdown, for example – and of estimating the age of many or all the components. One set of estimates has been prepared by Steve Worthington (1984) for the Friars Hole system of West Virginia, USA (Figure 2.5 and Table 2.2). This is an extensive system that is not well-endowed with formations and reflects episodes of cold climate conditions throughout its long history of some 4 million years. Limestone accounts for only 3% of its surface catchment, the remainder of which is underlain by shale, sandstone, and clay-rich limestone – these weather to produce abundant regolith that can be transported. Not surprisingly, 95–98% of the total mass flux through the cave was this allogenic fluvial sediment.

This result is probably representative of mid- to high-latitude caves where speleothem deposition has been restricted during glacial periods and may decline seasonally today.

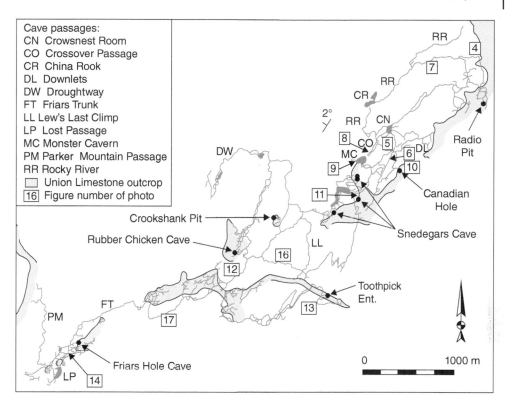

Cave passages:
CN Crowsnest Room
CO Crossover Passage
CR China Rook
DL Downlets
DW Droughtway
FT Friars Trunk
LL Lew's Last Climp
LP Lost Passage
MC Monster Cavern
PM Parker Mountain Passage
RR Rocky River
☐ Union Limestone outcrop
16 Figure number of photo

Figure 2.5 Plan of the Friars Hole system, West Virginia, USA. Several caves have been linked by exploration since 1960; 10 entrances are presently known (from Medville and Worthington 2017).

Older caves, long abandoned by their streams, might yield higher proportions of calcite speleothems. Caves in the tropics would probably yield higher proportions of organic matter, both from allogenic suspended sediment and authigenic bat guano and its by-products. The proportion of open cave to debris-filled cave may vary significantly: deep guano fills, many up to 20 m thick, occupy most of the cavity in Sarawak (Malaysia) and Australian caves, and the open space through which we can walk is but a small component. Consequently, cave histories gained solely from the examination of evidence in the open air-filled cave may neglect crucial information contained in or buried by the sediments.

2.3.2 Caves as Biological Systems

Caves have attracted the interest of biologists and ecologists for over a hundred years (Culver and Pipan 2009). Physiological and evolutionary adaptations to cave living have developed in many phyla of the animal kingdom, and also in some flowering plants and fungi. The principal changes are a loss of pigmentation, a partial or total loss of eyes, extension of sensory hairs or antennae, and changes in body-part proportions (Culver 1982). This often produces somewhat bizarre-looking animals that have thus attracted attention despite their cryptic habits. These adaptations are broadly classified as troglomorphic through a more detailed scheme of cave life – trogloxene, troglophile,

Table 2.2 Friars Hole Cave system, West Virginia (from Worthington 1984).

Length–68 km, depth – 188 m				
Total volume of known cave				**2700 × 10³ m³**
	– open cave			1800 × 10³ m³
	– debris-filled			900 × 10³ m³
Catchment area				85.7 km²
	– includes limestone area			2.6 km²
Mass flux	Total 10³ m³	%	Present 10³ m³	%
Dissolved limestone:				
1 from catchment	57 300	1.9	trace	—
2 from cave	2400	0.0008	trace	—
Cave breakdown	1000	0.0004	280	31
Allogenic fluvial sediment	3 000 000	98.0	600	67
Authigenic fluvial sediment	400	0.0001	20	2

troglobite – which relates to the proportion of its life that an organism spends in the truly dark zone.

The fundamental characteristics of any cave ecosystem and its constituent trophic levels are highly dependent on the availability of energy to sustain life. This relates back to the basic concepts of mass and energy flow in the karst system, especially carbon (Simon et al. 2007). External energy sources are absolutely critical for cave life, and these may be transported in as debris or organic sediment or derived from bat or bird guano. Cave invertebrates are dependent on periodic inputs of nutrients, usually swept in by floods. They are also severely disadvantaged by quite minor disturbances. Thus, they have low resilience in the face of a change to the cave ecosystem. The spectrum of impacts on caves have serious consequences for their biology and ecology, and adequately conserving cave biota is a major challenge for protected area management.

2.4   Where Are the Deepest and Longest Caves?

Caves provide some of the few unexplored places left on Earth. Speleologists are driven to seek out the deepest and longest caves. Such caves also reveal much about the Earth's geomorphic evolution. The search continues, so Tables 2.3 and 2.4 were correct in October 2019, but the listing of the deepest caves likely will change radically due to the fast pace of cave exploration and mapping. Currently, most of the world's really deep caves are located

Table 2.3 The 20 deepest caves of the world (as of October 2019).

Name	Country	Depth (m)
1 Veryovkina	Georgia	2204
2 Krubera (Voronja) Cave	Georgia	2197
3 Sarma	Georgia	1830
4 Illyuzia-Mezhonnogo-Snezhnaya	Georgia	1753
5 Lamprechtsofen Vogelschacht Weg Schacht	Austria	1632
6 Gouffre Mirolda/Lucien Bouclier	France	1626
7 Reseau Jean Bernard	France	1602
8 Torca del Cerro del Cuevon (T.33)–Torca de las Saxifragas	Spain	1589
9 Hirlatzhohle	Austria	1560
10 Sistema Huautla	Mexico	1560
11 Sistema Cheve (Cuicateco)	Mexico	1524
12 Shakta Vjacheslav Pantjukhina	Georgia	1508
13 Sima de la Cornisa–Torca Magali	Spain	1507
14 Cehi 2	Slovenia	1502
15 Sistema del Trave	Spain	1441
16 Sustav Lukina jama–Trojama	Croatia	1431
17 Evren Gunay Dudeni (Mehmet Ali Ozel Sinkhole) Peynirlikonu EGMA	Turkey	1429
18 Boj-Bulok	Uzbekistan	1415
19 Gouffre de la Pierre Saint Martin Gouffre des Partages	France/ Spain	1408
20 Sima de las Puertas de Illaminako Ateeneko Leizea (BU.56)	Spain	1408

Source: Data correct as of 22 October 2019, www.caverbob.com.

in Eastern Europe, especially in the Caucasus mountains, and in the European Alps. In these locations, the drained depth of the cave system will be greatest, and many of these caves are steeply descending stream canyons with shafts to be negotiated. They generally terminate in siphons or break down barriers close to the regional base level, often marked by springs. Thus, the entrance of Veryovkina and Krubera Caves are high in the Arabika Massif of the Caucasus, but the lowest levels of the caves – 2.2 km lower down – are close to the level of large springs close to the Black Sea.

The longest caves are dominantly in flat-lying or gently dipping limestones or gypsum and have been the scene of systematic exploration and mapping over many decades. It is unlikely that their ranking will change radically in the near future, but the prospect of it doing so is one of the most tantalising facets of speleology. Over four decades of cave exploration in the tropics is starting to yield very significant caves. The largest cave chambers in the world are being found in China, Sarawak, and Vietnam, while deep caves with impressive underground rivers are being explored in Papua New Guinea and Mexico.

Table 2.4   The 20 longest caves of the world (as of October 2019).

	Name	Country	Length (km)
1	Mammoth Cave System	USA	667.9
2	Sistema Sac Actun (Nohoch Nah Chich, Aktun Hu) (Underwater+Dry)	Mexico	371.9
3	Jewel Cave	USA	334.8
4	Sistema Ox Bel Ha (Underwater)	Mexico	271.0
5	Suiyang Shuanghe Dongqun	China	257.3
6	Optymistychna	Ukraine	257.0
7	Wind Cave	USA	241.5
8	The Clearwater System (Gua Air Jernih)	Malaysia	227.0
9	Lechuguilla Cave	USA	222.5
10	Fisher Ridge Cave System	USA	208.8
11	Hoelloch	Switzerland	201.9
12	Sistema del Alto Tejuelo	Spain	164.0
13	Siebenhengste-hohgant Hoehlensystem	Switzerland	157.0
14	Schoenberg-Hohlensystem	Austria	147.8
15	Sistema del Mortillano	Spain	145.0
16	Ozerna (Gypsum)	Ukraine	140.5
17	Schwarzmooskogelhoehlensystem	Austria	133.8
18	Bullita Cave System	Australia	120.4
19	Sistema del Gandara	Spain	116.7
20	Hiriatzhohle–Schmelzwasserhohle	Austria	112.9

Source: Data correct as of 22 October 2019, www.caverbob.com.

References

Culver, D.C. (1982). *Cave Life: Ecology and Evolution.* Cambridge, MA: Harvard University Press.

Culver, D.C. and Pipan, T. (2009). *The Biology of Caves and Other Subterranean Habitats.* Oxford: Oxford University Press.

Ford, D.C. and Williams, P.W. (2007). Karst hydrogeology. In: *Karst Hydrogeology and Geomorphology*, 103–144. John Wiley and Sons.

Gillieson, D.S. (1996). Caves: Processes, Development, and Management. In: *Processes Development and Management.* Oxford: Blackwell: 7.

Klimchouk, A.B. (2016). The karst paradigm: Changes, trends and perspectives. *Acta Carsologica* 44.

Medville, D.M. and Worthington, S.R. (2017). The Friars Hole system. In: *Caves and Karst of the Greenbrier Valley in West Virginia* (ed. W. White), 137. Cham, Switzerland: Springer.

Simon, K.S., Pipan, T., and Culver, D.C. (2007). A conceptual model of the flow and distribution of organic carbon in caves. *Journal of Cave and Karst Studies* 69: 279–284.

White, W.B. (1984). Rate processes: Chemical kinetics and karst landform development. In *Groundwater as a Geomorphic Agent*. Boston: Allen & Unwin Incorporated.

Worthington, S.R.H. (1984). *The paleodrainage of an Appalachian fluviokarst: Friars' Hole*, West Virginia. MSc thesis, McMaster University.

3

Cave Hydrology

3.1 Basic Concepts in Karst Drainage Systems

In most landscapes, small streams unite to form larger rivers that take water, solutes, and sediments to the sea in an obviously integrated drainage network. In karst landscapes, this surface integration is disrupted by the formation of small centripetal drainage basins, and the landscapes are punctuated partially or wholly by closed depressions. Below the surface, a complex network of fissures and conduits (Figure 3.1) carries the water and erosion products to springs where the karst drainage network re-emerges at the surface.

The karst drainage network, thus, includes surface elements such as vegetation, soil, regolith, and closed depressions that regulate the quantity and quality of water passing underground. It encompasses the subterranean elements of various sizes and shapes of pores, fissures, and conduits (including caves), which transmit and store water and act as repositories for chemical precipitates and organic and inorganic sediments. The output of the karst network is regulated by springs at the karst margin, which may be enterable caves, flooded tubes, or diffuse seepages in a stream bed. The most important attributes of the karst drainage system are the intimate connections between surface and underground drainage, and the higher fissure density that enhances the rate at which surface water may enter and percolate down through the karst rock.

Rock formations capable of storing large amounts of water are called *aquifers,* a term widely used in hydrology and economic geology. Aquifers are characterised by their thickness, area, water-storing or transmitting properties, and their mean water quality. In this text, aquifers in karst are referred to as karst drainage systems because this term encompasses other elements of the karst, such as vegetation, soil, regolith, recharge, and discharge, which generally fall outside those elements considered in aquifer behaviour. It also avoids any assessment of either the quantity or quality of the water involved. In the scientific literature, these terms are often used interchangeably.

Water movement through the karst drainage system can be divided into flow and storage. We need to understand the various pathways that water takes through the rock and how water is stored in various zones of the system. Driven by energy (potential or kinetic), water will always move from areas of high energy to areas of low energy. So, within a karst, water will move slowly from recharge areas, where the water table is highest, to discharge areas, where the water table is lowest. This movement occurs along a *hydraulic gradient,*

Caves: Processes, Development, and Management, Second Edition. David Shaw Gillieson.
© 2021 John Wiley & Sons Ltd. Published 2021 by John Wiley & Sons Ltd.

Figure 3.1 Stream passage in the Grotte Milandre, Switzerland.

which is the height difference (or head) divided by the distance. The rate of water movement will be limited by this gradient and by the porosity and permeability of the karst rock.

The development of a flow network in karst depends on the amount of energy available – potential, kinetic, and thermodynamic – to drive the process. The main factors involved are the total volume of water passing through the rock mass, the difference in elevation between sink points and springs, whether the recharge to the karst is evenly distributed (diffuse) or focused (point), and the aggressivity or solutional capacity of the circulating water.

Most of the available potential energy is dissipated as the water descends through the air-filled vadose zone, where the cave stream can do mechanical work as well as some solution. Below the water table in the phreatic zone, water is moving in conduits at varying speeds. Results from dye-tracing experiments in Slovenia (Milanovic 1981) show flow velocities between 0.002 and 0.5 m s^{-1}, with an average of <0.05 m s^{-1}. These are low rates, but they increase dramatically when a pulse of water from heavy rain injects water and sediment into a karst aquifer. Such a flood passes as a kinematic wave or pressure pulse, which can travel at speeds of tens to thousands of metres per hour.

Thermodynamic energy is best indexed by the rock and water temperatures. Water temperature affects its viscosity – lower at higher temperatures – with greater discharge possible through capillary-sized tubes at higher temperatures, enhancing the solutional process. This may be a significant factor in the development of tropical caves and in the formation of caves deep in the rock mass (Worthington 2001).

There are three main pathways for water flowing through a karst drainage system: diffuse, fissure, and conduit flow. All of them are dependent on the porosity and permeability of the karst rock, key concepts that are defined next.

3.2   Porosity and Permeability

All rocks are capable of transmitting water to some degree through the pore spaces between the mineral grains. The *porosity* of a rock is the volume fraction occupied by voids or pores. It is dimensionless and can be defined as:

$$\theta = \frac{V_{pores}}{V} = \frac{1 - V_{min}}{V} \tag{3.1}$$

where V is the bulk volume of the rock and V_{min} is the net volume of mineral grains.

Permeability is the ability of a rock to transmit fluid, usually water. A rock must be porous to transmit fluids, but the pores must be connected for this to occur. In a uniform porous medium, the fluid flow is governed by Darcy's law, which states that the flow through a porous medium depends on the hydraulic gradient:

$$q = \frac{Q}{A} = -K\frac{dh}{dl} \tag{3.2}$$

where q is the unit flow in m^3 s^{-1} across an area, A (m^2), of the medium. The hydraulic gradient, $\frac{dh}{dl}$ is dimensionless, so both K and q have units of velocity, m s^{-1}. The hydraulic conductivity, K, depends on both fluid and rock properties. Darcy's law is usually stated as:

$$q = \frac{-Nd^2\rho g}{\eta} \cdot \frac{dh}{dl} \tag{3.3}$$

where g = gravitational acceleration, ρ = fluid density, d = grain diameter, η = fluid viscosity, and N is a dimensionless shape factor.

The quantity Nd^2 is the effective permeability of the rock; this is difficult to calculate and so is usually based on laboratory measurements or on aquifer characteristics. It is usually expressed in area units as cm^2 or m^2. For limestones, the range of values is from 10^{-5} to 10^2 m^2 or more, while for sandstones it may range from 10^{-8} to 10^{-3} m^2 (Ford and Williams 2007, p. 109).

There are two forms of porosity that contribute to θ in Eq. (3.1). *Primary porosity* is the result of the packing of mineral grains during the formation of a rock (Field 2002). This may be modified by solution or deposition during diagenesis. Thus, primary porosity is rarely uniform in a rock. Through time the formation of fractures, enlarged joints or bedding plane partings produce connected openings through which water can circulate. Brittle but strong rocks, such as limestone, dolomite, sandstone, and granite, readily crack, and these fractures tend to remain open for a long time. Weaker rocks, such as shales, mudstones, and chalk, may crack, but the fractures often seal under the pressure of overlying strata. Together, the various fractures, joints, and bedding plane partings constitute *secondary or fissure porosity*. This porosity is highly variable in the rock mass and allows for the concentration of water flow in areas of higher permeability. Through time some fissures may, thus, enlarge in a karst to form conduits – producing the *conduit porosity*. These conduits may range in size from small solutionally widened fractures or fissures, 1–100 mm wide, to major trunk passages 30 m in diameter (Table 3.1).

3.2.1   Diffuse Flow

In a homogeneous porous rock, the flow of water can be characterised by Darcy's law which has already been stated in Eq. (3.2). The movement of water in a homogeneous rock mass

Table 3.1 Porosity types and karst aquifer properties.

	Primary porosity	Secondary porosity	Conduit porosity
Components	Intergranular pores Vugs Mineral veins	Linked joints and fractures Bedding plane partings Connected mineral veins	Open channels and pipes of variable size and shape
Homogeneity	Usually isotropic	Usually anisotropic due to fracture origins, often oriented	Highly anisotropic forming networks
Size of apertures	µm–mm	10 µm–10 mm	10 mm–10 m
Flow regime	Laminar	Laminar to just turbulent	Turbulent
Governing hydraulic law	Darcy	Hagen-Poseuille	Darcy-Weisbach
Water table	Well defined	Irregular surface	Often perched and at varying levels
Flow response to input water	Slow	Moderate	Rapid

can be regarded as being at right angles to the contours of the water table or piezometric surface, and this surface of hydraulic equipotential lines can be mapped from borehole or well level data. Thus, there is a flow net made up of streamlines that extend down the maximum slope or hydraulic gradient of the piezometric surface, from the inputs at the upper boundary of the catchment, to the springs or outlets at the lowest point. This flow net may descend to considerable depths below the spring level in any karst drainage system. If flow potential increases with depth, then at some point the flow will converge and turn up to an area of lower potential, usually below the spring or surface river. A large conduit in the saturated zone will also have lower potential and will thus serve as a target for the flow net. The rate of groundwater inflow per unit length along such a conduit can be estimated by

$$Q_i = \frac{2_\pi K h_z}{2.3 log\left(^{2h_z}/_r\right)} \tag{3.4}$$

where h_z is the depth of the conduit of radius (r) below the water table (Freeze and Cherry, 1979).

For Darcy's law to be applicable to karst systems, laminar flow must be occurring, and there must be conservation of mass (the principle of continuity), where water entering a unit cube of rock is balanced by water leaving it. At higher velocities beyond a certain threshold, turbulent flow occurs, and the velocity is no longer proportional to the hydraulic gradient. Turbulence also occurs in water flowing through porous media when the interconnected pore spaces are larger than fine gravels. Despite these problems, many groundwater hydrologists have applied the principles of Darcian flow to karst aquifers, and it remains the most commonly used method for modelling flow from borehole and well data.

For Palaeozoic aquifers, Worthington (1991) has noted that the velocity predicted from Darcian theory ($10^{-7}–10^{-5}$ ms^{-1}) is much less than that actually observed from dye-tracing experiments from well to well (average $0.02\,\mathrm{ms^{-1}}$). In part, this is due to the fact that conduits are the rapid-flow elements in the karst drainage system and are fed by the much slower movement of water through the rock mass. Thus, to understand the movement of water in karst, it is necessary to look at all components of the underground drainage network. This view is exemplified by Gunn (2010) for the karst of Buxton, Derbyshire, UK. In an $8\,\mathrm{km^2}$ block of carbonate rocks around the town, the degree of interconnectedness of the channel networks discharging at springs and intersected at wells varies to such an extent that there is clearly no single karst aquifer. It would be impossible to model the carbonate block as a simple porous medium aquifer dominated by diffuse flow; rather there is a complex network of conduits with flow paths varying with flow conditions.

3.2.2 Fissure Flow

Where laminar flow occurs in a jointed or fractured rock, the hydraulic conductivity K (in $\mathrm{m\,s^{-1}}$) of a fissure with parallel sides can be determined from the equation

$$K = \frac{\rho g w^2}{12\mu} \tag{3.5}$$

where w is the width of the fissure, ρ is the kinematic viscosity of the liquid, and μ is the dynamic viscosity (Ford and Williams 2007, p. 112).

The kinematic viscosity is heavily dependent on temperature, while it is apparent that a slight increase in fissure width will cause a dramatic increase in hydraulic conductivity. Thus, laminar flow in an enlarging karst fissure may quickly become turbulent, and conduit flow conditions obtain. Scaling up from a single fissure to a fractured aquifer is very difficult. Irregularities and variations in fissure width along the flow path make it difficult to arrive at a single representative width, and the dependence of flow on the cube of this value makes it a sensitive parameter (White 1988, p. 171)

$$Q = \frac{\rho g w^3}{12\mu} \cdot \frac{dh}{dl} \tag{3.6}$$

where Q is the flow rate through a single fracture with parallel sides and distance w apart, ρ is the fluid density, and $\frac{dh}{dl}$ is the change of hydraulic head with distance along the fracture.

Finally, the geometry of complex fracture sets is often unknown and is very difficult to model in three dimensions. Worthington (1991) established a relationship between flow path length and joint orientation (Figure 3.2).

3.2.3 Conduit Flow

Active conduits are major determinants in karst drainage system behaviour and function. They are the foci for the discharge of tributaries in the karst, and they are the main vectors for groundwater flow through and discharge from the karst system. Unfortunately, the flow of water in active conduits is very hard to observe directly and even harder to model satisfactorily (Figure 3.3). Significant recent advances have occurred in our understanding of the structure and function of conduits, due largely to the efforts of cave divers and by carefully designed dye-tracing experiments. Inferences can be drawn from observations of drained conduits or caves, with Lauritzen et al. (1985) warning that the study of fossil

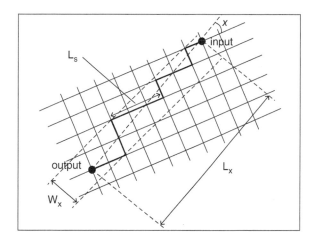

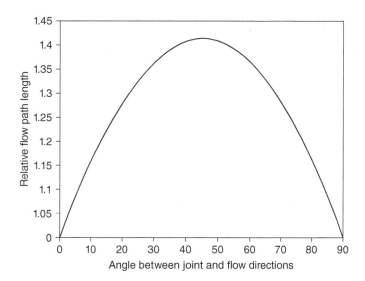

Figure 3.2 Relationship between flow path length and angle between joint and flow directions. L_x and W_x indicate the length and width of the flow zone, respectively. Source: From Worthington (1991).

conduits "is a direct parallel to the case of medieval anatomists dissecting a dead body rather than studying the physiology of the living organism." However, key questions, such as the magnitude of storage and the velocity of water movement, remain unresolved.

The flow of water through small tubes was studied by Hagen (1839) and later by Poiseuille (1847). They discovered that the water flow or specific discharge μ was proportional to the hydraulic head loss by friction Δh along the tube:

$$\mu = \pi r^2 \left(\frac{r^2}{8} \cdot \frac{\rho g}{\mu} \cdot \frac{dh}{dl} \right) = \frac{\pi r^4}{8} \cdot \frac{\rho g}{\mu} \cdot \frac{dh}{dl} \tag{3.7}$$

where r is the radius of the tube, ρ is the kinematic viscosity of the water, and l is the length of the tube. The unit head loss $\frac{dh}{dl}$ is the difference in head between one end of the tube and

Figure 3.3 In Atea Kananda, Papua New Guinea, the feature known as The Turbine marks the transition between the vadose and phreatic zones of the cave. Approximately 4 m^3 s^{-1} of water disappear into a vertical shaft that connects to lower levels of the cave.

the other, where the head at a point is the sum of the pressure head, the depth of the point below the water table, and the elevation head, the height of the point above the datum or base level.

Under laminar flow condition in small tubes, the discharge per unit length can be calculated using the Hagen-Poseuille equation (Vennard and Street 1976):

$$Q = \frac{\pi d^4 \rho g}{128 \mu} \cdot \frac{dh}{dl} \tag{3.8}$$

where $\frac{dh}{dl}$ is the head loss over a unit length and d is the diameter of the tube.

From this, large tubes are much more conductive than small ones, as the diameter is raised to the fourth power in the equation. A tube 1 mm in diameter will thus conduct the same flow as 10 000 capillaries 0.1 mm in diameter. This means that a single enlarging tube will rapidly capture the flow from smaller ones and is of great importance for an evolving cave network.

The Reynold's number, R_e, is used to help identify the critical velocity at which the transition from laminar to turbulent flow takes place and beyond which Darcy's law is inapplicable. This may occur when R_e is greater than values of 1–10, but full turbulent flow is attained at high velocities and R_e values in the range 10^2–10^3. Reynold's number is expressed as:

$$R_e = \frac{\rho v d}{\mu} \tag{3.9}$$

where v is the mean velocity of the fluid flowing through a pipe of diameter d.

When the flow in an enlarging tube becomes turbulent, a threshold is passed beyond which the Hagen-Poseuille equation becomes invalid, and discharge is better estimated using the Darcy-Weisbach equation (Thrailkill 1968):

$$u^2 = \left(\frac{2dg}{f}\right) \cdot \frac{dh}{dl} = \frac{Q}{a^2} \tag{3.10}$$

where f is a friction factor, from which

$$Q = \sqrt{\frac{2dga^2}{f}} \cdot \sqrt{\frac{dh}{dl}} \tag{3.11}$$

The relationship between f and Q has been studied by Lauritzen et al. (1985) for an active phreatic conduit in Norway. Apparent friction shows a rapid exponential decrease with increasing discharge until a constant value is reached. Friction is affected by tube dimensions, wall roughness, and the complexity of conduit geometry, including the presence of breakdown.

3.2.4 Understanding the Karst Drainage System

Early work on karst drainage systems emphasised two modes of flow: diffuse flow and concentrated or conduit flow. Atkinson (1977) demonstrated that both existed in the Mendip Hills, England, and that conduit storage made up only about 3.5% of the total baseflow storage in the karst drainage system. Conduits were, however, very important in draining the diffuse part of the karst system. From an analysis of spring water chemistry, Bakalowicz and Mangin (1980) concluded that this bimodal scheme was too simple to explain the observed variation, and from an evaluation of a much larger sample of karsts Atkinson and Smart (1979) suggested that an intermediate-sized void, the karst fissure, was important in some drainage systems. This led Atkinson (1985) to propose a three-member spectrum of water flow, which has found wide acceptance (White 2002; Figure 3.4). Depending on the nature of the karst host rock and its geomorphic history, any karst drainage system may be placed in this scheme.

Smart and Hobbs (1986) elaborated this approach by suggesting that the properties of recharge, transmission, and storage might help to position individual karst drainage systems in the spectrum (Figure 3.5). For recharge, the end members are concentrated and diffuse inputs; for transmission, conduit and diffuse flow; and for storage, unsaturated and permanently saturated stores. Later consideration by Worthington (1991) led to an appreciation of the importance of the role of spring aggradation in controlling drainage system behaviour, and of the role of fissure flow. The consequences of these three factors of recharge, flow, and storage for spring behaviour will be considered later. It is first necessary to define the three modes of water flow in the karst drainage system.

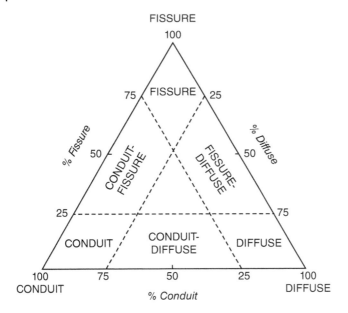

Figure 3.4 Conceptual types of karst aquifers and their mixtures. Source: From Smart and Hobbs (1986).

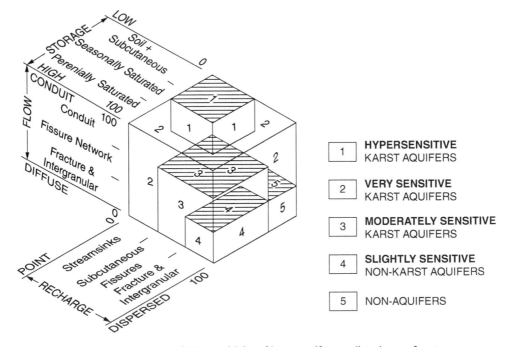

Figure 3.5 Conceptual scheme of the sensitivity of karst aquifers to disturbance, for storage, recharge, and flow types. Boundaries between fields 1 to 5 are approximate and intuitive. A crude measure of the diagonal from field 4 to field 1 is given by increasing values of the coefficient of variation (s.d./mean) of specific conductivity of the water. Source: Modified from Smart and Hobbs (1986).

Hence, any karst aquifer may be regarded as having elements of all three porosity types: diffuse, fissure, and conduit. It is unwise to analyse any single karst hydrologic system assuming that one type prevails in water movement. Analysis of a karst drainage system assuming diffuse flow – an equivalent porous medium obeying Darcian flow – will generally significantly underestimate flow velocities from recharge zone to spring by two or three orders of magnitude (Worthington 1991). This may have serious repercussions if the pollutant movement is being predicted (Gunn 2010).

The presence of all three flow types may be seen in many active stream passages. The water flowing in the cave stream is typical of low stage conduit flow. Water dripping from straws and stalactites in the roof, perhaps aligned along a fracture or joint, is obeying fissure flow. Water seeping through the intergranular pores and emerging as wet patches on the roof is governed by diffuse flow. The magnitude and relative importance of each type in an individual karst hydrologic system will depend on the lithology and secondary porosity of the host rock. Thus, in a porous rock such as chalk, diffuse flow may dominate, though there may be a significant amount of fracture flow. However, chalk is a mechanically weak rock and conduits may not survive. In contrast, marble has low primary porosity, and thus diffuse flow may be negligible. Water will tend to move quickly in fractures and conduits from recharge zones to the springs.

3.3 Zonation of the Karst Drainage System

The karst drainage system may be usefully divided into several zones, each of which has distinctive hydraulic, chemical, and hydrological characteristics. Water infiltrates into the karst through the soil and regolith, moving down under gravity until it reaches the zone in which all the pores are already filled with water. The surface separating air-filled from water-filled pores is called the piezometric surface or *water table,* and its contours can be mapped from elevations in boreholes. The saturated rock below the water table is called the *phreatic* zone, while the largely air-filled rock above the water table is called the *vadose* zone (Table 3.2) (Field 2002). If the vadose zone connects directly with the surface so that the water table can respond freely to rainfall and spatially varying infiltration, then the aquifer is *unconfined.* This is common in karstic rocks where there is not an overlying rock layer of lower permeability. If there is an overlying impermeable rock layer, then the water may be in a *confined aquifer* and may be under some pressure. The water table cannot respond to varying infiltration, and instead there is a potentiometric surface that corresponds to

Table 3.2 Zonation of the karst drainage system.

Broad zone	Functional zone	Components
Epikarst	Cutaneous zone	Surface and soil
	Subcutaneous zone	Regolith and enlarged fissures
Endokarst	Vadose zone	Water unsaturated, with free air
	Epiphreatic zone	space flood or temporary phreas
	Phreatic zone	Water-saturated, no air space

the varying pressure head developed under the confining rock formation. Any well or bore drilled into the rock will have a water level at this potentiometric surface. If this lies above the ground surface, water will flow freely from the well, and we will have an *artesian aquifer*. These zones are not mutually exclusive, and under radically altered flow conditions one zone may temporarily gain the properties of the one below it in the sequence.

The extent of groundwater through these zones has been the subject of intense scientific debate since the time of Martel and has been summarised by White (1988). Early workers considered that most flow occurred in the vadose zone rather than in the phreatic zone, but

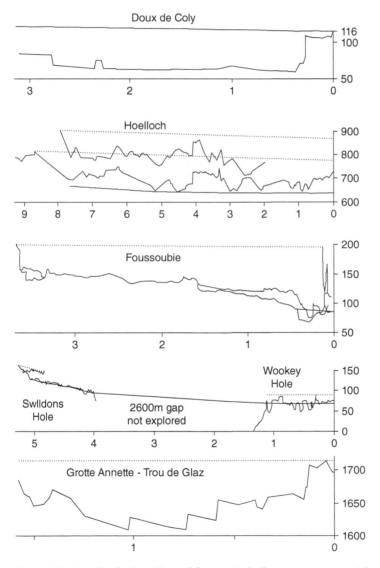

Figure 3.6 Longitudinal sections of flow paths in limestone cave conduits, indicating depth of circulation below the water table. Conduits in the Doux de Couly (France) and Swildons-Wookey (UK) systems have been mapped by divers; the remainder are air-filled and indicate fossil water tables. Source: From Worthington (1991).

the latter view eventually predominated. Subsequently there was debate as to whether the locus of flow was close to the water table (the shallow phreatic theory of Swinnerton [1932], Rhoades and Sinacori [1941], and Thrailkill [1968]) or some distance below it (the deep phreatic theory of Davis [1930] and Bretz [1942]). These various authors all tried to create a universal scheme applicable to all caves but based on the characteristics of the individual region that they studied. Subsequent approaches (Dreybrodt 1990; Ford and Ewers 1978) have used stochastic approaches to flow path location.

Most, if not all, of these approaches have been based on analysis of the morphology and context of drained conduits or caves. These cavities are now realised to be a small fraction of the total conduit length, and the sample afforded by them has itself increased fourfold in the last 20 years (Courbon et al. 1989). Detailed work examining the relationships between passage morphology and geological structure (Filipponi et al. 2009; Lowe and Gunn 1997; Maire 1990; Palmer 1987) has added greatly to our understanding of why caves form where they do. The exploration and mapping of flooded caves by cave divers (Figures 3.6 and 3.7) has yielded even more useful information about active conduits up to 10 km long (Caramanna and Giordani 2010; Ginès and Ginès 2007; Webb and James 2006).

Some evidence suggests that successive tiers of phreatic conduits are equally spaced in line with a Hagen–Poseuille flow net (Worthington 1991). An individual tier may be active for 0.3–10 million years (Worthington 1991, p. 130) based on radiometric and palaeomagnetic dating, but eventually it will be abandoned owing to intrinsic changes, such as blockage by sediments, or extrinsic changes, such as base level lowering or spring aggradation. During the relatively short initiation phase, conduits are of small diameter and experience

Figure 3.7 Improvements in cave diving technology have resulted in major advances in our understanding of flooded cave conduits and flow paths. Diving in Tank Cave, Mount Gambier, Australia, photo by Stefan Eberhard.

Figure 3.8 The main passage of Gua Tempurung Cave, Malaysia, shows a fossil phreatic tube in the roof breached by a huge vadose canyon that carries the present cave stream.

the transition from laminar to turbulent flow. Active enlargement to tubes of 3–5 m diameter occurs over a notional timespan of up to 100 000 years. After this, flooded conduits may enlarge until they collapse and other tiers take over, or may be drained (Figure 3.8). If base level lowering is active, then vadose canyons may result, or dry tubes may persist until the unsupported passages fragment or are removed by surface erosion. The likely timespan from initiation to decay is of the order of 1 to 10 million years. Partially or wholly infilled fragments of these conduits may persist in the geological record as palaeokarst.

3.4 Defining the Catchment of a Cave

The karst drainage system resembles that of a surface stream or river in that it is comprised of a network of channels great and small that transport water, solutes, and sediments from a single input or an array of inputs, often known as stream sinks (Figure 3.9) to single or multiple outputs, often known as springs. The surface analogues of these are the first-order streams in a drainage basin, where integrated flow commences, and the terminal basin of the catchment, where the channel debouches into a lake, an estuary, or the sea. Whereas the catchment of a surface stream may be readily defined using aerial photography

Figure 3.9 Stream sink of the Baia River, Karius Range, the Southern Highlands of Papua New Guinea.

or satellite imagery, the catchment of a karst drainage system may be very difficult to define unequivocally.

Although some of the inputs to the drainage network may include discrete stream sinks whose connections can be proven by dye-tracing experiments, many will be diffuse feeders in the subcutaneous zone whose flow paths will vary depending on individual storm intensity and antecedent rainfall. These diffuse sources are fed by soil water infiltration and percolation on the karst surface, and by soil water throughflow from adjacent non-karstic rocks and regolith. It is extremely difficult to determine the uppermost elements of a drainage network from diffuse sources, although some short-distance tracing of these inputs has been carried out (Trudgill 1987). There is also the possibility that high-level abandoned conduits or palaeokarstic features may be reactivated under exceptional rainfall conditions.

The catchment of a karst drainage system is usually more extensive than just the area of limestone outcrop and the obvious non-karstic contributing catchment. Karst drainage systems frequently breach surface drainage divides, as the network may have formed under an older surface topography that bears little resemblance to that visible today, or the evolving karst drainage network may have captured adjacent networks. The details of these connections can only be worked out by careful dye tracing and by mapping of cave passages.

The karst catchment boundary is rarely a single line that can be represented on a map, rather a zone that has a dynamic outer boundary dependent on local details of surficial geology and weather conditions. It is more useful to think of a core catchment area, within which flow will usually be directed to a particular cave network, and a peripheral or buffer catchment area that may be activated periodically. If the precautionary principle applies in karst research or management, then the larger catchment may provide a truer representation of the sources for the karst drainage network.

3.5 Analysis of Karst Drainage Systems

3.5.1 Water-Tracing Techniques

The technique of water tracing using artificial dyes or particles is very widely employed in groundwater hydrology and is the subject of several major works (Goldscheider and Drew 2007; Hotzl and Werner 1992; Milanovic 1981; Quinlan et al. 1987). The most common use has been the identification of flow routes from inputs to output springs or seepages. Repeated tracing of flow routes has shown that there is a strong relationship between discharge and tracer travel time (Stanton and Smart 1981), and that flow paths may vary with each stage as alternative routes become activated. It is, therefore, necessary to carry out dye-tracing experiments over the full range of discharges from low to high, and to monitor all possible springs, however unlikely the connection may seem to be.

Several substances have been used for tracing experiments. These include microorganisms, environmental isotopes, salts, dyes, and spores, and they are fully discussed in Ford and Williams (2007, p. 191). Because of their low cost, low toxicity, easy detectability, and high sensitivity, fluorescent dyes have been and will continue to be the most widely utilised in karst hydrology. The application of precision fluorometric techniques permits a definition and estimation of flow networks and hydraulics that are not readily available from other techniques. The principal dyes in use are outlined in Table 3.3. Smart (1984) provided a very thorough review of the toxicity of commonly used fluorescent dyes, but today only Tinopal CBS-X (an optical brightener) and Fluorescein have no demonstrated carcinogenic or mutagenic hazard (Goldscheider et al. 2008). Several others in common use are of unknown toxicity.

Two important conditions must be met for artificial dye tracers to yield reliable results. First, the tracer must be conservative, that is it must exhibit little or no loss within the karst drainage system. This is unfortunately not tenable due to the adsorption of fluorescent dyes onto clay-rich sediments or organic matter. Second, it is important that the tracer signature is unique or unequivocal. It must be clearly separable from other artificial dyes (Figure 3.10), and it must be distinct from naturally occurring compounds. Failure to meet these two criteria is often the cause of a false interpretation of a dye-tracing experiment.

Three specific problems need to be considered here: suppression, adsorption, and natural fluorescence. The presence of certain ions may cause a suppression or quenching of fluorescence (Smart and Laidlaw 1977). Large losses of Rhodamine WT fluorescence were noted in the presence of sodium and potassium chloride, with loss increasing through time. However, the study of Bencala et al. (1983) showed little suppression in solutions containing lithium, sodium, potassium, strontium, chloride, and nitrate. Dye adsorption is a more serious problem, although many studies assume that the tracer is conservative. Rhodamine WT has long been perceived as being conservative, though major losses can be expected in a

Table 3.3 Summary of some fluorescent dyes used in water tracing.

Tracer name (CAS RN)	Detection limit ($\mu g\, l^{-1}$)	General problems	Known issues
Uranine (518-47-8)	10^{-3}	Sensitive to light and strong oxidants	No concerns
Eosin (17372-87-1)	10^{-2}	Sensitive to light and strong oxidants	No concerns
Amidorhodamine G (5873-16-5)	10^{-2}	Sensitive to light and strong oxidants	No concerns
Sulforhodamine B (3520-42-1)	10^{-2}		Ecotoxicological concerns
Rhodamine WT (37299-86-8)	10^{-2}	Analytical interference with other fluorescent dyes	Genotoxic
Pyranine (6358-69-6)	10^{-2}	Analytical interference with other fluorescent dyes	Biodegradable
Napthionate (130-13-2)	10^{-1}	Analytical interference with other fluorescent dyes	No concerns
Tinopal CBS-X (27344-41-8)	10^{-1}	Analytical interference with dissolved organic carbon	No concerns

Chemical Abstracts Service Registry Numbers (CAS-RN) provide unambiguous identification of specific dyes.
Source: Modified from Goldscheider et al. (2008).

range of hydrological environments. Trudgill (1987) investigated adsorption in soils, while Bencala et al. (1983) recovered only 45% of the Rhodamine after a flow length of 300 m over sediment. Fluorescein is readily adsorbed by clays and probably by colloidal organic matter.

Natural fluorescence in soil and runoff water has been well-documented (Smart and Laidlaw 1977; Trudgill 1987). It varies with time, flow conditions, and source areas. Increased fluorescence is often noted under high-flow conditions owing to flushing of organic compounds from the soil. Fluorescence spectra for natural waters have broad bandwidth in the blue-green wavelengths, and this is largely owing to dissolved organic matter. The natural orange-red band (due to phytoplankton) is many times smaller than the blue-green band, and this in part explains the preference for Rhodamine over Fluorescein. Although natural fluorescence can be separated from that of dyes in the laboratory using a spectrofluorometer, this is not yet possible in the field.

The form of the time-concentration or tracer breakthrough curve depends on the structure of the conduit network, the prevailing flow conditions, and the nature of the tracer used (Smart 1988a). Consideration of these factors, in combination with the regional hydrogeology, can yield a great deal of information about the structure and dynamics of the karst drainage system (e.g. Crawford 1994; Smart 1988b). The form of the breakthrough curve has been used to indicate off-line storage in the conduit network (Atkinson et al. 1973).

Spores of the club moss *Lycopodium clavatum* were once widely employed for dye tracing, but their use has declined since the 1970s. The best overall review of their use is by

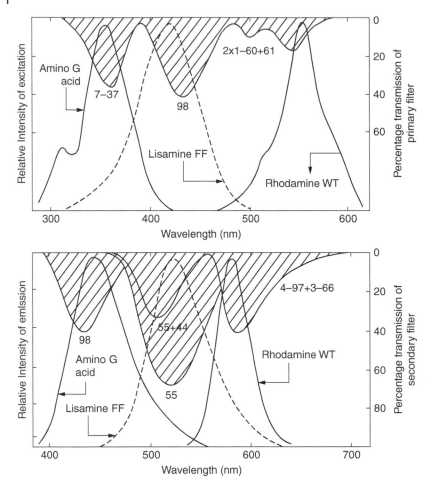

Figure 3.10 Absorption and emission spectra of blue (Amino G acid), green (Lissamine FF), and orange (Rhodamine WT) fluorescent dyes. Source: From Smart and Laidlaw (1977).

Smart and Smith (1976). The main disadvantages of spores are that operator error in colour identification may be considerable, that the aquifer is contaminated with spores for a long time, and that the technique is non-quantitative because the trap efficiency of the nets and laboratory recovery rates are unknown (Smart et al. 1986). More recently, coloured microplastic spheres have been used (Goldscheider and Drew 2007), but the environmental impact of microplastics is recognised as a global problem due to their incorporation into food chains.

Environmental isotopic techniques are based on measurements of variation on the isotopic composition of natural waters, and are most useful in problems related to the dynamics of drainage systems (using the radioactive isotopes tritium and carbon-14) and the origin of waters (using the stable isotopes deuterium and oxygen-18) (International Atomic Energy Agency 1983). In Crete, Leontiadis et al. (1988) determined the origin of recharge to coastal springs from karst plateaux and estimated the age of groundwater to range from

45 to >100 years using tritium input functions. The high variability in percolation input to a shallow cave in Israel was noted by Even et al. (1986). Using tritium, they determined the age of seepage water to be several decades. Enrichment of deuterium and oxygen-18 in the seepage water suggested that the water was derived from a relatively dry winter in 1962–1963 when westerly air masses dominated the region. In Australia, Allison et al. (1985) have investigated the rate of recharge of groundwater in calcretes using stable isotopes. Chlorine-37 concentrations in soil can increase dramatically through evaporation, and this saline water moves slowly to the groundwater table, and hence, to rivers. It is clear that, where the mallee vegetation has been cleared, recharge rates are increased. Recharge rates vary from 60 mm y^{-1} for secondary dolines to 0.06 mm y^{-1} for calcrete flats. Recharge rates for vegetated dunes are 0.06 mm y^{-1}, while for cleared dunes they are 14 mm y^{-1}. Thus, the highest rates of saline recharge are associated with secondary collapse dolines, where hydraulic conductivity is enhanced by fissures. There is considerable scope for further investigation of the authigenic percolation system in karst using these techniques (Perrin et al. 2003).

3.5.2 Spring Hydrograph Analysis

Karst drainage basins respond to rapid changes in recharge in a similar manner to surface basins. Analysis of the flood hydrographs from a spring can reveal much about the structure and function of the network feeding it. The most useful data come from the monitoring of intense storms that inject pulses of water into the karst. Stream sink hydrographs have a steep rising limb as overland flow feeds the input point, a peak flow (Q_{max}) and a long declining limb or recession to the baseflow unless another storm occurs (Figure 3.11). There is usually a lag time before the pulse is detected as a rise in discharge at the spring. The length of this lag will depend on the structure of the karst drainage system and the amount of stored water in the system. Thus, for a fairly direct flow path, with little stored water, most of the inflowing water will be stored temporarily in fissures and conduits, and the rise at the spring will be slow and of modest size. There is usually considerable damping of the spring hydrograph, with a broad peak in discharge. In contrast, if antecedent rainfall has filled the available conduit and fissure storages, then the pulse will be rapidly transmitted with a small lag and the spring hydrograph response may be dramatic, and the peak may resemble the inflow hydrograph with little damping. Often there will be flushing of older stored water from the conduits as the inflowing water acts like a piston.

If the spring is monitored for water chemistry and sediment load as well as discharge, then this flushing effect may be detected, and the size of the storage estimated (Ashton 1966). An idealised spring hydrograph is shown in Figure 3.12. The hydrograph can be separated into zones where a particular water source is dominant. On the initial rising limb, phreatic conduit water of long residence time may be displaced by incoming water. This is replaced by a mix of phreatic and subcutaneous water as the main flood pulse comes through, characterised by higher conductivity and increasing turbidity (Figure 3.13).

Finally, surface floodwater with a great deal of suspended sediment raises turbidity but has a lower solute concentration (lower conductivity). This water dominates the early part of the recession but is supplanted by baseflow from subcutaneous and phreatic storage. The magnitude of each water type can be roughly estimated by integrating under the separated hydrograph. This approach also permits some clarification of conduit network

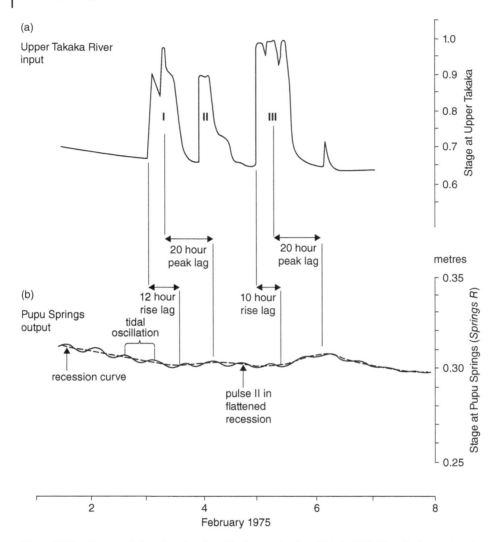

Figure 3.11 Stream sink and spring flood hydrographs from Takaka Hill, New Zealand, showing lag effects and peak flattening. Source: From Williams (1984).

geometry by interpretation in terms of combinations of spatially distributed inputs to produce complex output pulses (Figure 3.14). Since Ashton's theoretical work, several authors have considered pulse wave analysis purely in terms of discharge (Goldscheider et al. 2008), while others have combined discharge with water chemistry in pulse analysis (White 1988). An elaboration of the technique is the labelling of natural or artificial pulse waves with fluorescent dyes (Christopher et al. 1981).

Springs with low variance in flow and chemistry are often fed solely by autogenic recharge, which is water derived wholly from the epikarst. In contrast, springs with high variance in flow and chemistry are often associated with allogenic recharge (Jakucs 1959), where water is derived partially or wholly from non-karstic rocks. Rarely is a spring fed by an entire karst drainage system and can be termed a full-flow spring (Worthington 1991).

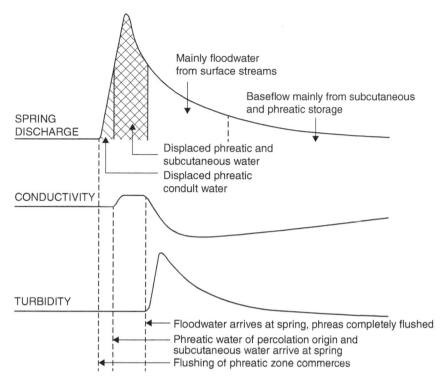

Figure 3.12 Interpretation of an idealised spring hydrograph and chemograph. Source: From Williams (1983).

More commonly, there is a hierarchy of springs draining a karst with underground distributaries feeding springs at differing levels in a valley. The lowest elements preferentially fed by baseflow are termed underflow springs (Figure 3.15), while upper springs preferentially fed by flood flow are called overflow springs. There are intermediate members in the hierarchy called overflow-underflow springs. Smart (1988c) studied about 80 springs over a 150 m vertical range near Castleguard Cave, Alberta, Canada. Steady and sustained flow came from the lowest underflow members, while short-lived variable flow came from the upper members, with the uppermost fossil springs showing no flow. Underflow springs may be hard to detect, as they may be aggraded by alluvium or sub-lacustrine or be seepages in a stream bed.

Where detailed discharge records are available, underflow and overflow springs may be distinguished by the ratio of maximum annual discharge to minimum annual discharge and the proportion of time that flow is present (Table 3.4). The shape of the flood hydrograph recession is a good diagnostic tool (Mangin 1975). Discharge usually decreases exponentially during the recession to baseflow such that

$$Q_t = Q_0 e^{-\alpha t} \tag{3.12}$$

where Q_0 is the baseflow at the start of the recession ($t = 0$), and α is the recession exponent. Q is usually expressed in m^3 s^{-1}, while t is expressed in days. There may be seasonal variation in the recession exponent depending on recharge and storage.

Figure 3.13 In Mamo Kananda, Papua New Guinea, the section of cave known as the Rinse Cycle floods each afternoon and shows typical phreatic features. This single tube has captured most of the water flow in the cave.

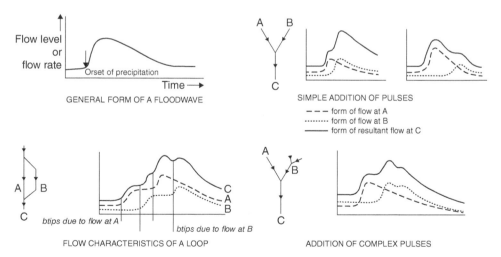

Figure 3.14 Flood pulse generation and hydrograph forms resulting from flow addition and complex flow paths. Source: From Ashton (1966).

The shape of the recession can be used to diagnose spring type (Worthington 1991, p. 52). For full-flow springs, α is constant (Figure 3.16 and Table 3.4). This is relatively rare, though many karst studies have assumed that it obtains in the Mammoth Cave area (Kentucky, USA), Quinlan and Ewers (1989) found that only one of the 21 larger karst drainage systems was drained by a full-flow spring.

Figure 3.15 The karst spring of Longgong Dong (Dragon Palace Cave), Guizhou, China, is harnessed for a hydroelectric plant as well as being used for tourism. Inside the cave, the water from a 600 m long lake falls 36 m to the entrance.

Table 3.4 Discharge characteristics for distinguishing spring flow types.

Spring type	Qx/Qn	Days with Q > 0
Full flow	High	All
Underflow	Low	All
Overflow	∞	Few – all
Underflow-overflow	∞	Few – all

Q_x is maximum annual discharge while Q_n is minimum annual discharge.

Overflow springs are characterised by convex log-normal baseflow recession (α increasing with time), with strongly seasonal or intermittent springs having a minimum discharge of zero and α tending to ∞. Underflow springs may be of two types. Losing or high-stage underflow springs are controlled by constrictions or aggradation close to the output, so at the high stage, the water backs up in the conduit until an overflow outlet is activated. This provides a constant head and a slightly decreasing or constant α depending on the magnitude of the recharge from the catchment relative to the output. Gaining or low-stage underflow springs have a concave recession, with α decreasing as a full-flow recession is

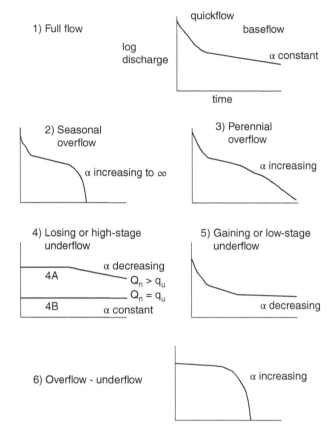

Figure 3.16 Karst spring types in relation to the exponent α of the recession curve. Source: From Worthington (1991).

supplemented by underflow from surface stream beds or other cave streams. Finally, many karst springs fall into the last category of underflow-overflow, where the spring is intermediate in a hierarchy and functions as an underflow spring for part of the time and then reverts to the intermittent overflow type.

In most karst drainage systems, each overflow spring has a complementary underflow spring. There is also some loss or gain from surface streams to groundwater in any karst. Finally, some groundwater flow will always emerge at the lowest point in the karst drainage system owing to the hydraulic head but may not do so as a discrete spring. Expect the unexpected!

3.5.3 Spring Chemograph Analysis

Where a spring has been monitored for both discharge and conductivity, it is possible to examine the seasonal and annual variation in solute load from the karst drainage system. There is usually a very good statistical relationship between conductivity and total hardness, and this can be used to calculate the solute load. Figure 3.17 is the 10-year record from the Argens Spring in Provence, southern France (Julian and Nicod 1989). This large underflow

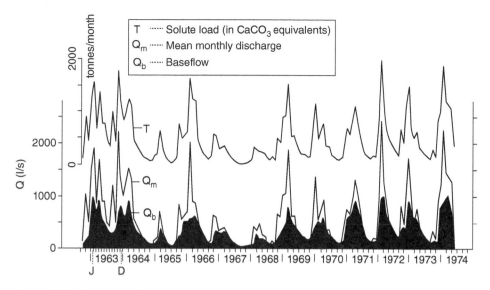

Figure 3.17 Ten-year spring hydrograph and chemograph from the Argens Spring, Provence, France. Note the seasonality of flow, the large proportion of baseflow discharge, and the dependence of solute load on flood discharges. Source: From Julian and Nicod (1989).

spring drains the karst plateau to the north of Ste. Victoire and shows strongly seasonal flow, with a maximum in winter. This is typical of karst springs whose recharge is due to snowmelt. There is a strong relationship between mean monthly discharge and solute load, with baseflow making a significant contribution in all months. Low recharge during the 1967–1968 drought was compensated by interannual carry over of karst water.

The frequency distribution of conductivity from such springs can be used to characterise their aquifers (Bakalowicz and Mangin 1980).

- Diffuse or porous aquifers – unimodal, high conductivity
- Fissured aquifers – unimodal, low conductivity
- Conduit or karstified aquifers – polymodal, wide range of conductivity.

Thus, in Figure 3.18, the Evian mineral water source is of the diffuse type, while the Source de Surgeint aquifer is probably more fissured. Most of the other examples are of the mixed conduit type, with several populations of unique geochemical evolution contributing to the final histogram.

3.6 Structure and Function of Karst Drainage Systems

3.6.1 Storage and Transfers in the Karst System

One consequence of the model of Smart and Hobbs (1986) is that spring hydrograph response can be affected by combinations of recharge, storage, and flow processes. In Figure 3.19, the effects of changing these factors are explored. Concentrated recharge (e.g. snowmelt) into a karst drainage system with low storage will produce a peaked spring

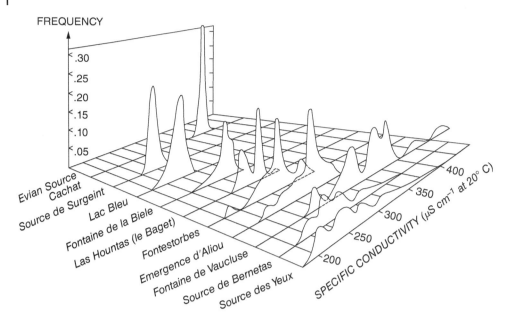

Figure 3.18 Frequency distributions of conductivity of karst spring waters in southern France. Source: From Bakalowicz and Mangin (1980).

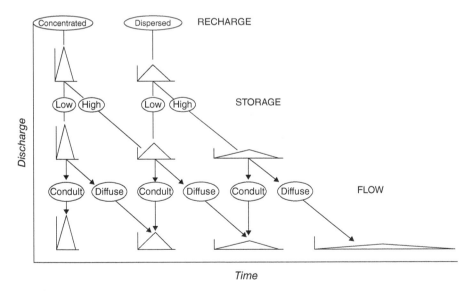

Figure 3.19 Effects of variation in recharge, storage and flow types on the form of flood hydrographs in karst drainage systems. Source: From Smart and Hobbs (1986).

hydrograph, while if storage is high the hydrograph will be flattened as water is gradually released. If diffuse flow is present as well, then the hydrograph is attenuated twice to produce a broad spring hydrograph more suggestive of baseflow dominance. With a dispersed recharge due to autogenic recharge in the epikarst, the nature of storage and the presence or absence of diffuse flow will radically alter the spring hydrograph to produce

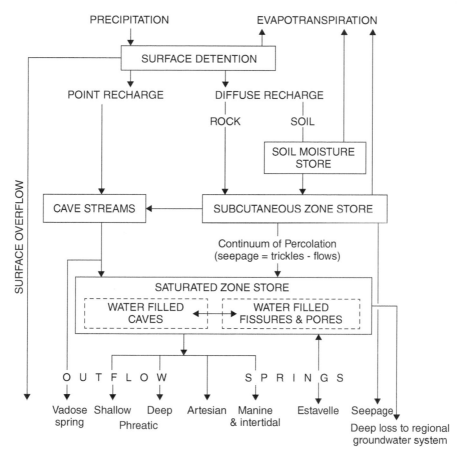

Figure 3.20 Scheme of storages and flow linkages in a karst drainage system. Source: From Ford and Williams (2007).

a wide range of responses. This complexity indicates that analysis of spring hydrographs alone is unlikely to produce sound interpretations unless some knowledge of the nature of storage and flow characteristics can be obtained.

Ford and Williams (2007, p. 139) have developed a model of karst drainage systems that incorporates these factors and serves as a basis for enhanced understanding of karst hydrology (Figure 3.20). If the major storages and fluxes in this scheme can be quantified, then the response of karst drainage systems to recharge events and to human-induced changes can be better estimated. Key elements of this scheme, which require better understanding are the subcutaneous zone store with its included percolation system, and the saturated zone store with the interchange between conduit (cave) storage and the fissure storage. The latter will require considerable input from empirical data and 3D modelling, but significant progress in this area has been made (Gabrovsek et al. 2004).

3.6.2 The Role of Extreme Events

The structure of a karst drainage system is dynamic in that certain linkages may be activated only once a limit is passed on a rising stage. A good example is provided by the Parker

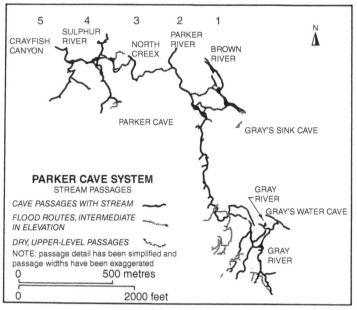

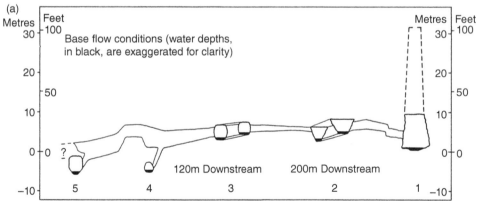

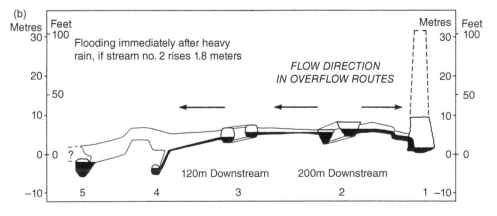

Figure 3.21 Flow paths in Parker Cave, Kentucky, USA, following rainfall events of different magnitudes. Numbered passages in the plan refer to passage cross-sections in A (baseflow) and B–D (varying flood flows). Source: From Quinlan et al. (1991).

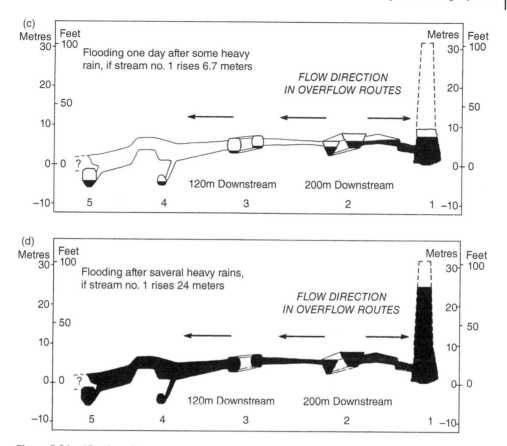

Figure 3.21 (*Continued*)

Cave system of Kentucky (Quinlan et al. 1991), which has five entry points to a 9 km long cave conduit (Figure 3.21). At baseflow conditions, each of the five conduits carries a small flow. If *stream 2* floods, then underground flow occurs in two directions, while if *stream 1* floods, the flow is unidirectional because of relative passage levels. However, a lagged stage rise in *stream 1* will also produce bidirectional flow, and a series of rainstorms will cause complete flooding of all five conduits and multidirectional flow. This complexity of flow, and dependence on antecedent rainfall, is typical of karst and creates unpredictability when pollutants become mobilised.

The nature of underflow-overflow springs in a hierarchy has already been described, and Smart (1988c) has provided an example of the dynamism of a conduit network from the Canadian Rockies. The extent to which fissure storage is filled by a single recharge pulse will determine the nature of the spring hydrograph, and there may be variable lags between the input pulse and the response at the spring. When we consider the response of the karst drainage system to extreme events, such as the one in a 100-year flood event, the prediction of behaviour is far less certain.

Each distributary link in a network will have its own unique characteristics of hydraulics and storage, and their combination may be variable in time and space. The friction within

Figure 3.22   The stream sink of the Tekin River Cave, Oksapmin, Papua New Guinea, shows large quantities of sediment and woody debris transported to the insurgence. When blocked this sink has a lake 10 m deep at the cave entrance.

each link will probably decrease with increasing discharge. In this context, the role of hydraulic semiconductors, described previously, becomes crucial to our understanding. Intrinsic changes in the nature of the conduit, due to the mobilisation of sediment banks and destabilisation of breakdown, may occur at long intervals during exceptional floods. This may explain the sudden backing up of water in caves, so that vadose conduits become temporary phreatic for periods of days to weeks, and active enlargement by solution may occur (Figure 3.22). This has been observed in the large caves of New Guinea by Checkley (1993) for the large sink of the Baliem River in Irian Jaya, and by Francis et al. (1980) for the passages of Atea Kananda in the Muller Range, Papua New Guinea. Palaeokarstic passages may become active and change the boundary of the karst drainage system (Kiernan 1993).

The radical changes in water level in the poljes of the Nahanni Karst, northern Canada, have been documented by Brook and Ford (1980). Although the spring thaw may be the annual hydrologic event of greatest magnitude in most of the arctic, in the Nahanni intense summer storms have profound effects. After 203 mm of rain over a week, the poljes filled as marginal sinks were incapable of coping with the inflow. At Raven Lake alone, the water level rose by 49 m. Alluviation of the poljes is held to account for the periodic flooding, and although the karst system can cope with annual snowmelt runoff, low-frequency spring and summer storms may have a greater role in karst development in the region than was hitherto realised.

3.7 Karst Hydrology of the Mammoth Cave Plateau, Kentucky

The Mammoth Cave System is located about 160 km south of Louisville, Kentucky, and about 56 km northeast of Bowling Green, Kentucky, USA. It is the longest cave in the world by a factor of three, with about 667 km of surveyed passage. Most of the cave lies within

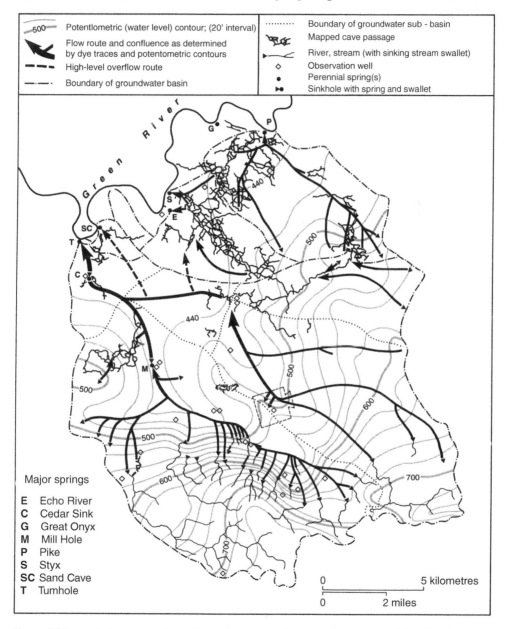

Figure 3.23 Hydrology, potentiometric surface, and underground flow routes defined by dye tracing in the Turnhole Spring karst basin of Mammoth Cave, Kentucky, USA. Source: From Quinlan et al. (1991).

Mammoth Cave National Park, a World Heritage Site and also a part of the United Nations program of International Biosphere Reserves. The cave system is formed in the Girkin, Ste. Genevieve, and St Louis limestone formations, which dip gently to the northwest and are wrinkled by a number of small anticlines. The limestone is in places overlain by insoluble sandstones and shales. To the south of Mammoth Cave National Park, these rocks have been largely eroded away to form the Pennyroyal Plateau, and limestone occurs just under the soil over a very large area punctuated by many sinkholes. To the north, these resistant rocks form the dissected ridges and small cuestas of the Chester Upland, with limestone in the deep valleys. Groundwater in the limestone is derived in part from drainage from these high areas, and there are numerous sinkholes along the edge of the sandstone uplands. The Green River lies in a deep trench and is fed by many springs from the limestone as it flows west to join the Ohio River.

The area is thus underlain by cavernous limestones and contains thousands of dolines, about a 100 stream sinks and about 200 springs. Generally, all surface and subterranean water within a groundwater basin flows to the same spring or set of springs. Water or pollutants from a point source can be dispersed through distributaries within the karst drainage system to as many as 53 springs along the Green River in the Bear Wallow groundwater basin. This is an extreme example, most being fewer in number, but the Mammoth Cave area is typical of the highly interconnected nature of karst drainage systems. The area is heavily used for agriculture, for grazing and for tourism, and thus, there is a wide range of potential pollutants for karst groundwater. Fortunately, this karst is quite well-known and has been the subject of many scientific studies (White and White 2013, 2017; Worthington 2009).

Figure 3.24 A passage in Flint Ridge Cave, Kentucky, USA, showing phreatic pendants and sediment banks. This is an upper part of the Mammoth Cave System.

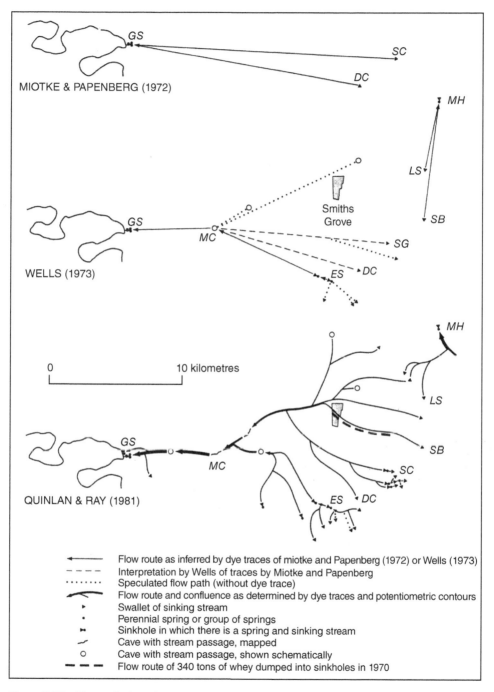

Figure 3.25 The evolution of knowledge about the Graham Springs karst basin, Kentucky USA, based on dye-tracing experiments, Doty Creek (DC); Elk Spring (ES); Graham Springs (GS); Little Sinking Creek (LS); Mill Cave (MC); Mill Hole (MH); Sinking Branch (SB); Sinking Creek (SC). Source: From Quinlan et al. (1991).

Our detailed knowledge of the hydrology of the Mammoth Cave region is the result of intensive study over the last three decades by Jim Quinlan, Ralph Ewers, Joe Ray, and other workers. This work has entailed more than 500 dye-tracing experiments, 2700 water level measurements, and mapping of more than 800 km of cave passages by the Cave Research Foundation, National Parks Service, and other speleological organisations. It is probably the most intensively investigated karst in the world. A detailed map of flow routes and the potentiometric surface (Quinlan and Ray 1981) is but one of the many products of this research. Any serious student of karst hydrology should study this map as an example of how to conduct a detailed investigation.

In Figure 3.23, the detailed hydrology of part of this map is shown for the Turnhole Springs Groundwater Basin. This includes Mammoth Cave and Roppel Cave in the northeast of the map, and Park City urban area in the centre. The flow of water is generally orthogonal to the water-level contours, and there is concentration of flow along troughs in the water-level contours. These troughs coincide with major conduits or caves, such as the Mill Hole or Roppel Cave conduit networks (Figure 3.24).

The evolution of understanding about flow paths in this basin is illustrated in Figure 3.25. Early dye-tracing work by Miotke and Papenberg (1972) indicated that water flowed directly from Sinking Creek (SC) and Doty Creek (DC) to Graham Spring (GS). Later work indicated that these flow paths actually went through Mill Cave (MC) and were joined by other caves to the north and flow from the Elk Spring (ES) complex. In addition,

Figure 3.26 Turnhole Spring, close to the Green River baselevel in the Mammoth Cave System (Kentucky, USA). For details of the flow paths, see Figure 3.23. Source: From Quinlan et al. (1991).

surface water from close to Sinking Creek entered two stream sinks (Sinking Branch [SB], and Little Sinking Creek [LS]) and emerged at the Mill Hole (MH) spring. The lowest diagram shows the complex pattern of groundwater flow mapped by Quinlan and Ray. Many more tributaries to the Graham Spring karst drainage system are evident, and there is a dense flow network that defines a groundwater basin boundary to the east of Smith's Grove. This divide is narrow, and under storm flow conditions, water could be diverted to either Graham Springs or Mill Hole.

Under flood conditions, many of these groundwater basin boundaries may be breached, and flow may occur to other springs. This has important ramifications for the flow of pollutants (Schindel 1984; Schindel et al. 1994), which inadvertently enter the karst drainage system of the Mammoth Cave area (Figure 3.26).

References

Allison, G.B., Stone, W.J., and Hughes, M.W. (1985). Recharge in karst and dune elements of a semi-arid landscape as indicated by natural isotopes and chloride. *Journal of Hydrology* 76: 1–25.

Ashton, K. (1966). The analysis of flow data from karst drainage systems. *Transactions of the Cave Research Group of Great Britain* 7: 161–204.

Atkinson, T.C. (1977). Diffuse flow and conduit flow in limestone terrain in the Mendip Hills, Somerset (Great Britain). *Journal of Hydrology* 35: 93–110.

Atkinson, T.C. (1985). Present and future directions in karst hydrogeology. *Annales de la Société Géologique de Belgique* 108: 293–296.

Atkinson, T.C. and Smart, P.L. (1979). Traceurs artificiels en hydrogéologie. *Bulletin du BRGM* 2: 365–380.

Atkinson, T.C., Smith, D.I., Lavis, J.J., and Wltaker, R.J. (1973). Experiments in tracing underground waters in limestones. *Journal of Hydrology* 19: 323–349.

Bakalowicz, M. and Mangin, A. (1980). L'aquifère karstique: Sa definition, ses characteristiques et son identification. *Mémoires hors série de la Societé géologique de France* 11: 71–79.

Bencala, K.E., Rathbun, R.E., Jackman, A.P. et al. (1983). Rhodamine WT dye losses in a mountain stream environment. *Journal of the American Water Resources Association* 19: 943–950.

Bretz, J.H. (1942). Vadose and phreatic features of limestone caverns. *Journal of Geology* 50: 675–811.

Brook, G.A. and Ford, D.C. (1980). Hydrology of the Nahanni karst, northern Canada, and the importance of extreme summer storms. *Journal of Hydrology* 46: 103–121.

Caramanna, G. and Giordani, M. (2010). Geomorphologic survey and hydrological measures of a karst spring by means of cave diving techniques (Amaseno, Italy). *Underwater Technology* 29: 95–99.

Checkley, D. (1993). Cave of Thunder: The exploration of the Baliem River Cave, Irian Jaya, Indonesia. *International Caver* 6: 11–17.

Christopher, N.S.J., Trudgill, S.T., Pickles, R.W.C.A.M., and Culshaw, S.M. (1981). A hydrological study of the Castleton area, Derbyshire. *Transactions of the British Cave Research Association* 8: 189–206.

Courbon, P., Chabert, C., Bosted, P., and Lindsley, K. (1989). *Atlas of the Great Caves of the World*. St Louis: Cave Books.

Crawford, S.J. (1994). Hydrology and geomorphology of the paparoa karst, North Westland, New Zealand. Unpublished PhD thesis. University of Auckland.

Davis, W.M. (1930). Origin of limestone caverns. *Bulletin of the Geological Society of America* 41: 475–628.

Dreybrodt, W. (1990). The role of dissolution kinetics in the development of karst aquifers in limestone: A model simulation of karst evolution. *Journal of Geology* 98: 639–655.

Even, H., Carmi, I., Magaritz, M., and Gerson, R. (1986). Timing the transport of water through the upper vadose zone in a karstic system above a cave in Israel. *Earth Surface Processes and Landforms* 11: 181–191.

Field, M.S. (2002). *A Lexicon of Cave and Karst Terminology with Special Reference to Environmental Karst Hydrology*. Washington, DC: United States Environmental Protection Agency.

Filipponi, M., Jeannin, P.-Y., and Tacher, L. (2009). Evidence of inception horizons in karst conduit networks. *Geomorphology* 106: 86–99.

Ford, D.C. and Ewers, R.O. (1978). The development of limestone cave systems in the dimensions of length and depth. *Canadian Journal of Earth Sciences* 15: 1783–1798.

Ford, D.C. and Williams, P.W. (2007). *Karst Hydrogeology and Geomorphology*. Chichester, England: Wiley.

Francis, G., James, J.M., Gillieson, D.S., and Montgomery, N.R. (1980). *Underground Geomorphology of the Muller Plateau*. Sydney: Speleological Research Council, University of Sydney.

Freeze, R.A. and Cherry, J.A. (1979). *Groundwater*. Englewood Cliffs, NJ: Prentice-Hall.

Gabrovsek, F., Romanov, D., and Dreybrodt, W. (2004). Early karstification in a dual-fracture aquifer: The role of exchange flow between prominent fractures and a dense net of fissures. *Journal of Hydrology* 299: 45–66.

Ginès, A. and Ginès, J. (2007). Eogenetic karst, glacioeustatic cave pools and anchialine environments on Mallorca Island: A discussion of coastal speleogenesis. *International Journal of Speleology* 36: 57–67.

Goldscheider, N. and Drew, D. (eds.) (2007). *Methods in Karst Hydrogeology*. London: Taylor & Francis.

Goldscheider, N., Smart, C., Pronk, M., and Meiman, J. (2008). Tracer tests in karst hydrogeology and speleology. *International Journal of Speleology* 37: 27–40.

Gunn, J. (2010). Is the term 'karst aquifer' misleading? In: *Advances in Research in Karst Media* (eds. B. Andreo, F. Carrasco, J.J. Durán and J.W. Lamoreaux). Berlin, Heidelberg: Springer.

Hagen, G. (1839). Ueber die Bewegung des Wassers in engen cylindrischen Röhren. *Annalen der Physik* 46: 423–442.

Hotzl, H. and Werner, A. (1992). *Tracer Hydrology*. Rotterdam: A.A. Balkema.

International Atomic Energy Agency (1983). *Guidebook on Nuclear Techniques in Hydrology*, Technical Reports Series. International Atomic Energy Agency.

Jakucs, L. (1959). Neue methoden der höhlenforschung in Ungarn und ihre ergebnisse. *Die Höbie* 10: 88–98.

Julian, M. and Nicod, J. (1989). Les karsts des Alpes du Sud et de Provence. *Zeitschrift für Geomorphologie, Supplement Issues* 75: 1–48.

Kiernan, K. (1993). The Exit Cave quarry: Tracing waterflows and resource policy evolution. *Helictite* 31: 27–42.

Lauritzen, S.E., Abbott, J., Arnesen, R. et al. (1985). Morphology and hydraulics of an active phreatic conduit. *Cave Science* 12: 139–146.

Leontiadis, I.L., Payne, B.R., and Christodoulou, T. (1988). Isotope hydrology of the Aghios Nikolaos area of Crete, Greece. *Journal of Hydrology* 98: 121–132.

Lowe, D.J. and Gunn, J. (1997). Carbonate speleogenesis: An inception horizon hypothesis. *Acta Carsologica* 26: 457–488.

Maire, R. (1990). La haute montaigne calcaire. *Karstologia – Mémoirés* 3: 731.

Mangin, A. (1975). *Insights into hydrodynamic behaviour of karst aquifers.* Dijon, France: Université de Dijon.

Milanovic, P.T. (1981). *Karst Hydrogeology.* Littleton, Colorado: Water Resources Publications.

Miotke, F.D. and Papenberg, H. (1972). Geomorphology and hydrology of the Sinkhole Plain and Glasgow Upland, central Kentucky karst: Preliminary report. *Caves and Karst* 14: 25–32.

Palmer, A.N. (1987). Cave levels and their interpretation. *The National Speleological Society Bulletin* 49: 50–66.

Perrin, J., Jeannin, P.-Y., and Zwahlen, F. (2003). Epikarst storage in a karst aquifer: A conceptual model based on isotopic data, Milandre test site, Switzerland. *Journal of Hydrology* 279: 106–124.

Poiseuille, J.L.M. (1847). Recherches expérimentales sur le mouvement des liquides de nature différente dans les tubes de très petits diamètres. *Acad. Sci. Paris Mem. sav. étrang* 9: 433–545.

Quinlan, J.F. and Ewers, R.O. (1989). Subsurface drainage in the Mammoth Cave area. In: *Karst Hydrology: Concepts from the Mammoth Cave Area* (eds. W.E. White and E.L. White). New York: Van Nostrand Reinhold.

Quinlan, J.F. and Ray, J.A. (1981). *Groundwater basins in the Mammoth Cave region, Kentucky.* Friends of Karst Occasional Publication.

Quinlan, J.F., Ewers, R.O., and Aley, T.J., and National Water Well Association & Association of Ground Water Scientists and Engineers (1987). *Practical Karst Hydrogeology with Emphasis on Groundwater Monitoring*, vol. 6, E1–E26. National Well Water Association.

Quinlan, J.F., Ewers, R.O., and Aley, T. (1991). *Practical Karst Hydrology, with Emphasis on Ground-Water Monitoring. Field Excursion: Hydrogeology of the Mammoth Cave Region, with Emphasis on Problems of Ground-Water Contamination.* Dublin, OH: National Ground Water Association.

Rhoades, R. and Sinacori, M.N. (1941). Pattern of ground-water flow and solution. *Journal of Geology* 49: 785–794.

Schindel, G.M. (1984). Enteric contamination of an urban karstified carbonate aquifer: the double springs drainage basin, bowling green, Kentucky. Master's thesis. Western Kentucky University.

Schindel, G.M., Quinlan, J.F., and Ray, J.A. (1994). *Determination of the recharge area for the Rio Springs groundwater basin, near Munfordville, Kentucky: An application of dye tracing and potentiometric mapping for determination of springhead and wellhead protection areas in carbonate aquifers and karst terranes. Project completion report.* Atlanta, Georgia: U.S. Environmental Protection Agency, Groundwater Branch.

Smart, P.L. (1984). A review of the toxicity of twelve fluorescent dyes used for water tracing. *The National Speleological Society Bulletin* 46: 21–33.

Smart, C.C. (1988a). Artificial tracer techniques for the determination of the structure of conduit aquifers. *Groundwater* 26: 445–453.

Smart, C.C. (1988b). Exceedance probability distributions of steady conduit flow in karst aquifers. *Hydrological Processes* 2: 31–41.

Smart, C.C. (1988c). Quantitative tracing of the Maligne karst system, Alberta, Canada. *Journal of Hydrology* 98: 185–204.

Smart, P.L. and Hobbs, S.L. (1986). Characteristics of carbonate aquifers: A conceptual basis. In: *Proceedings of Environmental Problems in Karst Terrains and Their Solutions*. Bowling Green, Kentucky: National Well Water Association.

Smart, P.L. and Laidlaw, I.M.S. (1977). An evaluation of some fluorescent dyes for water tracing. *Water Resources Research* 13: 15–23.

Smart, P.L. and Smith, D.I. (1976). Water tracing in tropical regions, the use of fluorometric techniques in Jamaica. *Journal of Hydrology* 30: 179–195.

Smart, P.L., Atkinson, T.C., Laidlaw, I.M.S. et al. (1986). Comparison of the results of quantitative and non-quantitative tracer tests for determination of karst conduit networks: An example from the Traligill Basin, Scotland. *Earth Surface Processes and Landforms* 11: 249–261.

Stanton, W.I. and Smart, P.L. (1981). Repeated dye traces of underground streams in the Mendip Hills, Somerset. *Proceedings of the University of Bristol Speleological Society* 16: 47–58.

Swinnerton, A.C. (1932). Origin of limestone caverns. *GSA Bulletin* 43: 663–694.

Thrailkill, J.V. (1968). Chemical and hydrologic factors in the excavation of limestone caves. *Geological Society of America Bulletin* 79: 19–46.

Trudgill, S.T. (1987). Soil water dye tracing, with special reference to the use of rhodamine WT, lissamine FF and amino G acid. *Hydrological Processes* 1: 149–170.

Vennard, J.K. and Street, R.L. (1976). *Elementary Fluid Mechanics*. New York: Wiley.

Webb, J.A. and James, J.M. (2006). Karst evolution of the Nullarbor Plain, Australia. In: *Perspectives on Karst Geomorphology, Hydrology, and Geochemistry – A Tribute Volume to Derek C. Ford and William B. White* (eds. R.S. Harmon and C.M. Wicks). Geological Society of America.

White, W.B. (1988). *Geomorphology and Hydrology of Karst Terrains*. New York: Oxford University Press.

White, W.B. (2002). Karst hydrology: Recent developments and open questions. *Engineering Geology* 65: 85–105.

White, W.B. and White, E.L. (2013). *Karst Hydrology: Concepts from the Mammoth Cave Area*. Berlin: Springer Science & Business Media.

White, W.B. and White, E.L. (2017). Hydrology and hydrogeology of mammoth cave. In: *Mammoth Cave* (eds. H.H. Hobbs III, R.A. Olson, E.G. Winkler and D.C. Culver). Cham: Springer International Publishing.

Williams, P.W. (1983). The role of the subcutaneous zone in karst hydrology. *Journal of Hydrology* 61: 45–67.

Williams, P.W. (1984). Karst hydrology of the Takaka valley and the source of New Zealand's largest spring. In: *Hydrogeology of Karstic Terrain: Case Histories* (eds. A. Burger and L. Dubertret). Hanover: Heise.

Worthington, S.R.H. (1991). Karst hydrogeology of the Canadian rocky mountains. Unpublished PhD. McMaster University.

Worthington, S.R.H. (2001). Depth of conduit flow in unconfined carbonate aquifers. *Geology* 29: 335–338.

Worthington, S.R. (2009). Diagnostic hydrogeologic characteristics of a karst aquifer (Kentucky, USA). *Hydrogeology Journal* 17: 1665–1678.

4

Processes of Rock Dissolution

4.1 Introduction

Karst landforms will form in a variety of rocks soluble in natural waters. The most soluble rocks are gypsum and halite, which will readily dissolve in drainage waters. The carbonate rocks (limestones, dolomites, and marbles) will dissolve in acidified waters. The acidification may be due to dissolved carbon dioxide, sulphur oxides, or organic acids derived from vegetation. Sandstones will dissolve slowly in natural waters due to a process termed arsenisation, in which the cement holding the quartz grains together is dissolved. The weakened rock is then readily eroded by flowing water yielding large amounts of sand. A similar process of granular disintegration also occurs in granites and related igneous rocks. All of these processes are enhanced in the warmer waters typical of temperate or tropical environments, past or present. The karst solution process may also be enhanced by the presence of high secondary porosity (fissures, fractures, and bedding planes) or high permeability (the capacity to transmit water). We first consider the various karst rocks, their composition and formation processes.

4.2 Karst Rocks

4.2.1 Limestone

Approximately 20% of the Earth's ice-free land surface is made up of carbonate rocks, with the major carbonate regions being in Europe, eastern North America, and East and South-East Asia (Figure 4.1), and the remnants of Gondwanaland being relatively depauperate in carbonates because of their extreme age. Because of overlying deposits, unsuitable climate, or low relief, only 10–15% of these areas display karst landforms (Ford and Williams 2007, p. 2). Despite this modest extent, about 25% of the world's population is dependent on karst groundwater. This dependence is increasing in the countries of Asia, where population growth is still rapid.

Limestones are generally regarded as being composed of more than 50% calcite or calcium carbonate, with pure cave-forming limestones generally being more than 90% calcite (Figure 4.2). Dolomites, the calcium magnesium carbonate rocks, follow a similar classification. Caves are less numerous in dolomite on account of its lower solubility under

Caves: Processes, Development, and Management, Second Edition. David Shaw Gillieson.
© 2021 John Wiley & Sons Ltd. Published 2021 by John Wiley & Sons Ltd.

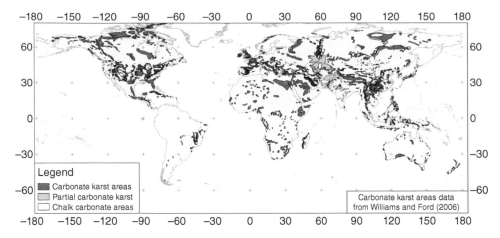

Figure 4.1 Global distribution of carbonate rocks. Carbonate karst category indicates that carbonates are relatively continuous; partial carbonate indicates areas in which carbonates are abundant but not continuous. Source: Modified from Williams and Ford (2006).

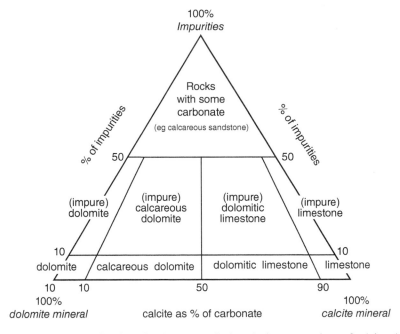

Figure 4.2 Classification of carbonate rocks by relative proportions of calcite, dolomite, and impurities. Source: After Leighton and Pendexter (1962).

Figure 4.3 Thinly bedded Tertiary limestone with stylolites at Punakaiki, New Zealand.

ambient conditions. Below 50% calcite or dolomite, it is unlikely that caves will form at all, though isolated examples form along tectonic fissures in such rocks. Limestones range in age from Precambrian to Holocene, with modes in the Ordovician, Carboniferous, Cretaceous, and mid-Tertiary (Figure 4.3). Limestones can form in both marine and freshwater environments, but most have formed in shallow warm to tropical seas. Limestones can also be divided into facies depending on the precise environment of their formation: for coral reefs, these are shown in Figure 4.7.

One of the most widely used classifications of limestone comes from the work of Folk (2014). This classification concentrates on the nature of the transported grains (allochems) and the calcite cements formed in situ (orthochems). Other components may include terrigenous material, such as sands and clays, and heavy minerals, such as zircon and rutile. Allochems may be subdivided as follows:

- Intraclasts: reworked fragments of older carbonates
- Oolites: rounded forms with a concentric structure
- Pellets: usually of faecal material
- Fossils: corals and shells, sometimes vertebrates

Orthochem cements may be either micrite (Figure 4.4), a fine opaque ooze of microcrystalline carbonate, or sparite, semi-translucent rhombic crystals of calcite. The detail is given in Table 4.1 and Figure 4.5. Depending on the proportions of allochems, sparry cement, or micritic cement, the limestone may fall into zones on the ternary diagram. Its only disadvantage is that it does not take account of diagenesis (chemical alteration after deposition) and is, therefore, difficult to apply in some instances. A massive concretion of fossil material, such as coral colonies or algal mats in growth position or as reef talus, may be called a biolithite (Figure 4.6). Another simple classification is based on the median grain size (Table 4.2). This classification is useful for the younger dune limestones or calcarenites,

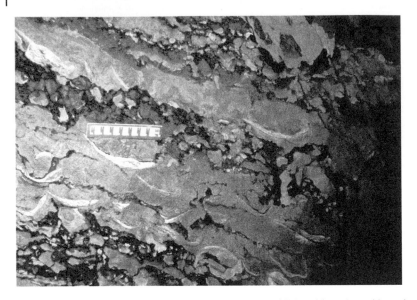

Figure 4.4 Massive fossiliferous micritic limestone with brachiopods and breccias of Silurian age at Yarrangobilly, New South Wales, Australia. The scale bar is 10 cm long.

Table 4.1 Classification of limestones.

| | *Orthochems* | |
Allochems	Micrite	Sparite
None	Micrite	Sparite
Intraclasts	Intramicrite	Intrasparite
Oolites	Oomicrite	Oosparite
Pellets	Pelmicrite	Pelsparite
Fossils	Biomicrite	Biosparite

Source: Based on Folk (2014).

which contain cemented allogenic clasts of sand or silt size, including fossil shell or bone material.

The genetic relationships between different limestone facies may be apparent in gorge sections and sometimes in caves. The classic example is the limestone of the Napier Range in the Kimberley of Western Australia, a Devonian reef complex formed along a shoreline shaped in Precambrian rocks. This reef complex included fine-textured inter-reef facies deposited in deep water, the tumbled blocks of the fore-reef slope or reef talus, the in situ bioherms of the growing reef rim and the muddier flat-bedded back-reef facies of the lagoon (Figure 4.7). Following uplift, Tertiary planation, and incision, a cross-section through this reef complex was exposed in the gorges of the Lennard and Fitzroy rivers. The karst morphology is affected by the facies type, with rugged "giant grikeland" terrain formed on the fore-reef and reef-rim facies contrasted with subdued relief on the muddier back-reef facies.

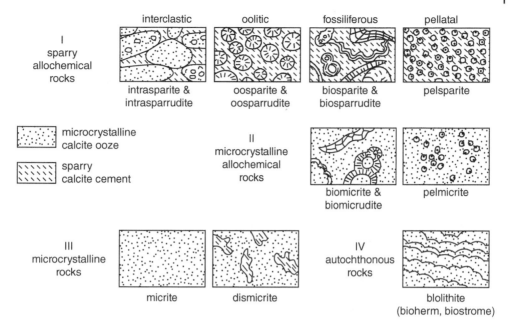

Figure 4.5   Major types of limestone. Source: According to Folk (1959).

Fissure caves and network mazes are numerous in the fore-reef facies, while caves in the lagoon facies are scarce and associated with fluvial incutting. Reef talus and lagoon facies limestones are well exposed in the walls of Rope Ladder Cave, near Townsville, Queensland, Australia; the jumbled blocks of fore-reef facies have been bevelled by solution under phreatic conditions (Figure 4.8).

Limestones are very susceptible to diagenesis because of their solubility at ambient temperatures and pressures. The circulation of acidified water through limestones creates endless opportunities for solution and recrystallisation in both small and large voids. Virtually all ancient limestones have experienced at least one such phase of diagenesis, which is both rapid and extensive. Following exposure of the limestone by uplift or sea-level change, circulating freshwaters dissolve the carbonate, especially the aragonite. This is succeeded by sparry calcite cements, often slowly by isomorphous replacement so that the "ghosts" of aragonite fossils are preserved. In contrast, rapid replacement removes the fossil traces, substituting sparry calcite throughout. Voids along bedding planes may be infilled by cements and sediments, including vein mineral deposits, such as those of the Derbyshire limestones. Under strong conditions of evaporation, calcrete nodules and crusts, gypsum, and pisolites may be deposited in the upper strata of the limestone.

Shallow water limestones may often contain interbeds of soluble minerals, such as halite or gypsum, which upon diagenesis are removed, leading to the partial collapse of the limestone. Minor fissures may also open in the rock because of uplift. These voids may be subsequently infilled by coarse material, including limestone fragments, creating breccias. The cavities surrounding the clasts are often infilled by sparry calcite or other minerals, including phosphates and limonites. Coarse breccias can also form as reef talus (fore-reef)

Figure 4.6 Gently dipping Miocene limestone in the Muller Range, Papua New Guinea, has pure limestone units 30–50 m thick separated by muddy facies that act as aquicludes.

Table 4.2 Classification based on the median grain size.

	Median grain size	
Calcirudite	>2 mm	e.g. coral or shelf fragments
Calcarenite	0.02–2 mm	e.g. sand-sized material
Calcilutite	<0.02 mm	e.g. fine-grained silts and clays

limestones with micritic cavity infillings. High temperature and pressure may also meta-morphose limestones to marble, a mosaic of interlocking large clear calcite grains. However, some limestones may also recrystallise by solution without high temperature and pressure to give coarse calcirudites in which the grains do not interlock.

Nodules or thin sheets of chert or flint will often form in lime-rich sediments by the dis-solution of quartz or silica from sponges, diatoms, and radiolarians. Nodules can be up to 1 m in diameter, especially in the chalky Wilson's Bluff limestone of the Nullarbor Plain (Australia) (Figure 4.9). Chert bands are usually less than 50 cm thick and, being relatively brittle, are fractured, allowing some water movement. Stylolites are pressure solution seams

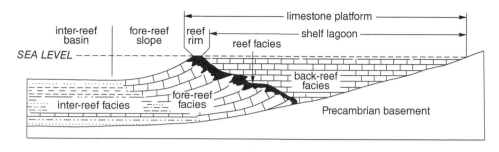

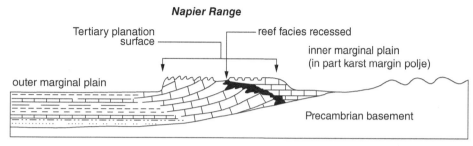

Figure 4.7 Facies types for carbonate reefs, with an example from the Napier Range, west Kimberley, Australia. Source: After Playford and Lowry (1966).

Figure 4.8 Reef talus limestone of Devonian age exposed in the walls of a formerly phreatic cave at Fanning River, northern Queensland, Australia.

Figure 4.9 Sea cliffs at the edge of the Nullarbor Plain, exposing Nullarbor Limestone overlying pale Wilson's Bluff Limestone, Australia.

formed under conditions of deep burial of the rock. They are often darker than the host limestone because of the concentration of insoluble minerals and organic matter along them (Figure 4.9). Stylolites are usually a few millimetres to a few centimetres thick, and individual forms are generally only a few tens of centimetres long.

4.2.2 Dolomite

At a global scale, dolomite is nearly as abundant as limestone. The formation of dolomite is a complex subject that remains a subject of intense research. Dolomite forms by the replacement of existing calcite and aragonite. There are four principal modes of formation: reflux, mixing, burial, and hydrothermal. Seawater concentrated by evaporation in shallow lagoons may reflux through limy sediments, exchanging ions with them. This process is enhanced where there is a high ratio of magnesium to calcium and abundant carbonate ions in the crystal lattice of calcite, breaking the bond of the strongly hydrated magnesium ions. This reflux process may occur very early in diagenesis. Under mixing conditions of fresh- and seawater, where calcite is soluble, but dolomite is not, the necessary magnesium to calcium ratio is met, and dolomite precipitates. Hanshaw and Back (1979) propose that this is occurring today in the mixing zone deep in the karst groundwater of Florida, USA. Deeply buried carbonates may be dolomitised by the expulsion under pressure of fluids, but the extent of this process is disputed. Hydrothermal dolomites tend to be located deep along fracture zones where warm fluids are circulating. Some dolomites are clearly of this type as they are locally present along fractures or pipes where hot waters have ascended.

The most common form of dolomite is a medium crystalline rock (grains 20–100 μm in diameter), while finer grain sizes or dolomicrites are found in association with the early replacement of aragonite. Coarsely crystalline dolomite is often found in association with silver-lead deposits where repeated cycles of dolomitisation have occurred.

Karst development on dolomites is generally restricted, though significant caves can be found in that host rock. In South Africa, there are extensive Proterozoic dolomite karsts in the Transvaal, with important hominid sites in the Sterkfontein, Swartkrans, and Kromdraai areas. The caves include Apocalypse Pothole, which is 12 km long, as well as tourist caves at Cango near Oudtshoorn. In Botswana, the Gcwihaba area (a tentative World Heritage site) contains cavernous dolomite hills rising above an arid sand plain. The caves contain windblown sand and are rich in fossils and speleothems, indicating humid conditions in the past.

Canada has approximately 600 000 km^2 of dolomite (Ford, 2004), but the legacy of multiple glaciations has meant that karst features are not well exposed in many areas due to glacial outwash and till deposits overlying the karst. In Ontario, dolomites are associated with Niagara Falls and the Bruce Peninsula on Lake Huron. There are karst pavements, coastal karren, and short caves. Doline karst on dolomite is found in Manitoba, between Lakes Winnipeg and Winnipegosis, while in the American states of Wisconsin and Minnesota there are many caves formed in the Prairie du Chien dolomite.

In Australia, the Barkly Tableland of Queensland and the Northern Territory contains extensive horizontal caves that intersect the regional water table (Figure 4.10). In Tasmania, the well-decorated Hastings Caves are formed in Precambrian dolomite, as are caves and karst at Mount Anne and Weld River in the Tasmanian Wilderness World Heritage Area (WHA). The Tasmanian dolomites exhibit interactions between karst landforms and glaciation as well as diverse endemic cave fauna. Humic acids derived from blanket peats may be significant in Tasmanian karst development and may account for the significant karst development in the less-soluble Precambrian dolomite carbonate rocks.

4.2.3 Evaporite Rocks – Gypsum and Halite

Evaporites are quite common, covering up to 25% of the continental surfaces. Gypsum karsts occur in China, Ukraine, and the USA, while halite karst is less common and confined to desert areas. The best-known halite karst is Mount Sedom in Israel (Frumkin 1994). Both gypsum and halite form by partial or complete evaporation of seawater or brines, with gypsum being the mineral first precipitated in most cases. Gypsum precipitates at concentrations around 3 times that of seawater, and halite at 11 times strength. These precipitation processes can be quite rapid: during the late Miocene, the tectonic collision between North Africa and Spain closed off the Mediterranean Sea, and the subsequent drying led to the precipitation of 1000–2000 m of gypsum and salt in the basin over less than 500 000 years (Ford and Williams 2007, p. 25).

Figure 4.10 Precambrian dolomite outcrop at the entrance doline of Nowranie Cave, Camooweal, Queensland, Australia.

Gypsum deposits tend to occur as coarsely crystalline massive beds up to 200 m thick and may be translucent (selenite), opaque white, or colored (alabaster: brown, pink, gray, yellow). Less commonly, gypsum occurs as interbeds or clusters of crystals in other rocks, such as dolomites, shales, or clay stones. The dehydration of gypsum to anhydrite usually occurs at depth, and most massive gypsum deposits have been subject to repeated cycles of dehydration and hydration. This creates complexity in the structure of the gypsum beds and makes the determination of the original depositional environment difficult.

Halite or salt can be present as thin interbeds in carbonates, sulphates, or shales, or can form massive deposits with economic significance. Buried salt deposits tend to have their cavities sealed by overburden pressure, and thus, have low permeability. Where salt outcrops at the surface, local vadose flow can form caves (Frumkin 1994; Frumkin et al. 1991).

4.2.4 Sandstone

Sandstones are essentially granular rocks comprised of silica grains cemented by calcite, clays, iron oxides, or silica. The mineral composition of the grains in the sandstone can provide much information about the source areas–granitic, metamorphic, volcanic, or sedimentary. Immature sandstones contain readily weathered minerals, such as feldspars and ferromagnesian minerals, and imply deposition close to their source. In contrast, sandstones dominated by quartz imply maturity and possibly long-distance transport. Four common types of sandstone are:

- Quartz sandstone (dominated by well-sorted and rounded quartz with little or no feldspar, mica, or fine clastic material)
- Arkose (dominated by angular quartz grains, but with more than 25% feldspar, bonded by calcareous cements, clays, or iron oxides)
- Graywacke (poorly sorted immature sandstones in a finer matrix of dark clays, silt, micas, and chlorite)
- Subgraywacke (sandstones in which better-sorted and rounded quartz grains sit in a matrix of not more than 15% fine cement in which feldspar is scarce)

Large caves may be found in quartz sandstones and arkoses but rarely in graywackes. Sandstones may be very fine-grained or quite coarse, containing a wide range of grain sizes depending on the energy and persistence of transport. The sedimentary structures, usually well-preserved, can give quite precise information of the environment of deposition–alluvial bars or fans, deltaic plains, turbidity currents, and shallow or deep water. The primary porosity of sandstones is a function of the mean grain size of the quartz grains and the nature of the cement. Sandstones are frequently highly fractured, and the joint networks in them often resemble those of limestone massifs. Silica is just soluble in meteoric waters and will dissolve, slowly; however, often the cement is more soluble and will rapidly dissolve, allowing granular disintegration termed *arenisation*. Once subject to metamorphism, they may be wholly or partially converted to finer-grained quartzites, in which the grains interlock. These hard rocks are less soluble, but very significant caves can form in them, primarily in Venezuela (Sauro et al. 2017) and Arnhem Land, Australia (Wray and Sauro 2017; Wray 2013).

4.2.5 Granite

Granite is an intrusive igneous rock that is granular in texture. By definition, it should contain between 20% and 60% quartz and at least 35% alkali feldspar (sodium and potassium forms), although the term *granitic* is often used to refer to a wider range of coarse-grained igneous rocks containing both quartz and feldspar. The remaining minerals are most often biotite and muscovite micas, and amphibole.

Chemical weathering of granite occurs when dilute carbonic acid and other acids present in rain and soil waters alter the feldspar in a process called hydrolysis. This causes potassium feldspar to form kaolinite clay, with potassium ions, bicarbonate, and silica in solution as by-products. An end-product of granite weathering is grus, which is often made up of coarse-grained fragments of disintegrated granite.

$$2KAlSi_3O_8 + 2H_2CO_3 + 9H_2O \rightarrow Al_2Si_2O_5(OH)_4 + 4H_4SiO_4 + 2K^+ + 2H_2O_3^- \quad (4.1)$$

The clay mineral products are then easily removed in suspension by running water. The quartz and micas remain as minerals more resistant to weathering and form a fine sand, which is easily transported by water or wind.

Physical weathering of granite involves the formation of exfoliation joints parallel to the rock surface. These are due to pressure release in the rock, the result of valley-side erosion,

Table 4.3 Simplified processes of solution of carbonates.

Process equation	Kinetics	Description
$CO_2 \leftrightarrow CO_2$ air dissol	Slow	Diffusion of CO_2 into water (4.2)
$CO_2 + H_2O \leftrightarrow H_2CO_3$ dissol	Slow	Hydration of dissolved carbon dioxide to form carbonic acid (4.3)
$H_2CO_3 \leftrightarrow H^+ + HCO_3^-$	Fast	Dissociation of carbonic acid into hydrogen and hydrogen carbonate ions (4.4)
$CaCO_3 \leftrightarrow Ca^{2+} + CO_3^{2-}$	Slow	Dissociation of calcite crystal lattice to ions (4.5)
$H^+ + CO_3^{2-} \leftrightarrow HCO_3$	Fast	Association of carbonate ions with hydrogen ions to form a hydrogen carbonate (4.6)

removal of overlying rock and soil, or expansion of water in fissures and crevices either as ice in winter or as steam after summer thunderstorms. The entry of water into the exfoliation joints facilitates chemical weathering.

4.3 Processes of Dissolution of Karst Rocks

4.3.1 The Solution of Limestone in Meteoric Waters

The slow dissolution of limestone is a superficially simple process in which two minerals, calcite and dolomite, are dissociated in acidified water. Although organic and mineral acids may be very important in certain circumstances, under most contemporary conditions, the dissolution of limestone is dominated by carbonic acid formed from dissolved carbon dioxide. The sources of this carbon dioxide will be considered later. There is a sequence of reactions, summarised in Table 4.3. The last Eq. (4.6), disturbs the equilibrium of (4.5) by the removal of CO_3^{2-}, so that more carbonate must dissociate to restore the balance. In addition, the association of $H^+ + CO_3^{2-}$ disturbs the equilibrium in (4.4), promoting further dissociation. This, in turn, disturbs the equilibrium in (4.3) and ultimately (4.2), causing more carbon dioxide to dissolve in water.

These processes continue until the forward and reverse reaction rates are equal, at which point the system is in equilibrium, and the solution is saturated with respect to calcite. Any acid (be it carbonic, organic, or inorganic) will add hydrogen ions to the system and will displace (4.4) and (4.6) in a forward direction, reducing the concentration of CO_3^{2-}, and thus permitting more dissolution of calcite.

If all these equations are combined, we get the commonly quoted dissolution equation for calcite:

$$CaCO_3 + H_2CO_3 \leftrightarrow Ca^{2+} + 2HCO_3^- \tag{4.7}$$

For dolomites, the presence of magnesium ions (Mg^{2+}) in Eq. 4.5 complicates the dissolution process, which is generally written as:

$$CaMg(CO_3)_2 + 2H_2CO_3 \leftrightarrow Ca^{2+} + Mg^{2+} + 4HCO_3^- \tag{4.8}$$

There are no thresholds in these reactions–they will occur in both static and moving water–in contrast to the mechanical erosion by water of other rocks, where there is a threshold for the transport of weathered material. But other factors, such as temperature and gas concentration, may affect the limestone solution.

The solubility of carbon dioxide gas in water decreases as temperature increases, in line with Henry's law, at a rate of about 1.3% per degrees Celsius. Although CO_2 is more soluble in cold water, there is much less soil CO_2 available in arctic regions, and the colder water makes the reaction proceed more slowly. The work of Ford and Drake (1982) shows that the gas concentration of CO_2 and the amount of water moving past the rock-water interface are more important than the absolute concentration of dissolved CO_2. Thus, limestone solution rates would tend to be greater in a tropical climate where there is a larger soil CO_2, concentration on account of bacterial activity, and the rainfall is much higher.

We can consider limestone solution to occur under two contrasting situations: an *open system*, where carbon dioxide gas, percolation water, and calcite-rich rock are continuously in contact, and a *closed system*, where carbon dioxide gas and water equilibrate, but the supply of gas is cut off before the water contacts the rock. From Eqs. (4.2) and (4.3) there is no replacement of CO_2 under closed conditions, and thus the total amount of calcite that can be dissociated is much less. This is shown in Figure 4.11. Note also the effect of temperature on the equilibrium solubility of calcite in an open system.

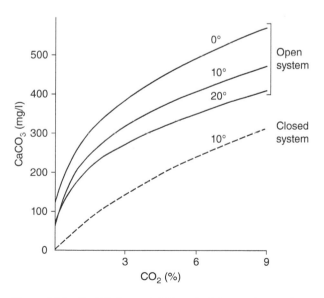

Figure 4.11 Equilibrium solubility of calcium carbonate in contact with air containing carbon dioxide for open and closed systems at varying temperatures. Source: After Picknett et al. (1976).

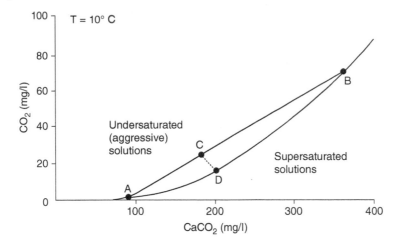

Figure 4.12 The "mixing corrosion" principle: solubility of calcite with respect to carbon dioxide in solution. The curve ADB shows the solubility of calcium carbonate with respect to dissolved carbon dioxide. Mixing of two saturated solutions A and B produces a solution (C), which is undersaturated. Solution C will then evolve to saturation along the line CD. Source: Based on Gunn (1986).

Thus, any body of percolation water contacting limestone will gradually reach saturation, beyond which no more dissociation of the rock will occur unless conditions change. However, one way in which this may occur is if two bodies of saturated water mix – as they may do at tributary junctions in a cave or below the water table in the phreatic zone. The process, first described by Bögli (1964), is termed mixing corrosion. If each body of water is saturated with respect to calcite at different partial pressures of CO_2 (PCO_2), then by mixing they will produce a new solution which is undersaturated (Figure 4.12). In extreme cases, an extra 20% more calcite could be dissolved, but the norm is more like 1–2%. When vadose seepage water at high PCO_2 meets a vadose stream at lower PCO_2, then a substantial increase in limestone solution is possible and may be expressed as dilation of the cave passage. At great depth, percolation water at high PCO_2 may encounter saturated water moving slowly. This form of mixing corrosion may be important for conduit enlargement in the early stages of cave development (Dreybrodt 1980).

4.3.2 Soil and Vegetation in the Limestone Solution Process

Most textbook discussions of the limestone solution process tend to concentrate on chemical kinetics and hydrology, especially those parts of the process occurring within active cave conduits. A notable exception is Trudgill's (1985) text on limestone geomorphology. Yet, the veneer of soil on most limestones has a central role in karst processes, through its control in water infiltration and storage, by acting as a CO_2 generator, and through the role of soil-buffering capacity in the solution process. The roles of soil and vegetation can be outlined using Figure 4.13. Healthy perennial vegetation acts to intercept rainfall and its dissolved gases, principally carbon dioxide (CO_2). Atmospheric carbon dioxide is currently 401 ppm (0.041%) of the total volume and is increasing at approximately 2 ppm per annum. To this can be added other aerosols, such as sulphuric acid from industrial

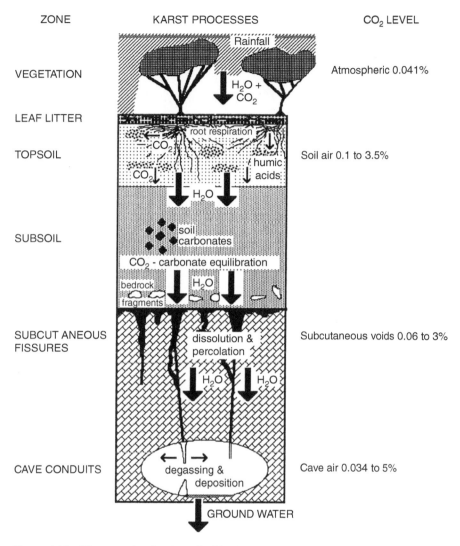

Figure 4.13 The cascade of carbon dioxide through the vegetation, soil, and subcutaneous zones of karst. Not all zones may be present in any single karst, and only the range of recorded carbon dioxide concentrations in temperate ecosystems is given.

sources, and often oxides of nitrogen formed in thunderstorms. The patchy cover of vegetation intercepts rain and protects the soil surface from rain and wind erosion. At the ground surface, the infiltration of rainwater is aided by leaf litter and hollows, which allow most water to percolate into the soil; the excess runs off to surface channels. The complex of decaying vegetation, soil fauna (including bacteria), and fungi in leaf litter provides a source of carbon dioxide and additional organic compounds, such as humic and fulvic acids. In the root zone, the respiration of up to 25% of the carbon dioxide taken up by plants occurs. This released gas may dissolve in percolating rainwater. Bacteria in topsoil also release copious quantities of the gas through their metabolism and are the primary producers. In acidic soils, bacteria are disadvantaged: decomposition of organic matter is achieved by fungi that are much slower yet also yield carbon dioxide. These processes result in soil CO_2 concentrations of between 0.1% and 3.5% in temperate regions, and up to 10% in the tropics. Much of this gas dissolves in soil water to give a weak acid, which is carried down by gravity to the subsoil along with humic and other organic acids.

In the subsoil, aeration is much less, and bacterial action is reduced, owing to this and lower organic matter content. The important process of carbon dioxide–calcium carbonate equilibration takes place in this zone. Carbon dioxide gas dissolves in water to give the weak carbonic acid, which dissociates to hydrogen and hydrogen carbonate ions. An increase in CO_2 concentration in the soil will allow more hydrogen ions to be released into drainage water, acidifying the soil (lower soil pH). This greatly enhances the ability of the percolating water to dissolve calcium carbonate. Some or all of the weak acid may be neutralised by exchangeable calcium ions released from clays, carbonate concretions, or bedrock fragments. Thus, bedrock erosion is much reduced where the soil has a high exchangeable calcium content released from leaf litter, or where it contains a high proportion of bedrock fragments. Changes in soil pH may be reduced by the reserve, or buffering, capacity of hydrogen ions derived from organic acids released from the decomposition of clay-humus particles and organic matter. The chemical balance of the karst solution process is thus very dependent on organic influences and especially the maintenance of vegetation, leaf litter, and a productive topsoil. Where native calcium-loving vegetation is replaced by plants adapted to more acidic soils, then the soil may become acidified to bedrock, and accelerated erosion of limestone occurs. This has been the case in Yorkshire, where more frequent burning has encouraged the growth of heather at the expense of the natural grassland and forest vegetation: there are now deep runnels formed in bedrock, down which soil can be lost (Trudgill 1976). A similar process has been observed in New Guinea, where slash and burn cultivation on limestone is accompanied by greatly accelerated soil loss (Gillieson et al. 1986).

4.3.3 The Zoning of Solution in the Unsaturated Zone

Below the soil zone in karst, there is a high porosity zone of weathered rock defined by (Williams 1983) as the subcutaneous zone. This zone has also been called the epikarst (Bakalowicz 1981; Friederich and Smart 1981; Mangin 1975). The epikarst stores large quantities of water and is the site of a high level of solutional erosion (Figure 4.14). Gunn (1981) measured the solutional load of the components of the epikarst water at Waitomo, New Zealand, and his study provides perhaps the most comprehensive set of data yet obtained on solutional processes in the upper skin of the karst. In this extensive polygonal karst subcutaneous flow, shaft flow and some vadose flow had hardness values in the same range as cave streams (Table 4.4).

Allogenic water flowing onto karst is highly capable of both chemical and mechanical erosion. Both allogenic and autogenic components of solutional denudation need to be separated if sense is to be made of rates of landscape development. This requires at least a partial solutional budget for the different components of the karst drainage system (Figure 4.15). In the Riwaka basin of New Zealand, karst rocks occupy just less than 50% of the basin, and the autogenic solution is about 79 mm ka^{-1}; from solute concentrations, much of which occurs

Figure 4.14 Enlarged joints in the epikarst are exposed in marble quarry walls at Wombeyan, New South Wales, Australia.

in the superficial zone. The large allogenic component increases the total basin solution by about 20%.

Both Gunn (1986) and Smith and Atkinson (1976) concluded that 15–50% of the total solutional erosion occurred in the deeper endokarst zone, which contains the cave conduits. However, a consideration of karst porosity by Worthington (1991) indicated that the effective void volume of the endokarst was around 1–1.4% only, while at the top of the epikarst it increased rapidly to 100% at the surface. Thus, only about 1% of the endokarst has been removed by solution processes, not the 15–50% cited earlier. Cave conduits formed by the solutional attack of circulating groundwater, therefore, occupy only a very small percentage of the total voids in the karst.

4.3.4 Limestone Solution in Seawater

In the marine environment, limestone dissolution is greatly enhanced owing to the common ion effect. Addition of large quantities of foreign ions, such as Na^+, K^+, and Cl^-, to bicarbonate-rich water decrease the activity of Ca^{++} and Mg^{++} ions, thus increasing the rate of dissolution of the solid phase carbonate or dolomite. The presence of 0–10% Mg^{++} reduces Ca^{++} solubility, while greater than 10% Mg^{++} increases Ca^{++} solubility (Picknett

Table 4.4 Total hardness measurements at Waitomo, New Zealand.

Hydrologic component	Number of samples	Mean water hardness mg l^{-1}	Standard deviation mg l^{-1}
Rainfall	26	5	4
Overland flow	6	21	20
Soil water	109	64	10
Throughflow	154	51	14
Subcutaneous flow	139	122	18
Shaft flow	58	122	18
Vadose flow (Mangapohue catchment)	84	130	15
Vadose flow (Glenfield catchment)	99	40	5
Vadose seepage	1714	96	19
Cave stream: Mangapohue	59	125	8
Cave stream: SP2	33	126	7
Cave stream: Glenfield	59	124	9

Source: After Gunn (1981).

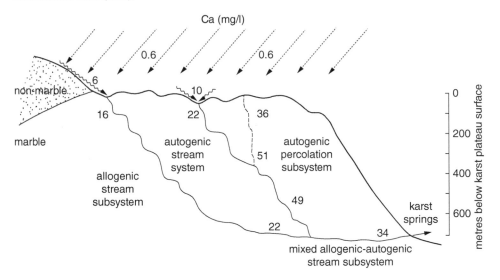

Figure 4.15 Dissolved calcium concentrations in the Riwaka karst, New Zealand. Source: From Williams and Dowling (1979).

et al. 1976). In seawater, the addition of large concentrations of sodium chloride greatly enhances limestone solution. In warm water, with high partial pressures of carbon dioxide, calcite solubility can be boosted to approximately 1000 mg l^{-1} (Plummer 1975). Where limestone coasts occur, a mixing zone between fresh and marine groundwaters exists. For the Yucatan peninsula of Mexico, Back et al. (1984) recorded the long-distance flow (100 km) of saturated karst groundwater with calcite concentrations of around 250 mg l^{-1}; in the final kilometre of flow, an additional 120 mg l^{-1} is added owing to mixing with a seawater lens.

This coastal mixing zone may be a site for intensive solution cavity formation, producing a narrow zone of maze caves that may invade pre-existing cavities in raised reefal limestone.

Careful examination of contemporary coral reefs shows that even very young limestones, actively undergoing diagenesis, have extensive rounded voids that are elongated along proto-bedding planes. In most cases, the corals are still clearly visible in either biohermal structures or in reef talus. These voids are interconnected to form networks through which seawater, and later freshwater, may circulate. These structures commonly survive uplift and may guide meteoric waters to produce mazes of solution cavities, some of which may be large enough for human entry. Further diagenesis may radically alter the original fossils and depositional structures, but the cavities tend to persist and enlarge.

The form of the freshwater lens on limestone islands is an important determinant of the types of cave that form there (Nunn 1994). The interface between fresh and salt water is a zone of enhanced limestone solution in a fresh–marine water mixing zone. Below the water table, phreatic passages form with a series of chambers linked by short passages, often complex in shape. The walls of these passages are frequently scalloped (Ollier 1975). Recently, caves of this type have been entered and mapped by divers (Palmer 1989). At the water table, epiphreatic caves form in a fluctuating zone and display characteristics of both phreatic and vadose caves. These may often connect with sea caves. Finally, vadose caves fed by rainwater may form, draining to lower phreatic systems. Thus, the caves in uplifted coral terraces combine primary reefal structures with solution features on account of rainwater inflow or inherited from previous phreatic conditions. Often stream invasion, lake, or shoreline incuts may further modify the form of the cave. Reactivation of entrenchment, wall collapse, or infilling may follow tectonic uplift or sea-level change. The occurrence of phreatic caves well above the water table is evidence of emergence, possibly uplift, while submerged caves may suggest island subsidence. Thus in Tuvalu, Gibbons and Clunie (1986) described a cave 46 m below present sea level with evidence of human occupation. These coastal caves are among the simplest in the spectrum of cave morphologies, yet from the preceding discussion their origins are diverse and complicated, and they may tell us much about relative movements of land and sea. The more familiar caves in the older limestones of major landmasses are an order of magnitude more complex to unravel.

4.4 Hydrothermal Solution of Limestone

Waters rising from deep sources (hypogenic recharge) may be partially enriched in dissolved CO_2. As they rise the pressure of overlying rock strata drops and water temperature drops, creating local conditions of increased solutional activity. This may be enhanced by mixing corrosion where water bodies of different origin, combine to create increased aggressiveness of the solutions.

Hypogenic recharge may be of several kinds:

- Basal injection of ancient meteoric water
- Rising thermal water with bicarbonate chemistry
- Rising thermal water with sulphuric acid chemistry

The effect of the increase in CO_2 solubility at depth (at high pressure) is very important, especially with regard to hypogenic CO_2 of volcanic or tectonic origin. An increase in hydrostatic pressure increases the solubility of CO_2, and consequently, the solubility of carbonate in water. Carbonate solubility increases at around $6\,mg\,l^{-1}$ for every $100\,m$ of increased water column depth.

In caves that occasionally become flooded (epiphreatic systems), air can remain trapped and compressed in niches in the roof, leading to CO_2 dissolution, and thus, to an increase in aggressiveness of the water, with the possible formation of cupolas and ceiling channels.

Sulphuric acid can be involved in dissolution reactions. This acid can be produced by the oxidation of hydrogen sulphide, commonly related to organic sources, such as hydrocarbon (oil and gas) deposits:

$$H_2S + 2O_2 \rightarrow 2H^+ + SO_4{}^{2-} \tag{4.9}$$

Sulphuric acid can also be produced by the oxidation of sulphides, such as pyrite:

$$2FeS_2 + 7O_2 + 2H_2O \rightarrow 2Fe^{3+} + 4SO_4{}^{2-} + 4H^+ \tag{4.10}$$

and

$$4Fe^{2+} + O_2 + 10H_2O \rightarrow 4Fe(OH)_3 + 8H^+ \tag{4.11}$$

Minor local concentrations of hydrochloric acid (HCl), can also occur:

$$CaCO_3 + HCl \rightarrow Ca^{2+} + HCO_3{}^- + Cl^- \tag{4.12}$$

Acidity can also derive from the oxidation of metals, such as manganese:

$$Mn^{2+} + \tfrac{1}{2}O_2 + H_2O \rightarrow MnO_2 + 2H^+ \tag{4.13}$$

and the oxidation of carbonate minerals, such as siderite:

$$2FeCO_3 + \tfrac{1}{2}O_2 + 5H_2O \rightarrow 2Fe(OH)_3 + 2HCO_3{}^- + 2H^+ \tag{4.14}$$

These reactions are occurring deep in a rock mass with no surface connection, as long as water is able to circulate in fissures and conduits.

4.5 Solution of Evaporites

All the evaporite rocks–gypsum ($CaSO_4.2H_2O$), anhydrite ($CaSO_4$), and halite (NaCl)–are soluble in pure water, producing little or no residue on dissociation, and they may be regarded as karst rocks. Evaporite karst will survive under arid- or semi-arid conditions only because of this high solubility. In addition, the large volume expansion of anhydrite on hydration to gypsum acts to close developing karst conduits (Jakucs 1977). Thus, evaporite karsts are rare on a global scale, but within them, true karst landforms of karren, dolines, stream caves, and springs may form (Wigley et al. 1973). Of more significance to the world of karst is the role of both gypsum and halite in the process of exsudation, where expansion on crystallisation breaks other karst rocks and enlarges cave passages. The deposition of both gypsum and halite speleothems is treated in Chapter 6.

4.6 Solution of Silicates in Meteoric Waters

Silicate rocks, such as sandstones and quartzites, are reasonably soluble in natural waters, with the solubility increasing markedly in alkaline conditions (Young and Young 2012, p. 62). Although under normal conditions of acidified drainage waters, quartz is less soluble (about $30\,mg\,l^{-1}$ or $0.5\,mmol\,l^{-1}$ at $25\,°C$), over a long time significant solution will occur. The weathering of silicate rocks involves both the dissociation of silica to silicic acid and the weathering of constituent minerals and cements to produce solutes and clays.

The dissolution of quartz can be written as follows (Young and Young 2012, p. 61):

$$SiO_2 + 2H_2O \leftrightarrow H_4SiO_2 \tag{4.15}$$

The weathering of a component mineral, such as a feldspar to gibbsite or kaolinite, can be simplified to:

$$NaAlSi_3O_8 + H^+ + 7H_2O \rightarrow Al(OH)_3 + Na^+ + 3H_4SiO_4 \tag{4.16}$$

$$NaAlSi_3O_8 + H^+ + 7H_2O \rightarrow \frac{1}{2}Al_2Si_2O_5(OH)_4 + Na^+ + 2H_4SiO_4 \tag{4.17}$$

In both cases, silica is mobilized in solution as weak monosilicic acid. This may precipitate as amorphous silica or opal-A, or combine with other compounds to form more complex clays.

$$2Al(OII)_3{}^+ + 2SiO_2 + H_2O \rightarrow Al_2Si_2O_5(OH)_4 + H^+ + H_2O \tag{4.18}$$

Amorphous silica is more soluble than the quartz: its solubility ranges from 60 to $80\,mg\,l^{-1}$ ($1.7–1.3\,mmol\,l^{-1}$) at $25\,°C$ to 100 to $140\,mg\,l^{-1}$ ($1.7–2.3\,mmol\,l^{-1}$) at $25\,°C$ over a wide range of pH. In general, silicate minerals weather more readily than amorphous silica, which itself is more soluble than quartz. Thus, the mineral composition and grain size of the silicate rock affects its weathering rate quite significantly.

Thus, in the weathering of silicate rocks, there is far more insoluble material produced, which may infill developing caves and dolines. These residues block the interstices between grains and infill small joints, reducing the permeability of the silicate karst. Surface solution features, such as solution pans and runnels, may be seen on many sandstones, especially in the tropics. The "beehive" terrain developed in tropical sandstones such as the Bungle Bungles (Pumululu National Park), in Western Australia, in some ways mimics polygonal karst (Young 1986), but subterranean drainage is restricted or absent. Large stream caves seem to be restricted to situations where prominent joints or faults combine with steep hydraulic gradients, such as near plateau edges. One example of this type is the Sima Aonde, in Venezuela, where very large dolines with solution pans (cinegas) combine with large stream caves showing strong joint control (Figures 4.16 and 4.17). The quartzite caves of Zimbabwe, including the 305 m deep Mawenge Mwena (Truluck 1994), extend the realm of deep caves in sandstone. Thus, Jennings (1983) concludes that this particular case is true karst and is deserving of further study. It would be instructive to compare karst forms and solution processes in sandstone and limestone existing in a similar tectonic setting and under a similar climate. Care must be taken to differentiate between true silicate karst and subjacent karst where limestone underlies the sandstone, and both solution and collapse in underlying cavities occur–for example, Big Hole, New South Wales, Australia (Jennings 1967).

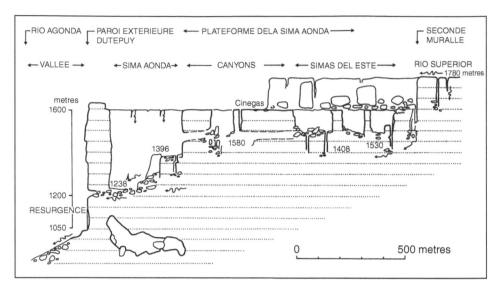

Figure 4.16 Underground circulation of the sandstone karst of the Sima Aonde, Venezuela. Simas are deep pits; cinegas are a type of solution pan. Source: From Courbon et al. (1989).

Figure 4.17 Sima Aonde on Auyantepui, Venezuela. Yellow circle indicates spring at base of shaft. Source: Based on Piccini and Mecchia (1986). © Piccini, L. and Mecchia, M. 2009. Solution weathering rate and origin of karst landforms and caves in the quartzite of Auyan-tepui (Gran Sabana, Venezuela) Geomorphology 106: 15–25.

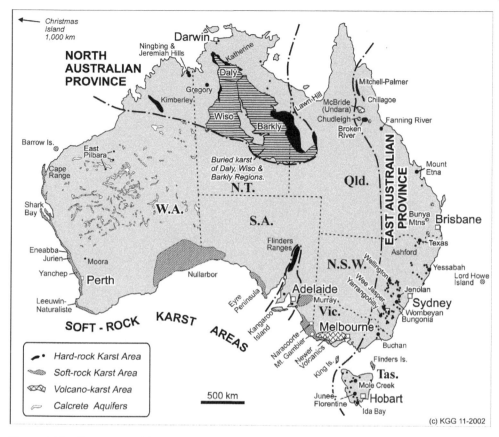

Figure 4.18 Karst areas of Australia. Map by K.G. Grimes. Source: From Webb et al. (2003).

4.7 Caves in Quaternary Limestone in Southern Australia

Quaternary limestones are widespread in southern Australia, including on Kangaroo and King Islands, and are also present on Lord Howe Island in the Tasman Sea (Figure 4.18). These limestones are cemented sand dunes that are porous, poorly cemented, and made up of well-rounded carbonate grains from fossil shells derived from shallow water environments along the southern Australian coast from the early Tertiary to the present. Broken up by wave action, the shell fragments were washed ashore onto the beaches, and then blown into dunes behind the beaches by the prevailing winds, which are predominantly strong westerlies across southern Australia. Additional quartz sand within the dune limestones was washed down to the coast by rivers and derived from erosion of granites and metamorphic rocks inland (Grimes 2002).

These carbonate dunes form ridges parallel to the coast and rise up to 40 m above the beach in Victoria, and to more than 100 m in Western Australia, where the prevailing winds are stronger. Sand was blown off the beaches during high sea levels to form dunes. The dunes contain well-developed steeply dipping cross-bedding. During their formation, sand

Figure 4.19 Cross-bedding in dune limestone 135 000 years old on south coast of Kangaroo Island, South Australia.

was blown up the upwind side of the dune and then avalanched down the lee side, burying the downwind face as a cross-bed. As the dune migrated downwind, it progressively formed these steeply dipping cross-beds. In many places, buried soil horizons reflect the stabilisation of the dunes by vegetation (Grimes 1994) (Figure 4.19).

In southeast South Australia and western Victoria, slow uplift continued throughout the Pleistocene, raising each dune system so it was not reworked by the waves and remained preserved as a long ridge. These dune systems now extend across the coastal plain as low ridges parallel to the shoreline, separated by swamps or lakes. Where the coastline is stable and not being uplifted in southwest Western Australia, the dune systems stacked up on top of each other, forming deposits of dune limestone more than 100 m thick.

Rain falling on the dunes dissolves calcium carbonate that is then precipitated as thin coatings of calcite cement around the grains within the dune. This cement does not totally fill the pore spaces, so the dune is poorly cemented and very porous; more cement may be precipitated closer to the dune surface to form a hard calcrete crust. Many of the carbonate dunes contain solution pipes, vertical tubes, or cones that open to the dune surface. These smooth walled pipes extend down into the limestone, often for up to 20 m. They can be soil filled or open and may open into cave chambers below (Figure 4.20). The origin of solution pipes is not fully known, but trees may be involved as follows: Rain falling on large trees such as eucalypts runs down the branches and trunk and infiltrates into the soil, carrying

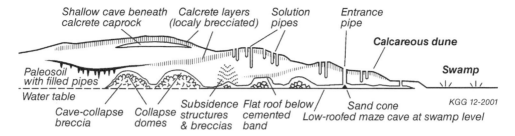

Figure 4.20 Features of syngenetic karst in a dune limestone result from solution occurring at the same time as the sand is being cemented into a soft rock. Source: From Webb et al. (2003).

Figure 4.21 Solution pipes in dune limestone of the Bridgewater Formation, Cape Bridgewater, Victoria, Australia.

carbon dioxide and organic acids that can dissolve the dune limestone under the trunk, forming a solution pipe directly below the tree.

The solution pipe (Figure 4.21) will fill with soil as it is slowly dissolved, and at least some of the dissolved calcium carbonate will precipitate to form the well-cemented walls of the pipe. The soil within the tube may later be washed away if a cave has developed beneath it.

Surface karst development is generally reduced on dune limestones, because of poor cementation and the presence a weathering residue of quartz sand left after limestone solution. There may be bare limestone pavements on more cemented caprock, with some

Figure 4.22 Collapse doline in dune limestone, Lake Cave, Leeuwin-Naturaliste Ridge, Western Australia.

solution pitting and weak solution rills. Shallow cracks or joints may also be present. Shallow dolines may be present, formed either by soil slumping down solution pipes or by collapse. Where present, collapse dolines can be 50–100 m across (Figure 4.22) and may have overhanging walls formed from projecting cap rock (Gillieson and Spate 1998).

If caves formed in dune limestones, several conditions must be met (White 2000). Firstly, the limestone must be sufficiently pure, at least 70–90% calcium carbonate. Less solution is needed to form a cavity due to the high porosity, but insoluble quartz sand left behind is less likely to clog the cavities. Secondly, for the solution process to be effective, there must be a continuous supply of aggressive water. This can come from streams or springs draining non-carbonate rocks, or from inter-dune swamps. Swamp water can be particularly aggressive due to carbonic acid derived from microbial degradation of organic matter and organic acids (mostly humic and fulvic acids). Thirdly, the limestone must be well-cemented so that solutional cavities do not collapse before they become very large; for shallow caves, the hardened cap rock often provides enough strength to support the roof.

Cave chambers and passages within dune limestone are often collapsed and solutional features may be hard to find (Figure 4.23). Entrances are either collapses or solution pipes. Dome-shaped roofs are common with rubble piles reaching near to the roof. One of the best examples is Kelly Hill Cave on Kangaroo Island, South Australia, which consists of several roughly circular collapse domes up to 60 m in diameter, with narrow fissures at the sides blocked by rubble. The rubble is masked by speleothems.

Figure 4.23 Speleothems developed on rubble pile, Ngilgi Cave, Leeuwin-Naturaliste Ridge, Western Australia.

In Western Australia, inclined fissure caves, representing one side of a collapse dome like Kelly Hill Cave, are common. They consist of passages with a rubble floor and steeply sloping roof, often representing a cross-bed within the dune. Where a stream is present, rubble may be dissolved, leaving quite large chambers. If there is a hydraulic gradient, then irregular mazes with sculpted walls may develop, occasionally with sinuous passages.

The Leeuwin-Naturaliste Ridge in southern Western Australia is a narrow strip of coastal dune limestone about 90 km long but never more than 6 km wide. It runs between Cape Naturaliste in the north and Cape Leeuwin in the south. The Leeuwin-Naturaliste Ridge features more than 100 caves and over 200 other karst features. The ancient dunes have blocked some of the valleys eroded into the Precambrian gneiss basement, forming lakes or swamps. The larger streams maintained their courses through the encroaching dunes, forming valleys with steep walls up to 60 m high. The smaller streams have formed cave systems, which can extend for hundreds of metres as sinuous stream passages oriented down the water table gradient (Figure 4.24).

In the southernmost part of the ridge, the caves are located in the oldest dune system, which is composed of thick relatively well-cemented limestone. The caves form multilevel complexes with network mazes; the Jewel-Easter Cave system is more than 8 km long (Figure 4.25). There is well-developed phreatic spongework in the cave walls, indicating dissolution by slowly moving groundwater; in Jewel Cave, a large chamber with a

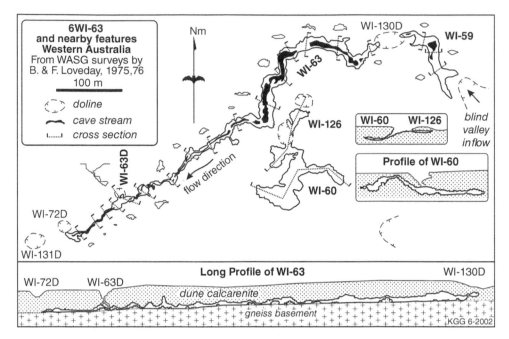

Figure 4.24 A linear stream cave (WI-63) follows the contact between dune limestone and impermeable gneiss. By contrast, the neighbouring cave system (WI-126 and WI-60) is more typical of caves in dune limestone, being a series of large collapse domes. Source: From Webb et al. (2003).

Figure 4.25 Aerial photo showing mapped cave passages in the Jewel-Easter and Labyrinth Cave subsystems on the inland flank margin of the dune ridge, Caves Road and Jewel Cave Reserve boundary (black). Source: Adapted from Eberhard (2005).

horizontal roof suggests that dissolution extended up to the water table. The groundwater flow paths may have extended deeper into the limestone than elsewhere in the region.

Caves in dune limestones often contain abundant speleothems. Because of the high porosity of the limestone, water percolation can occur over the entire ceiling, rather than being confined to joints as in Palaeozoic limestones. As a result, the cave roofs can be covered with thousands of straws and stalactites; this is particularly true of the caves in Western Australia. Individual straws in these caves may be up to 6 m long, and helictites are well developed.

References

Back, W., Hanshaw, B.B., and Van Driel, J.N. (1984). Role of groundwater in shaping the eastern coastline of the Yucatan peninsula, Mexico. In: *Groundwater as a Geomorphic Agent* (ed. R.G. La Fleur). Boston: Allen & Unwin.

Bakalowicz, M. (1981). Les eaux d'infiltration dans l'aquifere karstique. In: *Proceedings of the 8th International Congress of Speleology, 1981. Department of Geography and Geology, Western Kentucky University. Kentucky*, 710–712.

Bögli, A. (1964). Mischungskorrosion: dn Beitrag zum Verkarstungsproblem. *Erdkunde* 18: 83–92.

Courbon, P., Chabert, C., Bosted, P., and Lindsley, K. (1989). *Atlas of the Great Caves of the World*. St Louis: Cave Books.

Dreybrodt, W. (1980). Kinetics of the dissolution of calcite and its applications to karstification. *Chemical Geology* 31: 245–269.

Eberhard, S. (2005). Ecology and hydrology of a threatened groundwater-dependent ecosystem: The Jewel cave karst system in Western Australia. *Journal of Cave and Karst Studies*: 67.

Folk, R.L. (1959). Practical petrographic classification of limestones. *AAPG Bulletin* 43: 1.

Folk, R.L. (2014). Spectral subdivision of limestone types. *Memoir* 1: 62–84.

Ford, D.C. (2004). Canada. In: *The Encyclopedia of Caves and Karst Science* (ed. J. Gunn). New York: Taylor and Francis/Routledge.

Ford, D.C. and Drake, J.J. (1982). Spatial and temporal variations in karst solution rates: The structure of variability. In: *Space and Time in Geomorphology* (ed. C.E. Thorn). Boston: Allen & Unwin.

Ford, D.C. and Williams., P.W. (2007). *Karst Hydrogeology, Karst Hydrogeology and Geomorphology*. Chichester, England: Wiley.

Friederich, H. and Smart., P.L. (1981). Dye tracer studies of the unsaturated zone: Recharge of the Carboniferous Limestone aquifer of the Mendip Hills, England. In: *Proceedings of the 8th International Congress of Speleology, 1981. Kentucky Department of Geography and Geology, Western Kentucky University*, 283–286.

Frumkin, A. (1994). Morphology and development of salt caves. *National Speleological Society Bulletin* 56: 82–95.

Frumkin, A., Magaritz, M., Carmi, I., and Zak, I. (1991). The Holocene climatic record of the salt caves of Mount Sedom Israel. *The Holocene* 1: 191–200.

Gibbons, J.R.H. and Clunie., F.G.A.U. (1986). Sea level changes and pacific prehistory: New insight into early human settlement of Oceania. *Journal of Pacific History* 21: 58–82.

Gillieson, D.S. and Spate, A.P. (1998). Karst and caves in Australia and New Guinea. In: *Global Karst Correlation* (eds. Y. Daoxian and L. Zaihua). Beijing: Science Press.

Gillieson, D.S., Gorecki, P., Head, J., and Hope, G. (1986). Soil erosion and agricultural history in the Central Highlands of New Guinea. In: *International Geomorphology* (ed. V. Gardiner). London: Wiley.

Grimes, K.G. (1994). The Southeast Karst Province of South Australia. *Environmental Geology* 23: 134–148.

Grimes, K.G. (2002). Syngenetic and eogenetic Karst: An Australian viewpoint. In: *Evolution of Karst: From Prekarst to Cessation* (ed. F. Gabrovšek). ZRC Sazu, Postojna: Inštitut za raziskovanje krasa.

Gunn, J. (1981). Limestone solution rates and processes in the Waitomo District, New Zealand. *Earth Surface Processes and Landforms* 6: 427–445.

Gunn, J. (1986). Solute processes and karst landforms. In: *Solute Processes* (ed. S.T. Trudgill). Chichester: Wiley.

Hanshaw, B.B. and Back, W. (1979). Major geochemical processes in the evolution of carbonate–aquifer systems. *Journal of Hydrology* 43: 287–312.

Jakucs, L. (1977). *Morphogenetics of Karst Regions: Variants of Karst Evolution*. Budapest: Akademiai Kiado.

Jennings, J.N. (1967). Further remarks on the Big Hole, near Braidwood, New South Wales. *Helictite* 6: 3–9.

Jennings, J.N. (1983). Sandstone pseudokarst or karst? In: *Aspects of Australian Sandstone Landscapes* (eds. R.W. Young and G.C. Nanson). Wollongong: Australian and New Zealand Geomorphology Group.

Leighton, M.W. and Pendexter, C. (1962). Carbonate rock types. In: *Classification of Carbonate Rocks-a Symposium* (ed. W.E. Ham). Tulsa, Oklahoma: American Association of Petroleum Geologists.

Mangin, A. (1975). *Contribution à l'étude hydrodynamique des aquifères karstiques*. Des, University of Dijon.

Nunn, P.D. (1994). *Oceanic Islands*. Oxford, UK/Cambridge, MA: Blackwell.

Ollier, C.D. (1975). Coral island geomorphology-the Trobriand Islands. *Zeitschrift für Geomorphologie* 19.

Palmer, R.J. (1989). *Deep into Blue Holes: The Story of the Andros Project*. London: Unwin Hyman.

Picknett, R.G., Bray, L.G., and Stenner, R.D. (1976). The chemistry of cave waters. In: *The Science of Speleology* (eds. T.D. Ford and C.H.D. Cullingford). London: Academic Press.

Playford, P.D. and Lowry, D.C. (1966). Devonian reef complexes of the Canning Basin, Western Australia. *Geological Society of Western Australia Bulletin*: 118.

Plummer, L.N. (1975). Mixing of sea water with calcium carbonate ground water. In: *Quantitative Studies in the Geological Sciences* (ed. E.H.T. Whitten). Geological Society of America.

Sauro, F., De Vivo, A., Bernabei, T., and De Waele, J. (2017). La Venta association, 25 years of exploration projects and discoveries. In: *Proceedings of the 17th International Congress of Speleology, July 2017, Sydney*, 361–365.

Smith, D.I. and Atkinson, T.C. (1976). Process, landforms and climate in limestone regions. In: *Geomorphology and Climate* (ed. E. Derbyshire). London: Wiley.

Trudgill, S.T. (1976). The erosion of limestones under soil and the long term stability of soil vegetation systems on limestone. *Earth Surface Processes* 1: 31–41.

Trudgill, S.T. (1985). *Limestone Geomorphology*. New York: Wiley.

Truluck, T. (1994). The sandstone shafts of the Chimanimani Mountains. *Caves and Caving* 65: 15–18.

Webb, J., Grimes, K.G., and Osborne, A. (2003). Black holes: Caves in the Australian landscape. In: *Beneath the Surface: A Natural History of Australian Caves*. Sydney: University of NSW Press.

White, S. (2000). Syngenetic karst in coastal dune limestone: a review. In: *Speleogenesis: Evolution of Karst Aquifers* (eds. A.B. Klimchouk, D.C. Ford, A.N. Palmer and W. Dreybrodt). Huntsville: National Speleological Society.

Wigley, T.M.L., Drake, J.J., Quinlan, J.F., and Ford, D.C. (1973). Geomorphology and geochemistry of a gypsum karst near canal flats, British Colombia. *Canadian Journal of Earth Sciences* 10: 113–129.

Williams, P.W. (1983). The role of the subcutaneous zone in karst hydrology. *Journal of Hydrology* 61: 45–67.

Williams, P.W. and Dowling, R.K. (1979). Solution of marble in the karst of the Pikikiruna range, Northwest Nelson, New Zealand. *Earth Surface Processes* 4: 15–36.

Williams, P.W. and Ford, D.C. (2006). Global distribution of carbonate rocks. *Zeitschrift Fur Geomorphologie Supplementband* 147: 1–2.

Worthington, S.R.H. (1991). Karst hydrogeology of the Canadian Rocky Mountains. Unpublished PhD. McMaster University.

Wray, R.A.L. (2013). Solutional weathering and Karstic landscapes on quartz sandstones and quartzite A2. In: *Treatise on Geomorphology* (ed. J.F. Schroder). San Diego: Academic Press.

Wray, R.A.L. and Sauro, F. (2017). An updated global review of solutional weathering processes and forms in quartz sandstones and quartzites. *Earth Science Reviews* 171: 520–557.

Young, R.W. (1986). Tower Karst in Sandstone: Bungle Bungle massif, northwestern Australia. *Zeitschrift für Geomorphologie* 30: 189–202.

Young, R.W. and Young, A. (2012). *Sandstone Landforms*. Berlin: Springer.

5

Speleogenesis

5.1 Classifying Cave Systems

The basic definitions of caves were reviewed in Chapter 2 and can either be anthropogenic, based on size relative to the human body, or genetic, based on hydrodynamics. The sample of caves used to develop a typology is based on thousands of accurately mapped caves worldwide, a sample that grows each year due to the tireless efforts of speleologists. The top-ranked longest and deepest caves are listed in Chapter 2 (Tables 2.3 and 2.4). Caves can be classified according to internal characteristics (size, shape, or contents) or external context (topographic, geologic, or hydrologic) (Ford and Williams 2007, p. 210). This basic classification is further outlined in Table 5.1.

Neither of these basic classification schemes has any real genetic significance, and so they are of limited use in understanding speleogenesis. A more useful classification is based on the nature of the waters circulating in the rock mass (Klimchouk 2015). Most of the caves that have been explored and mapped (perhaps 80%) are due to the circulation of *meteoric* waters (rainfall) in karst rocks where the water is not confined above or below an impermeable rock layer or aquiclude. These are sometimes known as *epigene* caves. If impermeable layers are present, then maze caves may form as the water is confined. In this context, there may be some circulation of deep sourced water (several kilometres depth) which leads to the *hypogene* cave classification (Klimchouk 2013). Where the water is deeply sourced it may be enriched by carbon dioxide (usually in thermal waters) or hydrogen sulphide (basin waters or those associated with mineralisation) leading to acidification. Recent research has shown that many of the solutional features in drained caves can be attributed to past circulation of hypogene waters; this is probably the most exciting development in our understanding of speleogenesis in the last 20 years. The two defining properties of hypogene karst are the dominance of the deep-seated sources of acidified karst water and recharge of soluble rock formations by water from below, independent of any recharge from overlying strata (Klimchouk 2013).

Finally, on the coast, the mixing of marine and fresh waters causes enhanced solution of karst rocks. This causes a zone of increased porosity and permeability at the margin of the freshwater lens. The decomposition of organic matter, producing weak acids, may enhance dissolution. Water flow velocities are also faster at this margin, again increasing solution rates. *Flank margin caves* form as solutional voids in this flow regime but do so without

Caves: Processes, Development, and Management, Second Edition. David Shaw Gillieson.
© 2021 John Wiley & Sons Ltd. Published 2021 by John Wiley & Sons Ltd.

Table 5.1 Classification of solutional caves.

From internal characteristics	From external context
By size: length or depth or volume	By rock type and structure, e.g. limestone, gypsum, joint-guided, steeply dipping, flat-lying beds
By plan form (chambers, linear passages, shafts, mazes) or by passage cross-section, e.g. circular, canyon, breakdown, or some combination	By topographic context (montane, plateau, coastal) or position, e.g. cliff-foot cave, meander cut-off, sub-glacial, perched above valley
By relationship to present or past water table, e.g. vadose, phreatic, epiphreatic, relict	By relationship to current fluvial processes, e.g. active, episodic, abandoned, relict caves
By contents, e.g. calcite speleothems, gypsum, ice, guano, artefacts	By climate (humid tropical, arid, temperate, alpine, arctic)

Source: After Ford and Williams 2007, p. 210.

entrances (Lace and Mylroie 2013). Their size and complexity are dependent on how long the freshwater lens was stable at that position relative to sea level. Small flank margin caves are usually single chambers with a few short side passages, while the largest caves are ramifying collections of chambers with spongework (Mylroie and Mylroie 2007). They are only accessible to humans after erosional processes, such as cliff retreat or collapse, have opened them to the surface. They may incorporate elements of simple meteoric caves as well due to percolating rain, often in the form of shafts and short rainwater inflow passages.

Finally, a distinction needs to be made between caves and karst formed in older karst rocks where primary porosity is low due to diagenesis and compaction (telogenetic karst) and younger karst rocks where the porosity is quite high (eogenetic karst), such as occur at the coast (Figure 5.1). In the former water circulation is really confined to the network of fissures that make up secondary porosity.

One of the major controversies surrounding the formation of caves rests on the nature of the initial cavity or cavities which guided the flow of water dissolving the rock. It is not possible to form a cave from solid rock without any cavities. The divergent views can be broadly summarised into three classes:

The *kinetic* view, in which the size of tiny capillaries in the rock determines whether flow is laminar or turbulent; in the latter case, the helical flow characteristic of larger tubes permits accelerated solution with a positive feedback mechanism. Over time this results in the preferential enlargement of a single tube or proto-cave which dominates the array and becomes the principal cave conduit (Ford and Ewers 1978).

The *inheritance or inception horizon* view, in which a pre-existing small cavity (Figure 5.2) or chain of vugs, formed by tectonic, diagenetic (mineralisation), or artesian processes, is invaded and enlarged by karst groundwater to form a cave conduit (Figure 5.3; Lowe 1992).

The *hypogene* view, in which hydrothermal deep waters (charged with carbon dioxide, hydrogen sulphide, or other acids) result in the formation of heavily mineralised cavities which may be invaded by cooler karst waters to give larger, integrated cavities, or networks of passages (Bakalowicz et al. 1987; Klimchouk 2000; Klimchouk 2009).

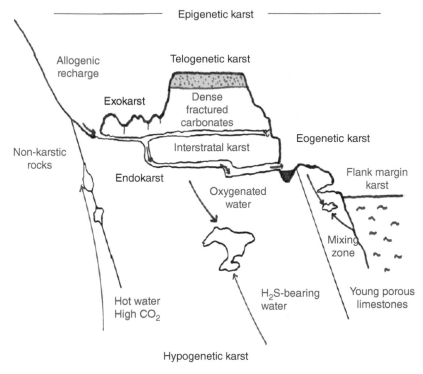

Figure 5.1 Various types of karst and modes of origin of caves. Source: From White and White (2013).

Figure 5.2 Proto-caves along a bedding plane in Clearwater Cave, Sarawak, Malaysia.

Figure 5.3 Elliptical passage in Clearwater Cave (Malaysia) showing the control of bedding and steeply inclined joints over passage shape.

It is possible that a single cave system may have formed under the operation of some or all of these processes during its very long history. One fundamental problem is that we can only see a partial sample of the cave passages in any system. The rest may be underwater, too small to enter, or sealed by breakdown or sediment. There is certainly a disjunct between the local features in a cave that we can observe and measure, and the more systemic observations and data that derive from a hydrologic analysis of a whole karst system. Clearly when thinking about cave development we must critically re-examine the evidence periodically, make careful observations, and keep an open mind about timescales.

5.2 Controls of Rock Structure on Cave Development

5.2.1 Role of Lithology

Quite often caves are guided by changes in lithology, with passages developing along or close to the junction of pure and impure limestone, limestone and underlying shales, or limestone and igneous rocks such as granite. There may be a stratum or horizon in which passage development occurs preferentially, even at the proto-cave stage; this is what Lowe (1992) calls the inception horizon. In the Forest of Dean (UK), the major cave inception horizons are associated with interbedded sandstones and unconformities in the Carboniferous limestone (Lowe 1992). In McBride's Cave (Alabama) flat-lying chert beds serve as aquicludes and result in a stepped long-profile with long, low epiphreatic tubes linked by waterfall shafts located at major joint intersections. Such an impure or impermeable bed need not be continuous: for example, discontinuous cherty layers in the Tertiary limestones of New Guinea act as aquicludes, and cave passages form just above them, as in Atea Kananda (Muller Range; Gillieson 1985). Strong's Cave (Western Australia) is formed at the junction of the Tertiary dune limestones and the underlying Proterozoic granite, which is partially exposed in the floor where speleothems are absent.

It is now very clear that there is an important role for impurities in limestone in guiding water flow in the karst. Insoluble beds in limestone act to confine water flow, especially in the vertical plane, and result in perched cave passages of varying sizes. This may occur at the earliest stages of cave formation – as with the inception horizons of Lowe (1992) – or much later in the cave's history, when cave passage incision owing to external valley downcutting is retarded by resistant or insoluble strata, producing stepped passage profiles or cave waterfalls, such as that of Barber's Cave, Cooleman Plain, Australia, or Gaping Ghyll, Yorkshire, UK (Jennings 1968). Impure muddy limestones occur every 20–30 m in the stratigraphy of the limestones of the Muller Range, Papua New Guinea, and act as aquicludes to direct cave passages to many levels over a vertical range of 525 m in the Mamo syncline (Francis et al. 1980). Stylolitic bands owing to pressure solution of impurities are widespread in the Ordovician limestones of Tasmania, Australia, and form waterfalls in Herbert's Pot and Khazad-Dum caves. Where the limestone strata have been tilted, stylolitic bands can force rapidly descending and ascending phreatic tubes which create inverted syphons, such as those in Murray Cave and River Cave, Cooleman Plain, or in Peak Cavern, Derbyshire, UK.

Although the development of caves is determined principally by hydraulic gradient and catchment size, much of the passage morphology and network architecture is determined by lithology and structure. Thus, karst systems tend to be more random in pattern in flat-lying porous limestones than in steeply dipping crystalline limestones. This randomness was investigated by Laverty (1987), who concluded that cave lengths exhibit fractal behaviour, with the fractal dimension in the range 1–1.5 over the compass of resolutions from 1 to 100 m for surveyed caves. The implication of this for speleology is that such relationships developed for individual caves may be applicable to a region, including unexplored or unenterable caves. Curl (1986) has investigated this possibility for the Appalachian karst and has developed regional area and volume equations of use to hydrology and biology.

5.2.2 Role of Joints, Fractures, and Faults

Water circulation within karst is greatly enhanced when a network of joints is present. Joints are simple fractures in rock with no, or minimal, movement of the strata. They are usually caused by tensional and shear forces and may be the result of diagenesis, uplift, folding, or valley-side unloading following erosion. Most joints run at right angles to bedding, but may be inclined or even sinuous. In plan view, they may intersect to give a rectangular or rhomboidal network, with cross joints penetrating only a few beds and master joints extending through many beds. Master joints may be several hundred metres long, and caves may form preferentially along them; for example, Lancaster Hole, Yorkshire is partly guided by vertical joints.

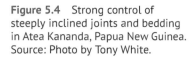

Figure 5.4 Strong control of steeply inclined joints and bedding in Atea Kananda, Papua New Guinea. Source: Photo by Tony White.

Small joints may be tight and relatively impermeable or may be filled with sediments or calcite. Large joints start as angular, irregular cavities (Figure 5.4) but become rounded by solution with time. Cave development is enhanced where the joint spacing is wide to very wide (100–300 m) because of the concentration of flowing water. Most caves will show some passages with joint network guidance (Figure 5.5). In extreme examples, the joint network dominates to give a maze cave, such as Wind Cave, South Dakota, USA, or Optimistich-eskaja, Ukraine.

Joint sets are not static in time; any rock mass may bear the imprint of many phases of joint development, as a result of tectonism or erosional unloading. Large, deep tensional joints, tens of metres deep, often form at the edges of plateaux or valleys owing to pressure release following scarp retreat (Figure 5.6). These may fortuitously intersect active or fossil phreatic passages in the limestone mass. Often the sequence of this joint development can be ascertained by careful examination of cave passage intersections.

Although in a single passage the role of bedding and joints may be reflected in the plan and cross-section of a cave, we need to consider major geologic structures when we attempt to explain the complex patterns of caves in length and depth. The increased circulation of water in and around faults is an obvious factor in karst system development. Faults are

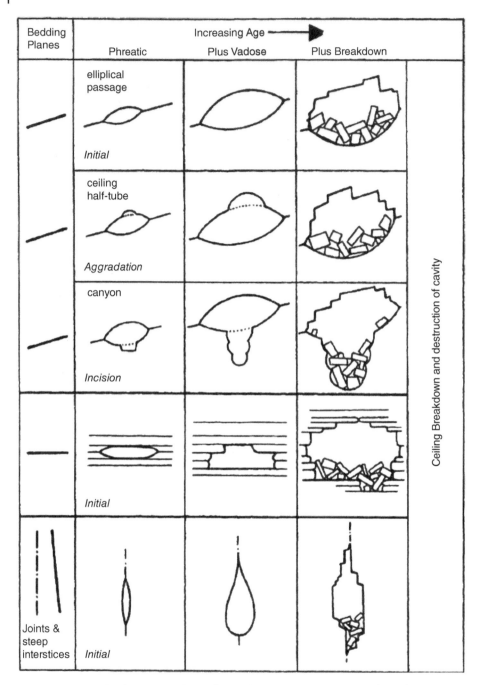

Figure 5.5 Variation in passage shapes in relation to bedding and joint orientation with increasing age of phreatic and vadose cave passages. Source: From Bögli (1980).

Figure 5.6 Joint guided shaft in Greftsprekka Cave, Gildeskal, Norway. The cave is developed in steeply dipping marble of Cambro-Silurian age.

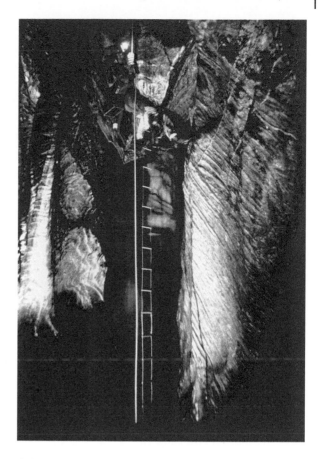

fractures with some displacement of the rock strata evident; a fault gangue of ground-up rock may be present to clarify the sense of movement. Such zones may act as inception horizons for proto-cave development and may continue to dominate the array of conduits.

Individual cave chambers may be guided by faults – for example, the main chamber of Gaping Ghyll, Yorkshire, UK; West Driefontein Cave, South Africa; P8 Cave, Peak District, UK; and Marakoopa Cave, Mole Creek, Tasmania.

Many major trunk cave passages that drain along the strike are within anticlines. The extension of joints in these locations promotes greater dissolution. In Kentucky, USA, Mammoth Cave has its principal passages formed within the noses of plunging anticlines (Ford and Williams 1989, p. 41). Both Ogof Fynnon Ddu and Dan yr Ogof in Wales, UK, show anticlinal control (Charity and Christopher 1977; Lowe 1989). Anticline Cave (Buchan, Australia) is a very small but clear example, while much of the 220 km of Clearwater Cave, Mulu, Sarawak, Malaysia, is formed in an anticline.

Some examples of synclinal guidance over karst drainage are the Sistema Purificacion (Mexico; Hose 1981) and the Mamo-Atea Kananda (Papua New Guinea; Francis et al. 1980). In both cases, streams sinking high on the flanks of the syncline have developed cave passages down dip which unite in the trough of the syncline.

Figure 5.7 The sheer size of Sarawak Chamber in Lubang Nasib Bagus (Good Luck Cave; Malaysia) defeats attempts to photograph its entirety. Source: Cavan / Alamy Stock Photo.

The largest known underground chamber (Figure 5.7) also owes its existence to a combination of folding and faulting. Sarawak Chamber in Lubang Nasib Bagus (Good Luck Cave) has a volume of approximately 12 million cubic metres, and floor dimensions of 600 by 415 m (Figure 5.8). The cave developed in upper Eocene to lower Miocene limestones which have partially recrystallised and folded. The large chamber has formed close to an anticlinal axis and to the underlying impermeable schists and slates (Gilli 1993) (Figure 5.9). There is also a fault on the western side of the chamber. Stoping of the chamber is aided by fretting and rotation of blocks in the arched roof, and the fallen blocks may be rapidly removed by stream action. The large passage of nearby Deer Cave has formed in a similar geologic context.

5.2.3 Cave Breakdown and Evaporite Weathering

Limestone is a strong but somewhat brittle rock which fractures readily. The creation of breakdown in caves will be considered further in Chapter 7. Here, the role of breakdown in shaping passage morphology will be reviewed. The role of hydrostatic support for cave cross-sections is evident when water-filled passages are drained; inevitably some wall and roof collapse occurs, causing a stoping process which continues through the beds until a stable arch is formed or a stronger interbed is reached. Valley-side unloading following hillslope erosion and fluvial incision may create a series of joints parallel to the valley axis. Close to cave entrances, this may produce a denser joint network which promotes collapse. Seismic shocks may also promote spalling of walls and opening of joints. Thus, in any cave

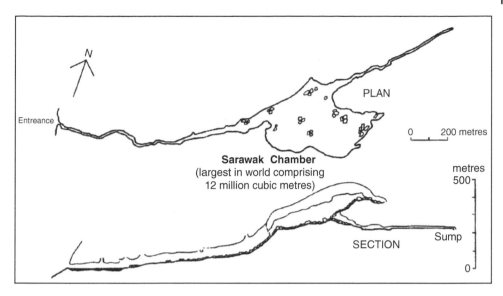

Figure 5.8 Plan and section of Lubang Nasib Bagus (Good Luck Cave), Mulu, Sarawak, Malaysia, showing Sarawak Chamber. Source: From Courbon et al. (1989).

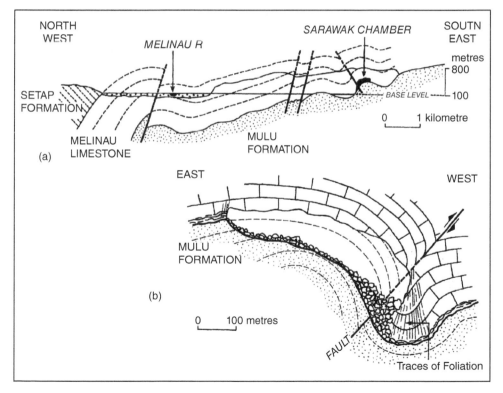

Figure 5.9 Geological context of Sarawak Chamber (Mulu, Sarawak, Malaysia). The chamber lies on a minor anticlinal axis and has been affected by faulting of the Melinau Limestone. Source: From Gilli (1993).

evolution sequence, ceiling and wall breakdown are an almost inevitable phase which may lead to partial or total infilling of the cavity.

The process of rock weathering by gypsum and halite crystallisation (known as exsudation) can be a powerful agent for passage modification. Seepage water charged with soluble salts evaporates upon reaching the cave wall, and the expansion of crystals in bedding or small fissures promotes dramatic spalling.

There are good examples from the Nullarbor Plain of Australia, where solutional conduits, formed under phreatic conditions and then drained, have been so modified by exsudation as to remove nearly all the initial solution features and speleothems (Gillieson and Spate 1992; Lowry and Jennings 1974).

5.3 Meteoric Speleogenesis, Unconfined and Confined

5.3.1 Formation of Caves in Plan

In Chapter 3, the factors governing the initiation of water circulation in a limestone massif were discussed. From this, the path of groundwater flow will depend on the balance between the direction in which resistance to flow is minimal and the direction of maximum potential energy (the shortest and steepest route from streamsink to spring). The long gradient of a karst hydrologic system will depend on the hydraulic head and the position of the springs in the system, the latter being controlled largely by geologic structure and base level (Audra and Palmer 2015). Once flow has been initiated, then a cave system will develop linking streamsinks and springs. The simplest situation is in the direction of steep dip. Initially a distributary pattern of solutional conduits develops, extending away from the water entry point down the hydraulic gradient (Figure 5.10). The precise pathways will depend on local lithology and structure, but in every case one tube will expand at the expense of the others and come to dominate the array. This will propagate downgradient until the streamsink is linked to the spring. Once this occurs, resistance to flow is drastically reduced, and a cave conduit exists. Ford (1965) termed these *dip tubes*, but they may not follow the dip precisely, and Ford and Williams (2007) employ the term *primary tubes* for these features. Where the limestone massif is fractured, then the primary tubes will extend along fractures or bedding planes, zig-zagging to minimise flow resistance. Additional water inputs at joint or joint-bedding intersections will enhance the development of the cave conduit through the process of mixing corrosion.

Thus, the plan form of the developing cave network will vary between a simple, meandering tube (in the absence of dominant structural control) to highly angular or linear conduits where the rock is highly fractured. A more common situation is where there are multiple streamsinks draining to a spring or series of springs located at a base level determined by valley incision or geologic structure. Under this circumstance, competitive development occurs, where one of the single input conduits links to and captures the flow from one or all of its neighbours. This capture occurs because flow resistance is not uniform in the rock mass, conduits form at different times, and pressure heads will not be equal. If one of the primary tube systems is close to the output boundary, then the local hydraulic gradient may be greater, and capture may not occur. Thus, there is complexity in plan form of karst hydrologic systems which depends on nuances of geologic structure and timing of events. Some

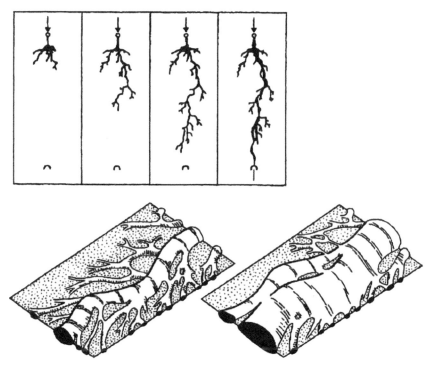

Figure 5.10 Development of cave conduits from a single input according to Ewers (1978). The conduit propagates from sink to spring by preferential development of one proto-conduit; the enlargement of this is shown in the lower detailed view. Source: From Gillieson (1996).

examples of simple down-dip systems with phreatic tubes are Friar's Hole Cave, West Virginia, USA, and at Peak Cavern, Derbyshire, the Kingsdale Master Cave and Lost John's, in the UK. The Subway in Castleguard Cave, Canada, is one of the best linear tube examples yet known. Finally, a classic high-dip example is provided by the Holloch, Switzerland.

5.3.2 Formation of Caves in Length and Depth

So far, the case of cave development along a single bedding plane has been investigated. Now, attention is focused on the more normal case where water entry occurs along multiple bedding planes or fissures, introducing the possibility of complex development in dimensions of length and depth.

Ford and Williams (2007) have proposed a four-state model (Figure 5.11) for the continuum of morphologies from deep phreatic to water table caves. In bathyphreatic or deep phreatic caves, there is a single downward loop below the water table owing to high hydraulic resistance. There may be several elements to the loop, following the linking of conduits developed in separate bedding planes. At the inflow end, there may be a drawdown vadose cave passage on account of the locally greater hydraulic gradient. As seen in Chapter 3, we now have good examples of water-filled bathyphreatic passages thanks to divers.

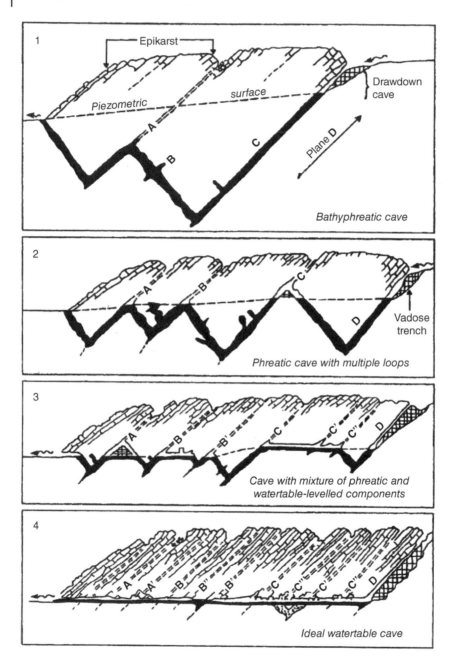

Figure 5.11 The four-state model differentiating the basic types of phreatic and water table caves. Source: From Ford and Williams (2007).

According to Ford and Williams (2007), where there is a higher fissure frequency, multiple loop phreatic caves occur. In such situations, the piezometric surface is initially higher in the rock mass, but as the network develops and enlarges it is lowered until it roughly coincides with the top of the loops. In the Holloch, where such an array of passages has developed along irregular strike passages, the amplitude of the looping is about 100 m, with a maximum value of 180 m. In the reduced relief of the Kentucky sinkhole plain, the amplitude is about 40 m.

With the diminished resistance to flow found in highly fissured rocks, a cave with a mixture of phreatic loops and water table levelled elements may be found. Commonly such systems develop along the strike or along major joints, and they are a very common type of cave. Multiple shallow loops, developed along individual bedding planes, are linked by short horizontal sections. In the Mendip Hills of England (UK), the Swildon's–Wookey system has at least 20 such loops, which are sumps, today. Upstream exploration is at present halted at a depth of 80 m in such a loop. The passages of Clearwater Cave, Mulu, Sarawak, provide a good example of such a system which has been partially drained. There are numerous phreatic tubes at varying levels in the cave – for example, the Revival series, linking horizontal passages developed along the strike. In some sections of the cave, deep river incuts mark former water table positions and may be temporarily invaded by floodwaters today (Figure 5.12). In Selminum Tem Cave, Papua New Guinea, abandoned phreatic tubes up to 30 m in diameter (Figure 5.13) are now perched several hundred metres above the piezometric surface feeding the Kaakil spring (Gillieson 1985).

Minor lowering of the piezometric surface will promote entrenchment of the top of the phreatic loops, giving short sections of vadose canyon (Figure 5.14). Bypass passages develop where blockage of a downward loop by detritus occurs (Farrant and Smart 2011). This phenomenon is very common in tropical caves subject to flash flooding, where large quantities of fine-grained sediment are washed into loops and create a large hydraulic head across the loop. Fissures are opened to form short bypasses, which may propagate downstream. Sometimes the downward loop may be re-excavated (as in Clearwater Cave) to create multiple flow paths dependent on flood stage.

Renault (1968) described a phenomenon termed *paragenesis* where the accumulation of sediment in a phreatic passage promotes solutional erosion at the top at the top of the

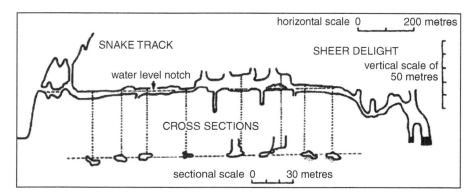

Figure 5.12 Wall notches in a section of Clearwater Cave, Sarawak, Malaysia, mark former stream levels in this multi-level phreatic and water table-levelled system. Source: From Smart et al. (1985).

Figure 5.13 The abandoned phreatic tube of Selminum Tem, Papua New Guinea, is 30 m in diameter and carried flood flows from the flanks of the glaciated Star Mountains at the height of the last Ice Age.

conduit (Figure 5.14) as the water is perched on top of the fill. The convergence of form between such features and normal vadose canyons makes determination of their origin difficult, and few examples cited so far are unequivocal. Lauritzen and Lauritzen (1995) developed a method based on a stereographic analysis of meander migration vectors developed from solutional scallop orientations. This innovation needs testing in a wider range of contexts but is very promising.

Finally, the ideal water table caves with long horizontal passages represent a fourth cave state developed where fissure frequency is high. Direct routes can be excavated between numerous input tubes developed down dip and, with enlargement, these passages can absorb all of the flow. The piezometric surface lowers, yielding canal passages penetrable for hundreds of metres. These may flood to the roof following heavy rain, at which times some localised dynamic phreatic development may occur. This commonly results in flat roof sections cut across the bedding. Many of the long river caves in tectonically mobile regions are of this type. Good examples are provided by the Chuan Yan Cave of Nanxu, Guanxi, China (Waltham 1985); the Caves Branch system, Belize (Miller 1982); and Tham Nan Lang Cave, Thailand (Dunkley 1985), which is so large as to be explorable on elephant back for the first few hundred metres! Many of these caves have deep sediment fills, tens

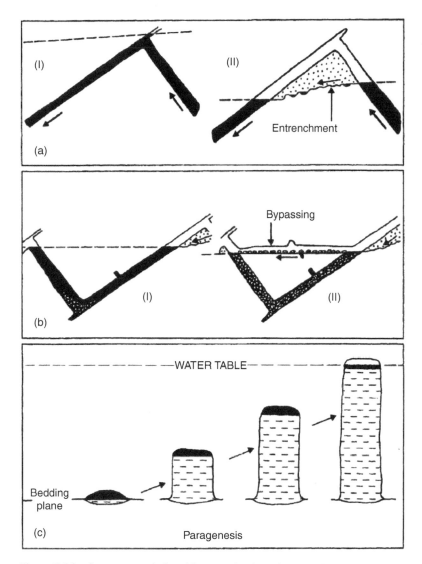

Figure 5.14 Common gradational features in phreatic caves: (a) isolated vadose entrenchment at the apex of a phreatic loop; (b) formation of a bypass tube across the top of a sediment-clogged phreatic loop; (c) upward development of a paragenetic passage to the water table. Source: From Ford and Williams (1989).

of metres thick, in both active and inactive passages. Cave streams may meander across the top of these sediment fills, eroding deep wall incurs which may subsequently be fully exposed by channel lowering.

It should be borne in mind that this simple four-state scheme of cave morphology represents a continuum, with individual caves often exhibiting more than one of the states at a given time. It does not apply as well to those caves where water flow is localised, on account of rainwater inflow in a relatively small impounded karst. Such caves may show both phreatic and vadose passage forms at the same stage of development, and there may be a temporary phreatic state following intense rainfall. Such is the case for the extensive caves in dissected towers of the Chillagoe–Palmer River karst of north Queensland (Australia), and for many small caves in other tropical tower karsts of Malaysia and Thailand.

According to Ford and Williams's (1989) model, fissure frequency exerts a dominant control over the long gradient of caves and the state of water flow. But fissure density evolves through time through the solutional enlargement of joints, providing a further possibility for changes in passage morphology. This is illustrated in Figure 5.15, for a steeply dipping

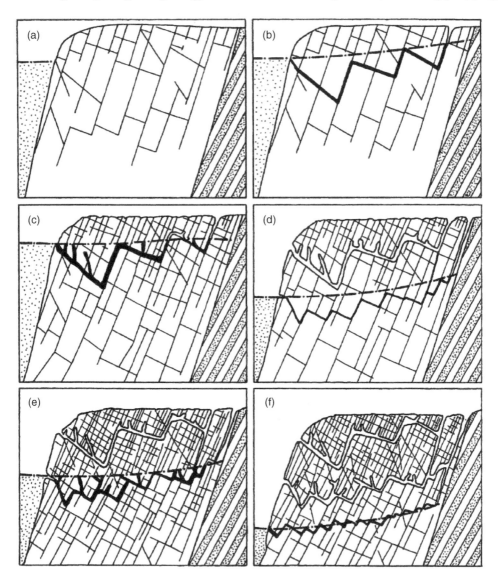

Figure 5.15 The geometry of successive caves in a multiphase system is affected by the increase in fissure frequency over time. First-generation caves are of the state 2, bathyphreatic type, while later caves tend to state 3 (mixed phreatic and water table), and 4 (water table-levelled) types. Source: From Ford and Williams (1989).

limestone. Fissure frequency diminishes markedly with depth in most limestone massifs, though exposed joints in cliffs, dolines, and quarries are but a poor sample of the total. From an initial low-frequency state, increased density of linked fissures allows for the development of a bathyphreatic (type 1) cave system with deep loops. This may develop multiple loops and connections of lesser amplitude along enlarged joints and fissures, giving a transition to types 2 and 3 as the spring lowers and upper passages are abandoned. Finally, in a highly fissured massif, a cave with a mixture of water table-levelled and phreatic passages may form. Such sequences are seen in the caves of Mulu, Sarawak, and in the caves of New Guinea (Farrant et al. 1995; Francis et al. 1980; Smart et al. 1985).

5.3.3 The Formation of Maze Caves

Maze caves often form where flat-lying well-bedded limestones are invaded by floodwaters (Palmer 1975). The juxtaposition of strike belts of limestone and river meanders creates a situation where water readily passes underground and erodes complex networks of intersecting passages, such as Sof Omar, Ethiopia, Bullock Creek, New Zealand (Crawford 1994), and the Atea Kananda, Papua New Guinea (Francis et al. 1980). These act as meander cut-offs and may be found at several levels in an outcrop. Careful inspection of lithology may help to resolve whether multiple levels reflect changing base level control or exploitation of only the more soluble beds in the limestone. When carrying capacity of the main cave passage is reduced by sedimentation (both inorganic and organic), backflooding may occur, creating temporary ponds. Large hydraulic heads may develop during floods, enhancing the enlargement of alternative flow paths. Figure 5.16 shows the flat floor of the Lavani Valley in Papua New Guinea. During the wet season lakes several metres deep form in this polje. A similar situation has been reported for the Baliem River sink of Irian Jaya (Checkley 1993).

Maze caves may also form by hydrothermal action where either carbon dioxide-enriched waters or corrosive sulphuric acid derived from pyrites excavate well-jointed limestones. The first case, where rising waters charged with carbon dioxide dissolve the limestone, is reasonably common and, according to Palmer (1991), accounts for about 10% of caves known and explored. Water circulating deep in the rock mass becomes heated and gains carbon dioxide gas from deep sources. This heated water allows strong thermal convection; as the water rises, it cools, and the carbon dioxide gas dissolves. Three distinct types of cave formed by this mode are recognised:

Rising hot water dissolves limestone upwards from a basal chamber along a branchwork of rising passages, which terminate in cupola-form dissolution pockets formed by slowly eddying waters. The best example is Satorkopuszta Cave in Hungary (Muller and Sarvary 1977).

Water ascending fissures in limestone is halted by an aquiclude which compels the water to spread laterally, forming a two-dimensional maze or network, often joint guided. In some cases, the hot water mixes with cooler groundwater and solution is enhanced by the mixing corrosion. One example is Crossroads Cave, Virginia, USA (Palmer 1991).

By far the most common type is a three-dimensional maze formed in a single phase by rising warm water. Usually, such caves are associated with modern hot springs discharging at stratigraphic boundaries with overlying or underlying non-karstic rocks, or along faults. The relict caves (Baradla system) of Buda Hill, Budapest (Hungary), are of this type. The

Figure 5.16 The Lavani Valley, a polje in the Southern Highlands of Papua New Guinea, is drained by two caves which mark a subterranean drainage divide.

longest examples are Jewel Cave and Wind Cave of South Dakota, where nearly 200 km of complex passages lie beneath the Black Hills (Ford 1985). They are formed in up to 140 m of well-bedded limestones capped by sandstones and shales.

Modern hot springs discharge where this shale cap is breached by valley incision. These caves have little or no relation to the modern topography, and consist of large passages up to 20 m high encrusted in calcite spar speleothems with silica overgrowths. Isotopic study of these formations has shown them to be clearly of thermal origin (Bakalowicz et al. 1987). Regionally heated water from the granitic core of the Black Hills converged and ascended through the limestone prior to discharging at palaeosprings now removed by denudation.

The second case, that of limestone dissolved by waters containing hydrogen sulphide, has produced some of the best-decorated caves in the world. Sulphuric acid can form by the oxidation of hydrogen sulphides derived from sedimentary basins, often associated with hydrocarbons. These sulphide-rich fluids can be released either slowly by the deep circulation of meteoric water, or rapidly by the migration of brines into marginal areas by tectonism, igneous activity, or sediment compaction (Ford 1988; Palmer 1991). The caves of the Guadalupe Mountains (USA), including the well-known Carlsbad Caverns (New Mexico, USA), display a network appearance with great rooms formed in the reefal limestone linked to blind shafts and higher maze passages. The large chambers reflect zones of enhanced solution near the water table and may have formed around existing vugs in the rock. Narrow fissures in their floors, clogged with speleothems, may be the original water entry points. The blind shafts represent the base of the mixing zone where sulphide-charged waters rose to meet meteoric waters. Flow of this acidified water into intersecting fissures has produced maze passages.

If the concentration of dissolved sulphate is great enough, then extensive gypsum formations may result, as in Lechuguilla Cave, New Mexico (Widmer 1998). This is arguably one of the most beautifully decorated caves in the world, containing a great diversity in forms and mineralogies of speleothems.

5.3.4 Tectonic and Eustatic Controls on Cave Development

Most caves in unconfined karsts show several levels or phases (multiphase systems) which relate to the shift of the outlet springs from the system. Usually, this shift is in a negative sense, for instance, due to lowering of the springs in response to base-level change. Positive shifts tend to result in infilling of the caves with sediment rendering them inaccessible and less likely to be explored and mapped. Examples can be found in oil well and water drilling logs as infilled voids or palaeokarst.

Lowering of springs at its simplest case results in cave development along the same fissure system to produce a vadose canyon, but more commonly a new array of passages develops and propagates headwards through the system, with its targets the primary tubes or proto-caves in the rock mass. Where the cave has formed in gently dipping strata the new array may form parallel to the existing passages along the strike and feeding to the new lowered spring.

In Switzerland, the Siebenhengste-Hohgant-St Beatus-Barenschacht system is a multiphase system (Figure 5.17) extending over a vertical range of 1340 m and 156 km of passages over 14 phases or levels (Jeannin and Häuselmann 2019). This multiphase development is attributed to Alpine uplift combined with stripping of cover rocks and valley entrenchment due to both fluvial activity and glaciation (Häuselmann et al. 1999). The earliest five phases of development drainage developed along the strike to the northeast at levels between 1950 and 1505 m asl, with springs in the Eriz valley (Figure 5.18). The lower eight levels including the present drainage fed springs along the Aare valley and Lake Thun to the south. These springs range from 1440 to 558 m asl, with five relict passages becoming reactivated during floods. Many of the passages in this system show deep phreatic loops with amplitudes of 220–250 m.

The effects of glacial erosion on the Siebenhengste system has been investigated by Häuselmann et al. (2007) using cosmogenic dating techniques on cave sediments. The transition from vadose to phreatic passage morphology at various phases of the system provide marker horizons for the altitude of the valley floor at the time of development of that passage. Samples of quartz sediment from different levels were analysed using cosmogenic ^{26}Al and ^{10}Be burial dating. These were deposited during interglacial episodes as inferred from the lack of carbonate rock flour derived from active glaciers. The results indicate burial ages ranging from 4 million years ago for a level at 1800 m to much younger sediments (140 000 years ago) in the St Beatus cave close to present spring levels. From these dates rates of valley incision can be calculated: early rates from the uppermost cave levels are around 120 m/million years while rates increased dramatically to 1200 m/million years around 0.8–1.0 million years ago. This corresponds to a major lowering of the

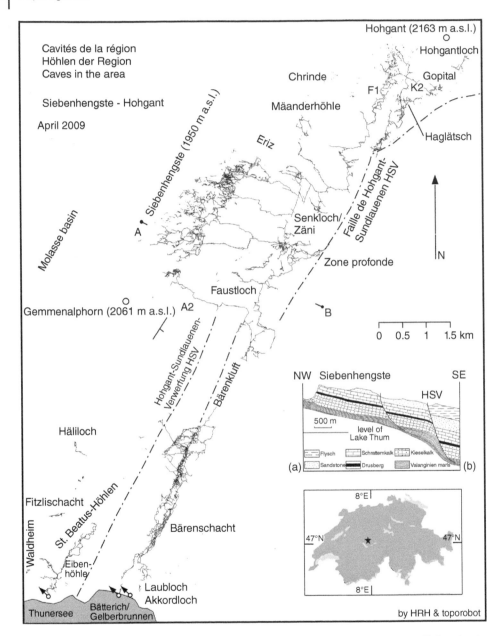

Figure 5.17 Plan of the Siebenhengste-Hohgant-St Beatus-Barenschacht system, Switzerland. Source: https://en.wikipedia.org/wiki/Siebenhengste-Hohgant-Hohle.

equilibrium line altitude and glacial coalescence leading to enhanced valley incision (Brocklehurst and Whipple 2004).

The second deepest cave in the world is Krubera (Voronja) in the Arabika Massif, Western Caucasus, Georgia. The depth of the cave is 2196 m, as of December 2017. It was displaced from the depth record by the nearby Veryovkina Cave (−2204 m) in December 2017. The

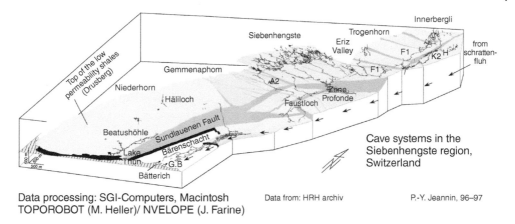

Data processing: SGI-Computers, Macintosh
TOPOROBOT (M. Heller)/ NVELOPE (J. Farine)

Data from: HRH archiv

P.-Y. Jeannin, 96–97

Figure 5.18 Three-dimensional view of the Siebenhengtse cave system (Switzerland). Source: From Jeannin and Häuselmann (2012).

exploration and mapping of the Krubera Cave since the early 1980s has been carried out by the highly dedicated Ukrainian Speleological Association and has been documented by Klimchouk (2012). Detailed hydrogeological studies in the Arabika Massif have shown that a deep and well-coordinated karst hydrological system exists, extending from the crest of the Berchil'sky anticline (Figure 5.19) steeply down through the limestone beds to substantial springs at and below the level of the Black Sea to the south-west (Klimchouk et al. 2009). The caves are dominantly combinations of vadose shafts and steep meandering passages over

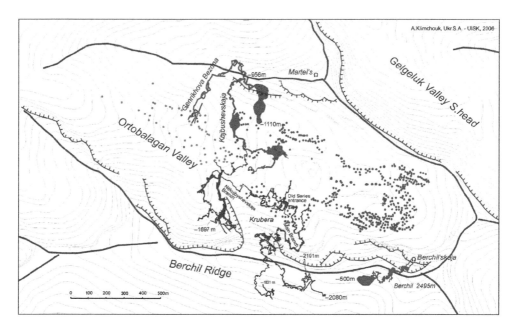

Figure 5.19 Major caves in the Ortobalagan Valley, Georgia. Red dots indicate dolines. Source: From Klimchouk (2012).

a depth range of 1660–1700 m, below which the passages become more inclined down-dip and show more evidence of phreatic conduits with some vadose downcutting elements.

The Arabika Massif is composed of Cretaceous and Jurassic limestones that dip southwest to the Black Sea. The Cretaceous cover is discontinuous over the area; in the Ortobalagan valley, the Jurassic rocks that dominate the upper 1000 m of the caves are sandy limestones. The tectonic structure of the Arabika Massif is dominated by a series of anticlines, with the Berchil anticline breached by the glacial Ortobalagan valley. A series of faults strongly control both cave development and groundwater flow from the central area to the Black Sea. These faults bound several large blocks of limestone which were uplifted in Plio-Pleistocene times.

Boreholes along the shore of the Black Sea yield karst groundwater from depths between 40 and 280 m. These and other submarine springs support the concept of deep karst groundwater circulation beyond the limits of Pleistocene sea-level fluctuations (around −125 m). Off the coast, the Arabika Submarine Depression (ASD) is around 400 m deep, and it is likely that it is fed by the karst system from the Arabika Massif. This flow of water is most likely guided by the fault systems and not constrained by the anticlinal structures; an idea proven by detailed dye-tracing experiments carried out in the 1980s. Its depth supports the hypothesis (Klimchouk 2012) that the earliest karst development in the massif originated in response to the Messinian salinity crisis of 5 million years before the present, when the Black Sea almost dried up. A sea-level drop of 1500 m is well established for the Mediterranean Sea, where deep karst circulation has been demonstrated. The development of karst systems deep below present sea level would be enhanced by the steep hydraulic gradients obtained. The early systems were flooded during the Pliocene marine transgression, after which differential uplift of the limestone blocks occurred. A number of zones of cavities with high conduit porosity due to the Messinian event were created at various depths (Figure 5.20) and acted as target zones for the developing extensive high-gradient

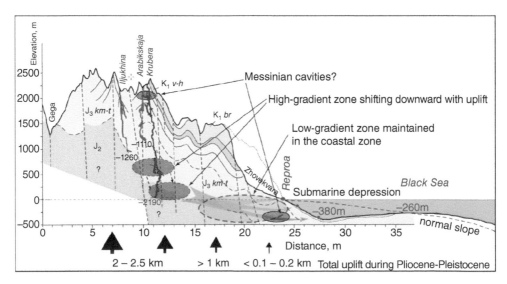

Figure 5.20 Geological and hydrological cross-section of the Arabika Massif, Georgia. Source: From Klimchouk (2012).

conduits. Stalagmite with dates of >200 000 years ago from depths of 1640 m and 1820 m support the idea of vadose conduits operating at depth during the middle to late Pleistocene. Two samples from fossil passages located at elevations of 2016–1906 m asl are dated beyond the limits (>500 000 years ago) of Uranium-Thorium series dating. The significant vertical development of Krubera Cave was influenced by the rapid uplift of the Arabika Massif during Pliocene-Pleistocene time (Figure 5.21). The total uplift was 2–2.5 km in the Arabika

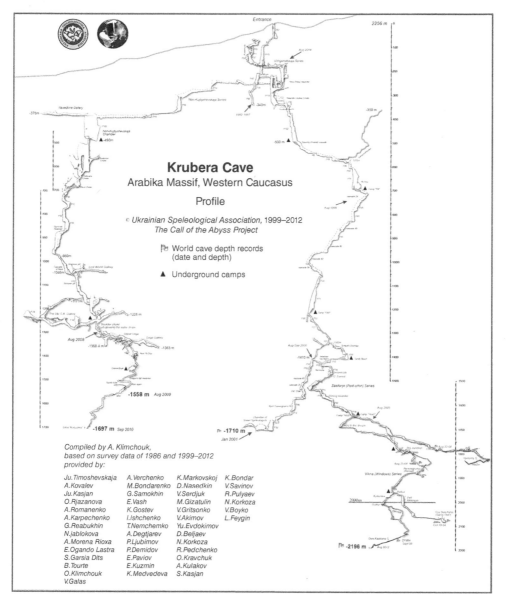

Figure 5.21 Profile of Krubera Cave, from Call of the Abyss project. Sourced from http://kruberacave.info/index.php/en/10-en/krubera-cave/24-topographic.

Massif, but much less close to the coast. Thus, the long uplift history of the Caucasus, couple with the Messinian event, resulted in the formation of a very deep system of caves, whose exploration continues today.

The Gunung Mulu karst in Sarawak hosts some of the longest caves in Southeast Asia. The Clearwater Cave system extends over 238 km of passages on many levels. Most of the cave passages are of phreatic origin and formed as a series of loops crossing the bedding planes. Drainage from the overlying sandstone also sank into the limestone down-dip, creating phreatic tubes with passages rising up to 100 m to intersect other bedding planes.

The geology of the area is comprised of three main lithologies. The Mulu Formation sandstones and shales are of Late Cretaceous – Eocene age and are 5–6 km thick. Overlying them are the shallow marine rocks of the Melinau Limestone, which are Late Eocene – Early Miocene in age. They are overlain in turn by the Setap Shales of Middle Oligocene – Early Miocene age. The Melinau Limestone outcrops as a spectacular line of karst hills reaching a maximum altitude of 1682 m at the summit of Gunung Api. The limestone is divided into four main blocks by gorges (Figure 5.22). From south to north, the Southern Hills have been separated from Gunung Api by the Melinau Paku valley. The spectacular Melinau Gorge divides Gunung Api from Gunung Benarat, while the Medalam Gorge divides Gunung Benarat from Gunung Buda. The limestones are pure and massively bedded, dipping 80° to the northwest, while faults and joints run to the northeast.

It is likely that there was pre-existing karstic relief on the limestone at the time of deposition of the Setap Shales, some 20 million years ago. Near the Mulu airport limestone pinnacles with relief of 5–10 m extend through the thin margin of the shale. Uplift occurred during the late Pliocene to Pleistocene, around 2 million years before present. This uplift was accompanied by folding and faulting and produced a synclinal trough along which the Melinau River runs today. The tension associated with this folding produced a dense network of conjugate joints which dissect the limestone and create the rapid infiltration resulting in features such as the Pinnacles on Gunung Api. A fault runs along the western edge of the limestone massifs, and another parallel fault is associated with the Clearwater Cave system. Another fault system is associated with the enormous Sarawak Chamber in Gua Nasib Bagus (Good Luck Cave). There is strong structural control over the cave passages in the Clearwater Cave system (Figure 5.23). The majority of cave passages propagated between sinks and springs as a series of loops guided along major joints crossing the bedding planes along which the passages formed. At a later stage, a large amount of gravel and mud was washed into the cave, blocking lower passages, and permitting paragenetic development to bypass the down-loops while bedrock canyons were cut though the up-loops. This sediment fill can be 50 m thick and is easily seen in Lagangs Cave, as well as in Stone Horse and Clearwater caves.

The high-level passages of Clearwater Cave are developed along the strike, forming an array of tubes developed on a few bedding planes. The hydraulic gradients involved, fed by water draining off the Mulu Sandstone, allowed phreatic tubes rising up to 100 m to intersect other bedding planes (Waltham 2004). Vadose development is present, for example, the 30 m high canyon in Clearwater Cave, but the original phreatic elements are clearly seen in the roof.

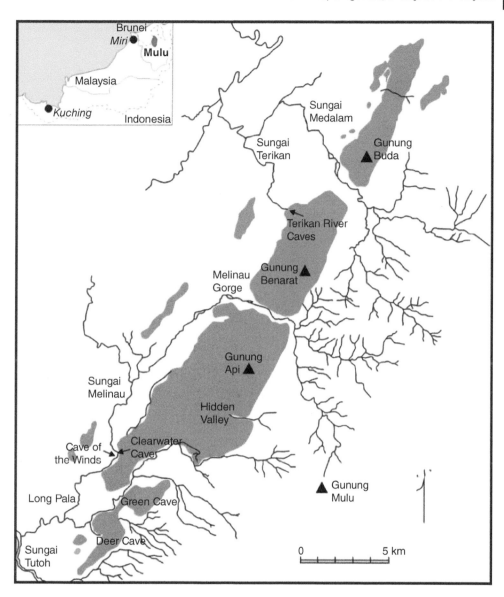

Figure 5.22 Gunung Mulu National Park with limestone massifs (Malaysia). Source: From Gillieson and Clark (2010).

As the overlying Setap Shale was stripped from the limestone, due to the incision of the Melinau valley, drainage from rain falling on the limestone massif created vertical shafts and vadose invasion passages, which intersected the drained phreatic tubes. There is, therefore, an age sequence from east to west, made more complex by reactivation of older passages when the lower cave passages became blocked by sediments. The stream sinks on the eastern side of the limestone massifs have also migrated over time, as alluvial fans formed in Mulu Sandstone sediments created a convex surface over which the surface streams could

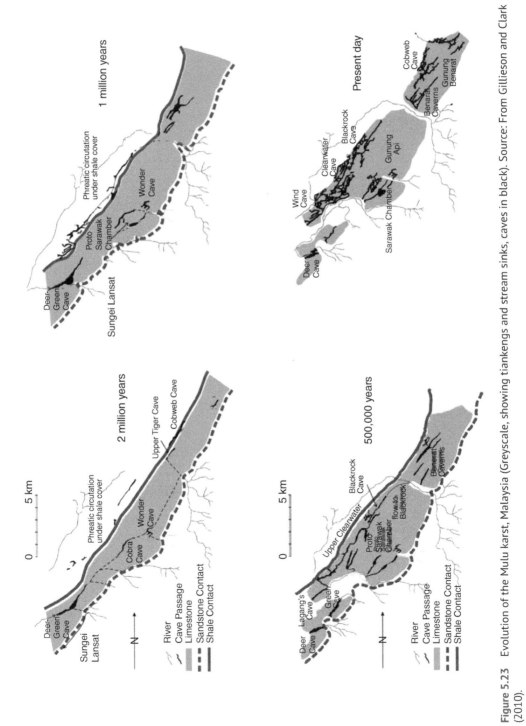

Figure 5.23 Evolution of the Mulu karst, Malaysia (Greyscale, showing tiankengs and stream sinks, caves in black). Source: From Gillieson and Clark (2010).

Figure 5.24 Fluvial solution notch in the wall of Clearwater Cave, Mulu, Malaysia.

migrate. Several caves show clear relationships with the headwards retreat and lateral shift of streamsinks along the flank of the Melinau Paku valley (Figure 5.23). The youngest caves in the sequence are the active springs from Wind Cave and the downstream parts of Clearwater Cave (Figure 5.23). Racer Cave is a former spring with a series of sediment-filled chambers leading towards a former sink in the Paku valley. The furthest downstream, and potentially oldest, is Lagangs Cave. This cave has large phreatic tubes with roof pendants, abundant speleothems, and both finely laminated clays and gravels.

The widespread cave sediments are still intact at all altitude levels, affording the opportunity for important scientific investigations of climatic change (Farrant et al. 1995). Paleomagnetic dating revealed a Plio–Pleistocene uplift rate of 190 mm/1000 years. Reversal of the earth's magnetic field, recorded in the sediments at 1.8 million years before present, indicates that the caves are at least 2 million years old, possibly as much as 3 million years. Notches in the walls of the caves at various levels over a range of 300 m vertically can be correlated with interglacial periods. Interglacial gravel aggradation created local base levels at which notches, typically 2 m high and 4–5 m deep, formed. Several of these can be traced over a distance of 2–3 km (Figure 5.24). At the lowest cave levels, at grade with the Melinau River, these solutional notches are forming today. Thus, the Mulu caves show interplay between uplift, glacial, and interglacial cycles and local catchment changes, which combine to produce some of the most spectacular and interesting caves on the planet.

Table 5.2 Deepest shafts of the world as of September 2019.

Rank	Name	Country	Depth (m)	Source
1	Vrtoglavica	Slovenia	603	Toni Palcic
2	Ghar-e-Ghala	Iran	562	Saadat Lotfi Zadeh
3	Patkov gust	Croatia	553	Toni Palcic
4	Da Keng	China	519	Descent N194P32
5	Lukina jama	Croatia	516	Descent N194P32
6	Velebita	Croatia	513	Velebit Speleological Society
7	Brezno pod velbom	Slovenia	501	Toni Palcic
8	Miao Keng	China	491	Erin e-mail
9	Melkboden-Eishohle	Austria	451	Descent N194P32
10	Hollenhohle Hades	Austria	450	Atlas – Great Caves of the World

Most of these are associated with major fault systems in tectonically active areas.
Source: www.caverbob.com.

5.3.5 Deep Shafts of the World

The current deep shafts in caves are listed in Table 5.2. The world's deepest individual roped pitch is currently in Vrtiglavica Cave in Slovenia at −603 m. Vrtiglavica (from the Slovene vrtoglavica 'vertigo') is located on the Kanin Plateau, in the Western Julian Alps, on the Slovene side of the border between Slovenia and Italy. The cave formed in a karst landscape that was subjected to Pleistocene glacial activity. The cave also contains one of the tallest cave waterfalls in the world; the estimated height of the falls is 400 m to 440 m. It was discovered in summer 1996 by Italian speleologists, and the bottom was reached on 12 October 1996, by a joint Slovene–Italian expedition.

The Kanin Plateau is dissected by large glaciokarst valleys. The Dachstein limestone is hundreds of meters thick and is underlain by dolomite. The area is tectonically active with deep fractures and fissures. Drainage flows to the southeast towards the Boka and Glijun springs, which are part of the Soča river basin. The combination of soluble carbonates and a series of Pleistocene glaciations created an exceptional glacial landscape in which meltwater flowed into the karst, creating many high-mountain abysses of exceptional depths. By 2010, more than three hundred caves had been explored in the wider area of the Kanin, two of which are deeper than 1000 m. The deep caves comprise a series of abysses connected by narrow canyons, the so-called meanders. At depth these shafts may intersect horizontal galleries created in completely different hydrologic and climatic conditions.

In the Zagros Mountains of western Iran, a large team of cavers led by Yousef Sorninia have descended the second deepest shaft in the world. It is located in the Parau massif near Kermanshah, and less than 6 km north of Ghar Parau. A single pitch descent, broken only by a traverse on a blockage at −358 m, descends to a large chamber where there is a sump which has yet to be dived (Figure 5.25). At 562 m, Ghar-e-Ghala is just nine metres deeper

(a)   (b)

Figure 5.25 Ghar-e-Ghala entrance and section, Zagros Mountains, Iran. Note caver on the far side of entrance. Source: https://darknessbelow.co.uk/wp-content/cache/all/iranian-cavers-discover-one-of-the-worlds-deepest-shafts/index.html.

than the 553 m pitch in Patkov Gust, in Croatia. The shaft was rigged with 700 m of rope, and includes three traverses and 40 rebelays! The entrance, at an altitude of 2794 m, is partially blocked with snow and ice. Surveying was completed in September 2016.

5.4 Hypogene Speleogenesis

In the last 20–30 years, there has been a minor revolution in our thinking about the formation of caves. Traditionally most cave formation has been attributed to epigenetic processes, where meteoric water percolates through soil and rock, gains aggressivity as carbonic acid, and then dissolves the bedrock to form cave conduits between sinks and springs. The formation of caves by water rising from below – as hydrothermal sources – was regarded as only accounting for 10% of the known caves (Palmer 2007). These cavities were poorly known, difficult to document and study, and probably grossly underestimated. Improvements in cave exploration techniques, cave diving, and the isotopic analysis of cave minerals have all led to the realisation that caves formed in deep-seated situations by water rising from

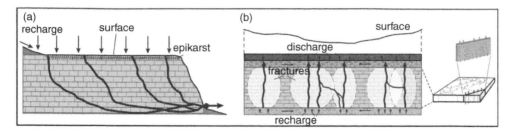

Figure 5.26 Conceptual representation of epigene (a) versus hypogene (b) speleogenesis. Source: From Klimchouk (2009).

depth are more common than hitherto realised (Figure 5.26). The chemical mechanisms involved included warm water charged with carbon dioxide, the oxidation of hydrogen sulphide to sulphuric acid, and the oxidation of organic matter derived from hydrocarbons (Dublyansky 2000).

Deep-seated sources of acidity may be associated with volcanism, where the production of gases such as carbon dioxide and hydrogen sulphide are common. Some carbon dioxide may also be produced during metamorphism, while the chemical degradation of organic matter in mudstones and marls may also produce hydrogen sulphide. This process is aided by sulphur reducing anaerobic bacteria (Ford and Williams 2007). Hydrogen sulphide and its by-products become more effective as agents of rock solution as the water rises to the surface. The mixing of hydrogen sulphide with oxygenated water produces sulphuric acid, especially near the water table where large cavities can form.

Even small amounts of sulphide minerals, such as pyrite (FeS) and marcasite (FeS$_2$), can be readily oxidised to form sulphuric acid. These minerals are quite common in carbonate rocks and may segregate in discrete layers, creating inception horizons in which proto-caves can form (Lowe and Gunn 1995). The oxidation of sulphide beds within Proterozoic dolomite has been held to be the cause of the development of the Toca da Boa Vista and nearby Tocada Barriguda caves in Brazil (Auler and Smart 2003). Mine drainage waters often contain sulphidic waste which oxidises to sulphuric acid. The oxidation of siderite (FeCO$_3$) in drainage waters can produce carbonic acid and iron-rich residues, as seen in small caves in Slovakia (Bosak et al. 2002).

H$_2$S can be generated by volcanic activity, reduction of sulphates, such as gypsum or anhydrite, in the presence of hydrocarbons (petroleum, methane) or organic carbon from marls, and is brought to the surface via deep tectonic structures. The origin of the sulphur can usually be ascertained using its stable isotope signature and that of its possible sources (Onac et al. 2011). The oxidation of H$_2$S produces sulphuric acid that reacts instantaneously with the carbonate host rock producing replacement gypsum and carbon dioxide. CO$_2$ can dissolve in water again and increase its aggressiveness even more. Also the local oxidation of sulphides, such as pyrite, often present in carbonate sequences, can generate sulphuric acid, boosting rock dissolution (D'Angeli et al. 2019).

Cave development by hypogene processes does not necessarily imply development at great depth. It is defined as '*the formation of solution-enlarged permeability structures by water that recharges the cavernous zone from below, independent of recharge from the overlying or immediately adjacent surface*' (Ford 2006). Hypogene caves can form by a variety of geochemical mechanisms, as briefly outlined above, including those that do not rely on acidification of groundwater (such as the dissolution of evaporites, such as gypsum). Hypogene caves formed in different lithologies and by different geochemical processes show great similarity in their passage morphology, small scale solution features, structural, and stratigraphic settings, which suggests a common hydrogeologic context.

5.4.1 Solutional Mesoforms as Indicators of Hypogene Origin

The following cave patterns are commonly seen in hypogenic caves, though they are not necessarily exclusive to that mode of origin (Klimchouk 2009):

- Zones of cavernous porosity
- Network maze
- Spongework maze
- Isolated passages or small clusters of passages
- Irregular isolated chambers
- Rising, steeply inclined passages or shafts
- Collapse shafts over large hypogenic voids

Multiple storey 3D patterns are also quite characteristic of hypogene caves. Where horizontal maze caves are involved, rectilinear networks of passages in the same stratigraphic unit are frequently seen. A good example is provided by the 118 km long Bullita Cave in the Northern Territory of Australia (Figure 5.27).

The various storeys may be connected by rising shafts or chimneys. Vertically extended 3D patterns (Figure 5.28) are comprised of chambers, clusters of networks or spongework mazes, and inclined shafts connecting them.

The most common features associated with hypogenic origin are cupolas or solution pockets, blind smooth solutional forms which may be a few metres across. These solution pockets are also found in phreatic caves, but in the hypogene setting, they are usually grouped where fluid flow paths can be discerned. These take the form of rising inlet conduits (risers or feeders) through wall and ceiling pockets and half-tubes. They feed cupolas and dome pits, large shafts with no discernible inlet or outlet. The combination of these features is termed the morphologic suite of rising flow (MSRF) and provides diagnostic evidence for hypogene speleogenesis (Figure 5.29).

At the base, a crack or feeder supplies water from the underlying aquifer. Similarly, at the top the hypogene conduit terminates in a cupola or blind chimney at the contact with overlying strata. These points of exchange may be small cracks or zones of higher porosity. Between feeder and outlet there may be rising wall channels and cupolas, which may extend over a significant height range and may be replicated in several storeys. The largest maze caves on the planet, such as Optimisticeskaya gypsum cave in Ukraine, Jewel Cave in the Black Hills of South Dakota, or Lechuguilla in the Guadalupe Mountains of New Mexico are of this type.

5.4.2 Condensation and Corrosion in Passage Enlargement

Condensation-corrosion (Audra et al. 2007) is an active speleogenetic process in thermal caves where a high thermal gradient drives air convection currents. Wall retreat rates are

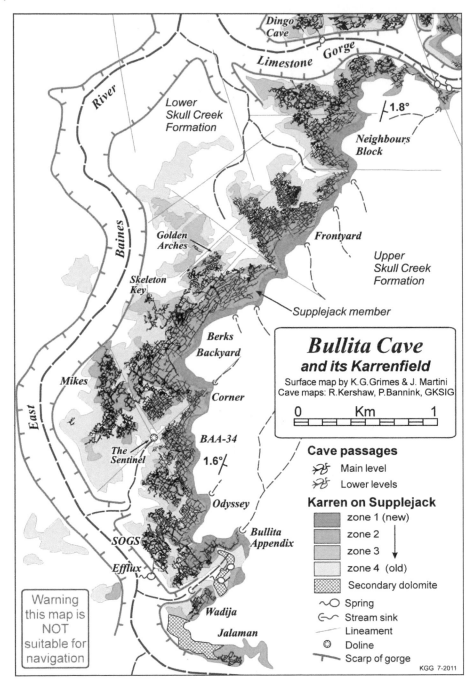

Figure 5.27 Bullita Cave (Norther Territory, Australia) and its karrenfield. Compilation from cave maps by Kershaw and Bannink, and personal field observations. Source: From Martini and Grimes (2012).

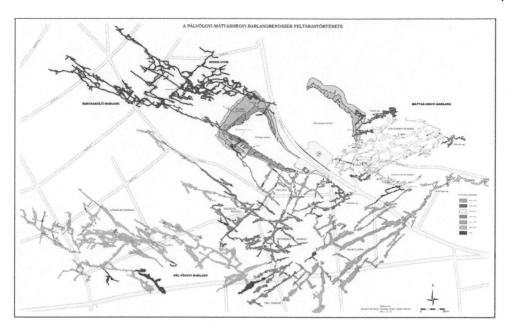

Figure 5.28 Caves of the Pál Valley, Budapest, Hungary. Source: https://dailynewshungary.com/3-caves-vicinity-budapest-adventure-seekers.

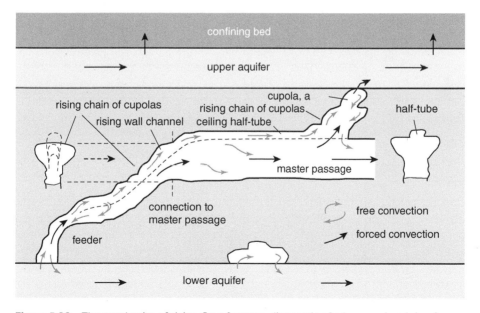

Figure 5.29 The usual suite of rising flow features, diagnostic of a hypogenic origin of caves. Source: From Klimchouk (2009).

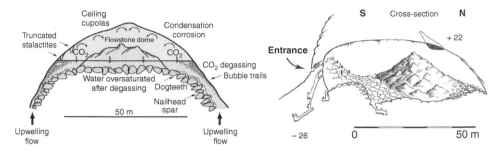

Figure 5.30 The Champignons Cave, Provence, France, is an isolated chamber. Thermal hypogenic flow degassed CO_2 at shallow depth. Thermal convection enhanced condensation-corrosion to develop a large isolated chamber more than 50 m wide, which tends towards a hemispherical shape. Simultaneously, massive calcite deposits occurred in the lake, which was supersaturated with calcite because of CO_2 degassing. Source: From Audra et al. (2002).

much greater than in caves dissolved by meteoric water. Condensation occurs at the contact of rising air with the cool walls of large spheres or cupolas (Osborne 2004); conversely, evaporation occurs at the output of narrow passages where air sinks down from the upper sphere. Condensation is greatly enhanced where there are thermal differences between the upwelling waters and the cave walls and atmosphere, even where the temperature contrast reaches only a few degrees. The corrosion distribution in the developing void varies according to thermal rock conductivity and causes the sphere to develop upwards. CO_2 and H_2S degassing greatly enhance the aggressivity of the condensation water. The void may enlarge laterally and may diverge into a series of stacked cupolas or spheres, as is seen in the Sátorkö-puszta Cave of Buda, Hungary (Szunyogh 1990). There a series of stacked spheres, isolated by narrow necks, can be seen arranged in a dendritic pattern. The development of two neighbouring spheres will be divergent which may provide the greatest potential heat transfer. This mode of development is clearly active in the vadose zone above a thermal water table (Figure 5.30). Conversely, evaporation produces depositional processes by replacement of limestone by gypsum and by aerosol decantation leading to the formation of cave popcorn.

5.5 Flank Margin Speleogenesis

Earlier in this chapter, we considered the formation of caves in very young limestone at the coast, termed eogenetic karst. This limestone has not been buried or compacted and has high porosity. This is very common in raised coral reefs around the world, but especially in the Pacific and Indian Oceans, and the Caribbean Sea (Taboroši et al. 2003). The water table in this limestone is largely determined by sea level. Fresh water sits on top of saline water which extends under the coast forming a halocline. This is complicated over time by the fact that sea levels have been fluctuating over Quaternary times between +6 and − 125 m relative to the present sea level.

While calcareous sand dunes and beach ridges accumulate, a calcrete caprock can form, up to several tens of centimetres thick. This can be strong enough to support a cave roof while unconsolidated material is washed out from underneath. These shallow syngenetic

Figure 5.31 Shallow caves developed under caprock in syngenetic karst, West Bay, Kangaroo Island, South Australia.

caves (Figure 5.31) were first described by Jennings (1968) and are quite common in young limestone on the coastal fringe of southern Australia and South Africa (Marker 1993). Where this limestone occurs next to coastal swamps, shallow spongework caves can be quite extensive, as at Kelly Hill Caves, Kangaroo Island, South Australia. If the outwashing of material is limited, shallow pits or shafts can develop, often guided by tree roots. Finally, allogenic streams may sink into the eogenetic karst creating stream cave conduits, as at Margaret River, Western Australia (Eberhard 2005).

It is worth noting that short inclined caves also occur in living fringing reefs, often on the edge of lagoons. These drain the lagoon at low tide and exit onto the reef front at depths of 10–15 m. If the reef is raised by tectonic activity, then these may become part of the assemblage of coastal caves.

Caves can develop very quickly along the coast where the freshwater table meets the sea. An enhanced solution of limestone in the mixing zone of fresh and salt water creates flank margin caves (Jensen et al. 2002; Mylroie and Carew 1990). The mixing dissolution front recedes rapidly into the rock leaving wide, low roofed cavities between pillars of more resistant rock (Figure 5.32). Subsequent coastal retreat or slumping may expose these cavities and allow human entry. The ramifying cavities terminate abruptly, but in some circumstances may be invaded by small tubes fed by rainwater inflow. One of the best examples of a flank margin cave is Lirio Cave, Isla de Mona, Puerto Rico (Figure 5.33) (Mylroie and Jenson 2002). This is an exceptionally large network of chambers which extend 250 m into the rock.

Figure 5.32 Flank margin caves in Pleistocene dune limestone, Ravine des Casoars, Kangaroo Island, South Australia.

5.6 Caves Formed in Gypsum

Two of the world's longest caves are formed in the Upper Tertiary gypsum of western Ukraine. Each is well over 100 km long, and the two caves are separated horizontally by only 750 m. They are both maze caves (Figure 5.34) with a very high passage density. The caves are established in gypsum beds ranging from 10 to 40 m thick and have formed at two or three levels controlled by insoluble clay layers; in plan, all the passages are controlled by joints which are perpendicular to each other (Figure 5.35). This jointing is a consequence of block faulting on a transition slope between the East European platform and the Pre-Carpathian foredeep. They are uniformly elliptical tubes, elongated along the bedding, and are typically 3 m to 10 m wide and 3–8 m high.

Two theories have been advanced to account for their formation. Originally it was thought that intermittent surface streams sank into the gypsum and formed caves of shallow phreatic and water table origin controlled by the jointing (Dubljansky 1979; Jakucs and Mezosi 1986). A theory proposed by Klimchouk (1992) suggests the upward development of the tubes from phreatic passages at the base of the gypsum, via discharge outlets for groundwater linking lower and upper passages. There are thousands of these outlets distributed uniformly along the master passages, and the evidence supports the theory.

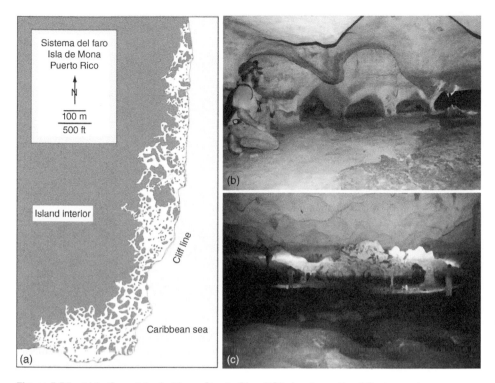

Figure 5.33 Lirio Cave, Isle de Mona, Puertp Rico, USA. A: schematic of flank margin cave development B: Oblique air photo of the cave entrances C: Plan view of the entire cave illustrating typical flank margin cave passages. Source: From Mylroie and Jenson (2002).

5.7 Lava Tubes, Weathering Caves, and Pseudokarst

The formation of lava tubes, weathering caves, and sandstone caves is another aspect of the subject where the dividing line between true solution and disintegration of the rock may be hard to define. Caves in rocks other than limestone are widespread and sometimes are very large. There may be gross similarity in form, although the processes leading to this are quite different.

5.7.1 The Formation of Lava Tubes

Volcanic lava has a number of flow styles which depend on the viscosity of the molten rock. Basaltic lavas flow readily, and an internal network develops which conveys the liquid rock from the erupting vent to the advancing toe of the lava flow. This may be likened to an arterial system, with tributary branches, distributaries, and anastomoses. This flow style is known as tube-fed pahoehoe (Swanson 1973), and it is identified in solidified flows by the ropy surface and lobate structure leading to surface mounds or tumuli. Lava caves, the drained segments of lava tube networks, are part of this landform suite. The precise mode of formation of the lava caves was a matter of some controversy until observations were made during the 1969–1974 eruptions of the Mauna Ulu vent of Kilauea volcano, Hawaii

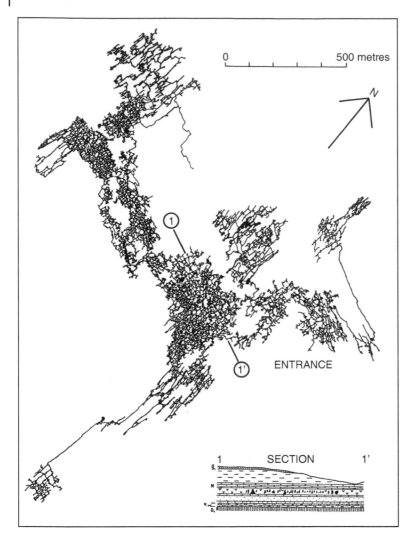

Figure 5.34 Plan of Optymistychna Cave, Ukraine. This 257 km long maze cave is formed in gypsum strata separated by aquicludes of clay. Source: From Courbon et al. (1989).

(Peterson and Swanson 1974). They confirmed that many of the large feeder tubes start as open lava canyons (similar to stream channels). Although the molten lava is at a temperature between 1000 and 1200 °C, the surface of the flow loses heat rapidly to the air and forms a surface crust adhering to the banks. This crust may extend to bridge the flow completely, or fragments may detach and travel downflow until they jam across the channel. In another mode, liquid lava splashing and spattering near the vent may build inward-leaning levees which eventually meet to bridge the flow. Irrespective of the precise mechanism of bridging, these roofs are strengthened by further channel overflows until they stabilise. Tube-fed pahoehoe flows advance by the overlapping of successive small tongues, fed from the main axial mass of the molten lava. Smaller tubes or canyons may form as distributaries, from

Figure 5.35 Typical passage in Ozernaja gypsum maze cave, Ukraine. Source: Photo by John Gunn.

which small lava tongues develop at the edges of the flow or on its surface. Thus, in plan, there will usually be a main axial or feeder tube, many tens of kilometres long, which has a large diameter. These large tubes are sinuous and may be braided; there are numerous terrestrial examples as well as extra-terrestrial examples from Mars (Baker 1981). At the distal end of the flow, there is usually a network of braided distributary tubes and channels, similar to a delta. Here the rapid heat loss causes clogging of the distributaries, frequent overflows and development of new tubes and channels.

Complex and extensive caves are usually associated with the drainage of the axial feeder tube. These tubes are full of lava only during periods of high discharge, as lava delivery from the vents is pulsatory. Long continued flow may cause channel erosion, in a manner akin to water in stream channels, lowering the liquid level. The insulating properties of surrounding basaltic rock will permit some flow after vent activity has ceased, allowing lava flow into lower tubes and channels in a layered process. This sub-crustal drainage (Ollier 1977) may produce isolated cave segments, sealed above and below, which may be entered only after roof collapse has taken place. Liquid elements in a largely cooled and solidified flow may exploit junctions or laminae between successive flows and enlarge them by remelting, producing further tube systems. Examples of this type are the multi-level lava caves of Mount Hamilton and Byaduk, western Victoria, Australia (Joyce 1980). Sagging of the roof may also produce surface depressions, linking cave entrances, which in plan give an idea of the form of the original axial feeder tube.

Currently, the longest and deepest lava tube in the world is Kazumura Cave in Hawaii, USA (Halliday 1995), which has a depth of 1101 m and a length of 65 500 m. It extends over

28 km downflow and is a complex multi-level cave of the type described above. Elsewhere, its nearest competitor is Cueva del Vento in the Canary Islands (Spain), which is 560 m deep and 17 032 m long.

Rarely lava flows may invade solution cavities in limestone. In Australia, examples of in situ lava infilling have been described at Timor caves, in the upper Hunter valley (Connolly and Francis 1979; Osborne 1986) and at Bunyan, in southern New South Wales (Osborne 1979). These palaeokarsts are given a minimum age by the basalts, which date from the mid-Tertiary.

5.7.2 Weathering Caves and Pseudokarst

Pseudokarst processes involve the non-solutional erosion of rock to produce cavities, which may be isolated voids or connected passages or tubes, or surface karst features such as enclosed depressions and minor sculpturing of rock surfaces to resemble *karren*. Solutional features in non-carbonate rocks, such as sandstone, gypsum, and halite, are excluded from this definition. However, solution can assist pseudokarst processes, so the distinction should be made according to the dominant process suite.

There are many examples of weathering caves in granite, laterite, and other granular rocks. Most of these are simple rock shelters, though many have small tubes fed by seepage moisture and developed along bedding planes or joints. The dominant process in these is granular disintegration in which the interstitial cement is removed by solution or by corrasion. Thus, caves formed in sandstone may be considered to be true karst or pseudokarst, depending on the dominant mode of passage enlargement. The large cave systems formed in the Proterozoic sandstones of Venezuela attain depths of 370 m and lengths of 2500 m. In these caves, there are clearly solutional features alongside features due to granular disintegration. Under ever-wet tropical conditions even the low equilibrium solubility of silica, perhaps aided by acidic drainage waters, can over a long time produce substantial solutional conduits. On a more modest scale, the ancient sandstones of northern Australia have many stream caves and a type of 'beehive terrain' which superficially resembles cone karst (Jennings 1985; Young 1986). Finlayson (1982) has described small stream caves developed in granite in southern Queensland, while Greenhorn Cave (USA) is a much larger granite cave (Courbon et al. 1989).

5.8 Life History and Antiquity of Caves

A major change in thinking about the origin of caves has resulted from the discoveries made in the last decade by a growing number of cave divers who have described cavities below the water rest level. In air-filled caves with sediments, we may be looking only at a very small proportion of the evidence; excavations and seismic analysis demonstrate that the depth of sediment may be tens of metres. In addition, the realisation that palaeokarstic cave forms and infills are widespread has radically revised thinking about the age of individual cave systems. Thus, the antiquity of caves is no longer constrained into the timeframe of the Quaternary, and it is clear that many cave systems have existed in some form throughout both the Quaternary and the Tertiary. In some cases, relict cave passages may be as old as the Mesozoic (Osborne and Branagan 1988). Thus, consideration of cave development must take account of tectonic and climatic change over timescales where our knowledge of the conditions obtaining is very fragmentary. Reliable age estimates from radiometric and microfossil analysis may help to constrain the age of the speleothems and sediments, but not necessarily the cavities which they infill.

Given these conceptual revisions, it now seems likely that any cave system may have formed in a variety of ways at various times. Elucidation of the sequence of cave formation must rest on careful examination of morphology above and below the present water rest level.

The life history of a single cave may be envisaged as passing through a sequence of stages, not necessarily irreversibly, from an initial state of no conduits to a long inception state during which certain horizons or inhomogeneities in the rock mass act to channel water into preferred pathways. Following this, a developmental phase, in which conduits form and enlarge, will persist and will respond to external factors of base-level change, rock mass evolution, and tectonism. Finally, in a multi-level system, passages will be abandoned and will collapse, leading to a final state of no connected, enterable caves at a particular level. The sequence may be idealised to:

- Inception
- Gestation or nothephreatic
- Turbulent threshold
- Conduit and cave development
- Decay and abandonment.

Gestation implies a phase of slow growth prior to the creation of a conduit. During timescales of speleogenesis which span millions, tens of millions, or even hundreds of millions of years, processes similar to those of today have operated, and their products can be seen – either wholly preserved or in fragments (Lowe 1992). From these, process suites can be reconstructed. Equally other processes have operated to produce relict features which have no contemporary analogues. There is a problem of convergence of forms from differing processes as well. The only way through this speleogenetic maze is careful observation, appropriate dating, and healthy scepticism.

5.9 Geological Control and the World's Longest Cave

Mammoth Cave, Kentucky, is currently the longest cave in the world – 667 km – and elegantly displays the role of geology in determining passage shape and orientation. The cave system is located at the south-eastern edge of the broad Illinois structural basin, in limestone of Mississippian (early Carboniferous) age which dips gently to the northwest. The cavernous limestone is underlain by impure limestone of similar age and overlain by Mississippian and Pennsylvanian (late Carboniferous) sandstone, shale, and thin limestone.

The limestone forms a broad plain of low relief, the Pennyroyal Plateau, punctuated by dolines and dissected by several sinking streams draining to the Green River to the north. The border of the limestone plain is an upland (the Chester cuesta) composed of limestone ridges capped by sandstones and other insoluble rocks. The Mammoth Cave system lies under these ridges, which are separated by broad valleys. Mammoth Cave occupies a 110 m thick sequence of limestones (Figure 5.36), with the lowest levels in the St Louis Limestone containing chert beds and some gypsum. The Ste. Genevieve Limestone in the middle of the sequence contains most of the cave passages and is made up of interbedded limestone and dolomite with some silty beds in the upper half. The Girkin Formation at the top of the sequence is 20 m to 40 m thick and consists of limestone with thin interbedded shales and siltstones.

Most canyons and tubes in Mammoth Cave are highly concordant with the bedding of the limestone (Palmer 1981). Thus, the main passage leading in from the Historic Entrance follows the same beds of the Ste. Genevieve Limestone for several kilometres. Passages tend to parallel the strata because there are very few joints, and they rarely intersect more than one bed. In the south-eastern area of Mammoth Cave, abrupt changes in passage level are associated with faults and major joints. This continuity in passages makes it possible to correlate passages and their sediments on either side of valleys. For example, Great Onyx Cave is now isolated from the rest of the Mammoth Cave system (Figure 5.37), but its stratigraphic position and passage forms suggest that it is genetically linked to Salts Cave and Dyer Avenue in the Flint Ridge system, also located in the Paoli Member of the Girkin Formation. The large canyon of Kentucky Avenue can be traced stratigraphically in the Joppa Member of the Ste. Genevieve Formation and correlates with Broadway and Gothic Avenue in the Historic Tour section of the cave (Palmer 2017).

The three most common passage types in Mammoth Cave are canyons, low gradient tubular passages and vertical shafts. The tubular passages are generally elliptical, reflecting the control of bedding (Figure 5.38). They trend along the strike, and their sinuosity is controlled by local variations in dip and strike of the controlling bedding plane. The vadose canyons (Figure 5.39) are much modified by collapse, but can be seen to drain down the dip.

The numerous vertical shafts in the cave are controlled by major joint intersections, with inflowing water perched on bedding planes or relatively resistant beds.

The dip tube of Cleaveland Avenue in Mammoth Cave shows geological control very well. It is about 1500 m long with almost no slope (4 m km^{-1}), and the limestone beds exposed in its walls show little variation along its length; it is clearly running almost exactly along the strike. Broad bends in the passage are caused by two gentle folds that push the tube to the west on the nose of an anticline and to the east into the trough of a syncline. It formed in the phreatic zone, and solution proceeded evenly at the junction of the Joppa and Karnak

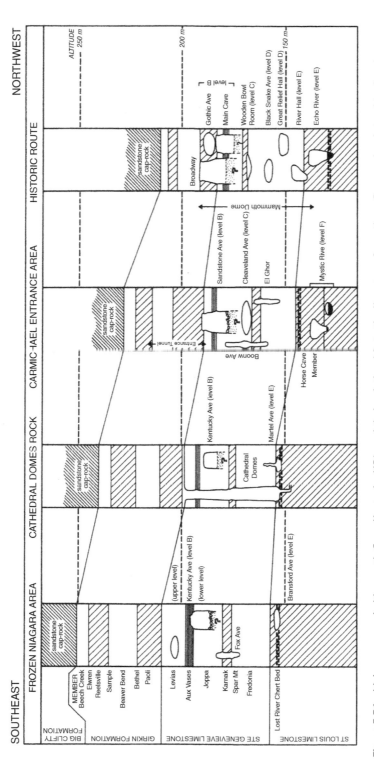

Figure 5.36 Major passages in Mammoth Cave, Kentucky, USA, and their relationship to the limestone formations. Passage levels tend to rise along a gradient away from the Green River. Source: From Palmer (1981).

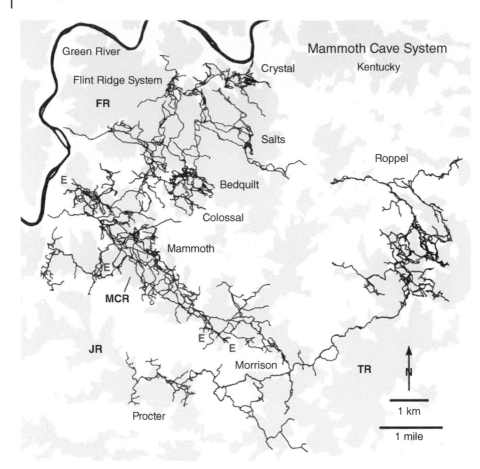

Figure 5.37 Map of the Mammoth Cave System (USA) and its relation to the Green River and local topography, simplified from map by the Cave Research Foundation. Map of the Roppel Cave section by the Central Kentucky Karst Coalition. Names of major sections of the system are shown. Green sandstone-capped ridges; yellow limestone-floored valleys. MCR: Mammoth Cave Ridge, JR: Joppa Ridge, FR: Flint Ridge, TR: Toohey Ridge. E: major entrances to tour routes of Mammoth Cave. Source: From Palmer (2017).

members of the Ste. Genevieve Limestone to produce an elliptical tube with gently scalloped walls (Figure 5.40).

The concentration of large passages at certain levels is a combination of this stratigraphic control and the progressive downcutting of the Green River. Thus, the passages cluster at three distinct elevations, 180, 168, and 152 m above sea level. Most passages at other elevations are shafts, narrow canyons, and tubes (Figure 5.41). These passages have been backflooded by the river at the time when their levels were accordant, leaving the underground equivalents of floodplains and terraces. These deposits, and the speleothems that cap them, have provided a rich source of data for Quaternary scientists.

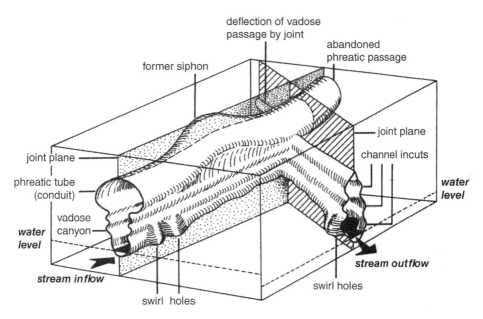

Figure 5.38 Principal morphological types of passages in Mammoth Cave, Kentucky, USA, showing the role of joint intersections. Source: From Quinlan et al. (1983).

Figure 5.39 Vadose canyon heavily modified by collapse processes at Dyer Avenue, Crystal Cave, Kentucky, USA.

Figure 5.40 Cleaveland Avenue in Lower Mammoth Cave, Kentucky, USA.

Figure 5.41 The main passage of Indian Cave, Mammoth Cave National Park, Kentucky, USA, an elliptical phreatic tube controlled by bedding subsequently modified by collapse. Source: Photo by Gareth Davies.

References

Audra, P. and Palmer, A.N. (2015). Research Frontiers in Speleogenesis. Dominant processes, hydrogeological conditions and resulting cave patterns. *Acta Carsologica* 44: 315–348.

Audra, P., Bigot, J.-Y., and Mocochain, L. (2002). Hypogenic caves in Provence (France). Specific features and sediments. *Acta Carsologica* 3: 33–50.

Audra, P., Hobléa, F., Bigot, J.-Y., and Nobécourt, J.-C. (2007). The role of condensation-corrosion in thermal speleogenesis. Study of a hypogenic sulfidic cave in Aix-Les-Bains. *Acta Carsologica* 2: 185–194.

Auler, A.S. and Smart, P.L. (2003). The influence of bedrock-derived acidity in the development of surface and underground karst: evidence from the Precambrian carbonates of semi-arid northeastern Brazil. *Earth Surface Processes and Landforms* 28: 157–168.

Bakalowicz, M.J., Ford, D.C., Miller, T.E. et al. (1987). Thermal genesis of dissolution caves in the Black Hills, South Dakota. *Geological Society of America Bulletin* 99: 729–738.

Baker, V.R. (1981). Pseudokarst on Mars. In: *Proceedings of the 8th International Congress of Speleology*, 63–65. Kentucky: Department of Geography and Geology, Western Kentucky University.

Bögli, A. (1980). *Karst Hydrology and Physical Speleology*. Berlin: Springer.

Bosak, P., Bella, P., Cilek, V. et al. (2002). Ochtina aragonite cave (Western Carpathians, Slovakia): morphology, mineralogy of the fill and genesis. *Geologica Carpathica* 53: 399–410.

Brocklehurst, S.H. and Whipple, K.X. (2004). Hypsometry of glaciated landscapes. *Earth Surface Processes and Landforms* 29: 907–926.

Charity, R.A.P. and Christopher, N.S.J. (1977). The Ogod Ffynnon Ddu cave system, South Wales, in relation to the structure of the Carboniferous Limestone. In: *Proceedings of the 7th International Congress of Speleology, 1977 Sheffield*, 108–109. British Cave Research Association.

Checkley, D. (1993). Cave of thunder: the exploration of the Baliem River cave, Irian Jaya, Indonesia. *International Caver* 6: 11–17.

Connolly, M. and Francis, G. (1979). Cave and landscape evolution at Isaacs Creek, New South Wales. *Helictite* 17: 5–24.

Courbon, P., Chabert, C., Bosted, P., and Lindsley, K. (1989). *Atlas of the Great Caves of the World*. St Louis: Cave Books.

Crawford, S.J. (1994). Hydrology and geomorphology of the Paparoa karst, North Westland, New Zealand Unpublished PhD thesis. University of Auckland.

Curl, R.L. (1986). Fractal dimensions and geometries of caves. *Mathematical Geology* 18: 765–783.

D'Angeli, I.M., Parise, M., Vattano, M. et al. (2019). Sulfuric acid caves of Italy: a review. *Geomorphology* 333: 105–122.

Dubljansky, V.N. (1979). The gypsum caves of the Ukraine. *Cave Geology* 1: 163–183.

Dublyansky, Y.V. (2000). *Hydrothermal Speleogenesis in the Hungarian Karst. Speleogenesis: Evolution of Karst Aquifers*. Huntsville, Alabama: National Speleological Society.

Dunkley, J.R. (1985). Karst and caves of the Nam Lang-Nam Khong region, Thailand. *Helictite* 23: 3–22.

Eberhard, S. (2005). Ecology and hydrology of a threatened groundwater-dependent ecosystem: the Jewel cave karst system in Western Australia. *Journal of Cave and Karst Studies*: 67.

Ewers, R.O. (1978). A model for the development of broad scale networks of groundwater flow in steeply dipping carbonate aquifers. *Transactions of the British Cave Research Association* 5: 121–125.

Farrant, A.R. and Smart, P.L. (2011). Role of sediment in speleogenesis; sedimentation and paragenesis. *Geomorphology* 134: 79–93.

Farrant, A.R., Smart, P.L., Whitaker, F.F., and Tarling, D.H. (1995). Long-term Quaternary uplift rates inferred from limestone caves in Sarawak, Malaysia. *Geology* 23: 357–360.

Finlayson, B.L. (1982). Granite caves in Girraween National Park, Southeast Queensland. *Helictite* 20: 53–59.

Ford, D.C. (1965). The origin of limestone caverns: a model from the Central Mendip Hills, England. *National Speleological Society Bulletin* 27: 109–132.

Ford, D.C. (1985). Dynamics of the karst system; a review of recent work in North America. *Annales de la Société Géologique de Belgique* 108.

Ford, D.C. (1988). Characteristics of dissolutional cave systems in carbonate rocks. In: *Paleokarst* (eds. N.P. James and P.W. Choquette). New York: Springer New York.

Ford, D.C. (2006). Karst geomorphology, caves and cave deposits: a review of north American contributions during the past half century. In: *Perspectives on Karst Geomorphology, Hydrology, and Geochemistry: A Tribute to Derek C. Ford and William B. White* (eds. R.S. Harmon and C.M. Wicks). Boulder, Colorado: Geological Society Of America.

Ford, D.C. and Ewers, R.O. (1978). The development of limestone cave systems in the dimensions of length and depth. *Canadian Journal of Earth Sciences* 15: 1783–1798.

Ford, D.C. and Williams, P.W. (1989). *Karst Geomorphology and Hydrology*. London: Unwin Hyman.

Ford, D.C. and Williams, P.W. (2007). *Karst Hydrogeology and Geomorphology*. Chichester, England: Wiley.

Francis, G., James, J.M., Gillieson, D.S., and Montgomery, N.R. (1980). *Underground Geomorphology of the Muller Plateau*. Sydney: Speleological Research Council, University of Sydney.

Gilli, E. (1993). Les grandes volumes souterrains du massif de Mulu (Borneo, Sarawak, Malaisie). *Karstologia*: 1–14.

Gillieson, D.S. (1985). Geomorphic development of limestone caves in the Highlands of Papua New Guinea. *Zeitschrift für Geomorphologie* 29: 51–70.

Gillieson, D.S. and Clark, B. (2010). Mulu: the World's Most Spectacular Tropical Karst. In: *Geomorphological Landscapes of the World* (ed. P. Migon). Springer.

Gillieson, D.S. and Spate, A.P. (1992). The Nullarbor karst. In: *Geology, Climate, Hydrology and Karst Formation: Field Symposium in Australia* (ed. D.S. Gillieson). Canberra: Department of Geography and Oceanography, University College, Australian Defence Force Academy.

Halliday, W.R. (1995). Puna emergency road proposal – Kazumura Cave. http://www.halcyon.com/samara/nssccmslpunaI.html (Accessed March 1995).

Häuselmann, P., Jeannin, P.-Y., and Bitterli, T. (1999). Relationships between karst and tectonics: case-study of the cave system north of Lake Thun (Bern, Switzerland). *Geodinamica Acta* 12: 377–388.

Häuselmann, P., Granger, D.E., Jeannin, P.-Y., and Lauritzen, S.-E. (2007). Abrupt glacial valley incision at 0.8 ma dated from cave deposits in Switzerland. *Geology* 35: 143–146.

Hose, L.D. (1981). Fold development in the Anticlinorio Huizachal-Peregrina and its influence on the Sistema Purificacion, Mexico. In: *Proceedings of the 8th International Congress of Speleology*, 133–135. Kentucky.

Jakucs, L. and Mezosi, G. (1986). Genetic problems of the huge gypsum caves of the Ukraine. *Acta Geographica* 16: 15–38.

Jeannin, P.-Y. and Häuselmann, P. (2012). Siebenhengste Cave System, Switzerland. In: *Encyclopedia of Caves*, 2e (ed. D.C. Culver). Amsterdam, The Netherlands: Elsevier/Academic Press.

Jeannin, P.-Y. and Häuselmann, P. (2019). Siebenhengste cave system, Switzerland. In: *Encyclopedia of Caves*, 3e (eds. W.B. White, D.C. Culver and T. Pipan). Academic Press.

Jennings, J.N. (1968). Geomorphology of Barber Cave, Cooleman plain, New South Wales. *Helictite* 6.

Jennings, J.N. (1985). *Karst Geomorphology*. Oxford: Blackwell.

Jensen, J.W., Mylroie, J.E., Mylroie, J.R., and Wexel, C. (2002). Revisiting the carbonate island karst model. In: *Proceedings of the Denver Annual Meeting*, 226. Denver, Colorado: Geological Society of America.

Joyce, E.B. (1980). Origin of lava caves. In: *Proceedings of the 13th Biennial Conference of the Australian Speleological Federation*, 40–48. Melbourne.

Klimchouk, A.B. (1992). Large gypsum caves in the Western Ukraine and their genesis. *Cave Science* 19: 3–11.

Klimchouk, A.B. (2000). Speleogenesis of the great gypsum mazes in the Western Ukraine. In: *Speleogenesis: Evolution of Karst Aquifers* (ed. A.B. Klimchouk). Huntsville, Alabama, USA: National Speleological Society.

Klimchouk, A.B. (2009). Morphogenesis of hypogenic caves. *Geomorphology* 106: 100–117.

Klimchouk, A.B., Samokhin, G.V., Cheng, H., and Edwards, R.L. (2009). Dating of speleothems from deep parts of the world's deepest cave – Krubera (Arabika Massif, western Caucasus). In: *Proceedings 15th International Congress of Speleology, Kerrville, Texas*, 1032. United States of America: National Cave & Karst Research Institute.

Klimchouk, A.B. (2012). Krubera (Voronja) Cave. In: *Encyclopedia of Caves* (eds. W.B. White and D.C. Culver), 443–450. San Diego: Elsevier Science & Technology.

Klimchouk, A.B. (2013). Hypogene speleogenesis, its hydrogeological significance and the role in karst evolution, Simferopol, DIP.

Klimchouk, A.B. (2015). The karst paradigm: changes, trends and perspectives. *Acta Carsologica* 44: 289–313.

Lace, M.J. and Mylroie, J.E. (eds.) (2013). *Coastal Karst Landforms*. Netherlands: Springer.

Lauritzen, S.E. and Lauritzen, A. (1995). Differential diagnosis of paragenetic and vadose canyons. *Cave and Karst Science* 21: 55–60.

Laverty, M. (1987). Fractals in Karst. *Earth Surface Processes and Landforms* 12: 475–480.

Lowe, D.J. (1989). Limestones and caves of the Forest of Dean. In: *Limestones and Caves of Wales* (ed. T.D. Ford), 106–116. Cambridge: Cambridge University Press.

Lowe, D.J. (1992). A historical review of concepts of speleogenesis. *Cave Science* 19: 63–90.

Lowe, D.J. and Gunn, J. (1995). The role of strong acid in speleo-inception and subsequent cave development. *Acta Geographica* 34: 33–60.

Lowry, D.C. and Jennings, J.N. (1974). The Nullarbor karst, Australia. *Zeitschrift für Geomorphologie* 18: 35–81.

Marker, M.E. (1993). Syngenetic karst in the southern cape, South Africa. *Cave Science* 20: 51–54.

Martini, J.E.J. and Grimes, K.G. (2012). Epikarstic maze cave development: Bullita cave system, Judbarra/Gregory karst, tropical Australia. *Helictite* 41: 37–66.

Miller, T.E. (1982). Hydrochemistry, hydrology and morphology of the caves branch karst, Belize. PhD Thesis. McMaster University.

Muller, P. and Sarvary, I. (1977). Some aspects of developments in Hungarian speleology theories during the last ten years. *Karszt-es Barlang*: 53–59.

Mylroie, J.E. and Carew, J.L. (1990). The flank margin model for dissolution cave development in carbonate platforms. *Earth Surface Processes and Landforms* 15: 413–424.

Mylroie, J.E. and Jenson, J.W. (2002). Karst flow systems in young carbonate islands. *Hydrogeology and biology of post-Paleozoic carbonate aquifers, Karst Waters Institute Special Publication* 7: 107–110.

Mylroie, J.R. and Mylroie, J.E. (2007). Development of the carbonate island karst model. *Journal of Cave and Karst Studies* 69: 59–75.

Ollier, C.D. (1977). Lava caves, lava channels and layered lava, Proceedings of Atti de Seminario sulle Grotte Laviche. pp. 149–158. Catania: Gruppo Grotte Catania.

Onac, B.P., Wynn, J.G., and Sumrall, J.B. (2011). Tracing the sources of cave sulfates: a unique case from Cerna Valley, Romania. *Chemical Geology* 288: 105–114.

Osborne, R.A.L. (1979). Preliminary report on a cave in Tertiary basalt at Coolah, New South Wales. *Helictite* 171: 25–29.

Osborne, R.A.L. (1986). Cave and landscape chronology at Timor caves, New South Wales. *Journal and Proceedings of the Royal Society of New South Wales* 119: 55–76.

Osborne, R.A.L. (2004). The troubles with cupolas. *Acta Carsologica* 32: 9–36.

Osborne, R.A.L. and Branagan, D.F. (1988). Karst landscapes of New South Wales, Australia. *Earth-Science Reviews* 25: 467–480.

Palmer, A.N. (1975). The origin of maze caves. *National Speleological Society Bulletin* 37: 56–76.

Palmer, A.N. (1981). *A Geological Guide to Mammoth Cave National Park*. Teaneck, New Jersey: Zephyrus Press.

Palmer, A.N. (1991). Origin and morphology of limestone caves. *Geological Society of America Bulletin* 103: 1–21.

Palmer, A.N. (2007). *Cave Geology*. Dayton, Ohio: Cave Books.

Palmer, A.N. (2017). Geology of Mammoth Cave. In: *Mammoth Cave – A Human and Natural History* (eds. Hobbs III, H.H., Olson, R.A., et al.), 97–109. Switzerland: Springer.

Peterson, D.N. and Swanson, D.A. (1974). Observed formation of lava tubes during 1970–1971 at Kilauea volcano, Hawaii. *Studies in Speleology* 2: 209–222.

Quinlan, J.F., Ewers, R.O., Ray, J.A. et al. (1983). Ground-water hydrology and geomorphology of the Mammoth Cave Region, Kentucky, and of the Mitchell Plain, Indiana. In: *Field trips in Midwestern geology*. Bloomington: Geological Society of America and Indiana Geological Survey.

Renault, P. (1968). Contribution à l'étude des actions mécaniques et sédimentologiques dans la spéléogenèse. *Annales De Spéléologie* 23: 530–596.

Smart, P.L., Bull, P.A., Rose, J. et al. (1985). Surface and underground fluvial activity in the Gunung Mulu National Park, Sarawak: a palaeoclimatic interpretation. In: *Tropical*

Geomorphology and Environmental Change (eds. I. Douglas and T. Spencer), 123–148. London: Allen & Unwin.

Swanson, D.A. (1973). Pahoehoe flows from the 1969–1971 Mauna Ulu eruption, Kilauea volcano, Hawaii. *GSA Bulletin* 84: 615–626.

Szunyogh, G. (1990). Theoretical investigation of the development of spheroidal niches of thermal water origin – second approximation. In: *Proceedings of the 10th International Congress of Speleology, 1989*, 766–768. Budapest: Hungarian Speleological Society.

Taboroši, D., Jenson, J.W., and Mylroie, J.E. (2003). Zones of enhanced dissolution and associated cave morphology in an uplifted carbonate island karst aquifer, northern Guam, Mariana Islands. *Speleogenesis and Evolution of Karst Aquifers* 1: 1–16.

Waltham, A.C. (ed.) (1985). *China Caves '85*. London: Royal Geographical Society.

Waltham, A.C. (2004). Mulu – the ultimate cavernous karst. *Geology Today* 13: 216–222.

White, W.B. and White, E.L. (2013). *Karst Hydrology: Concepts from the Mammoth Cave Area*. Springer Science & Business Media.

Widmer, U.F. (1998). *Lechuguilla: Jewel of the Underground*. Basel: Speleo Projects.

Young, R.W. (1986). Tower Karst in Sandstone: Bungle Bungle massif, northwestern Australia. *Zeitschrift für Geomorphologie* 30: 189–202.

6

Cave Interior Deposits

6.1 Introduction

Cave formations or speleothems make up one of several kinds of cave interior deposits, the others being in situ breakdown materials and clastic sediments transported mechanically into and deposited within the cave (White 1976). Cave sediments, both coarse and fine, will be further discussed in Chapter 7. Speleothems are secondary chemical precipitates derived from cool or hot water circulating in the karst. They display successive parallel bands or laminations which represent progressive stages of their growth.

These cave formations may be composed of only one or a combination of over two hundred known minerals which range in form from thin crusts a few millimetres thick to vertical columns 20 m high (Figure 6.1).

There is an enormous literature on the subject of speleothems. They have been the subject of review chapters (or parts thereof) in some key karst texts (e.g. Fairchild et al. 2007; Ford and Williams 2007; Jennings 1985; Palmer 2007; White 1976; White 1988) and have been described in great detail by Hill (1976) and Hill and Forti (1997), the latter of whom cite over 2000 references. The part-chapter in Ford and Williams (2007) book devoted to cave formations is broader in scope but less detailed. While this global literature is principally concerned with speleothems formed from calcite and other carbonate minerals, exploration and documentation of caves in other karst rocks (such as sandstone and evaporites) has revealed many new minerals and associated forms. This is especially true in the humid tropics, where organic compounds derived from guano may combine with inorganic chemicals to produce a complex range of carbonate and phosphate minerals.

Speleothem shape does not depend wholly upon the properties of the component mineral but is controlled exclusively by the type of motion of the water that generates them (Table 6.1). All speleothems are precipitates forming at boundary layers between rock and air, rock and water, or air and water.

Speleothems forming from dripping water occur where supersaturated water enters an air-filled cave. They develop on the roof where water drips or runs, where the droplets land, or where splashes hit the wall or floor. The most common speleothems on a roof are stalactites or soda straws, both of which have an internal feeding channel. A single droplet forming on the roof can precipitate crystals within itself, and as long as the drip rate remains high, then a straw will form, which for a given drop size will be a constant diameter of about 5 mm (Figure 6.2). The minimum diameter of a stalactite depends exclusively on

Caves: Processes, Development, and Management, Second Edition. David Shaw Gillieson.
© 2021 John Wiley & Sons Ltd. Published 2021 by John Wiley & Sons Ltd.

Figure 6.1 A massive speleothem column dwarfs the caver in Khan Hall, Kubla Khan Cave, Tasmania, Australia.

Table 6.1 Modes of water motion and resulting speleothem shapes.

Motion of water	Main resulting speleothem shapes
Droplets	Fall – stalactites, straws, draperies, shields
	Impact – stalagmites, conulites, splash circles
Running water	Flowstones, rimstone pools, crusts, columns, moonmilk
Submersion	Cave pearls, moonmilk, concretions, pool deposits
Evaporation	Rafts, cave coral, trays
Capillary	Helictites, cave coral, anthodites
Condensation	Rims, boxworks, cave coral, moonmilk

Source: Based on Forti (2009).

the chemical-physical characteristics of the water (density and surface tension) and on the force of gravity. The longest such straws recorded are 4.2 m in the Grotte de la Clamouse, France, and 6.2 m in Strong's Cave, Western Australia. Soda straws are usually monocrystalline, and often of calcite, with the c-axis parallel to the growth direction (Onac 1997). Initially, the crystal orientation is totally random because the crystals grow epitaxially on the

Figure 6.2 Straw speleothems about 1 m long in Jewel Cave, Augusta, Western Australia.

mineral grains forming the cave roof. However, crystals oriented with their c-axis parallel to the growth direction prevail after only a short time due to competitive selection. Not all crystals have the same opportunity to develop, given that, during growth, some crystals hinder the development of neighbouring crystals (selective competition). Only those crystals that have their major growth axis parallel to the direction of development will survive (Frisia et al. 2000).

If the drip rate slows, then crystals may deposit on the outside, forming a conical tube or stalactite (Figure 6.3). Stalactites grow under the influence of two different mechanisms: along the growth direction, like straws, and radially, because of the influence of water that flows along their external boundaries. Due to the effects of selective competition, the c-axis orientation of the crystals will be vertically downwards along the zone surrounding the feeding tube and radial within the external parts of the stalactite. Selective competition is exclusively active if the level of supersaturation is neither too low nor too high.

Another widespread speleothem shape is a drapery, curtain or shawl (Figure 6.4). This can form if water droplets flow along the roof for a significant distance before detaching. In this way, a 'track' of precipitate is deposited, with the average width of a water droplet (5.1 mm). The tracks follow each other vertically, forming a drapery. Lateral enlargement occurs only if water starts flowing along the outside of its vertical walls. Variations in the chemistry of the feed water and rates of flow can cause banding and some variation in

Figure 6.3 Stalactites and columns reflected in a pool, Lake Cave, Margaret River, Western Australia.

thickness. A change in the saturation level of the feed water can cause draperies to partially dissolve, leading to small holes or slots in the drapery.

Stalagmites are convex speleothems that grow upward, commonly fed by a corresponding overhead stalactite. They show flat, rounded, or slightly hollow tops, but lack a central feeding channel. Calcite and aragonite stalagmites form through degassing of a saturated film of water (about 0.1 mm thick) which slowly flows down the flanks of the speleothem from its tip. The form of stalagmites is mainly determined by the drip rate (Dreybrodt and Romanov 2008), and in particular by the interannual variation in the drip rate and the distance from the tip of the feeding stalactite. Given similar levels of saturation. Drips with low interannual variation in drip rate produce candle-shaped stalagmites with near-constant diameter (Figure 6.5), while drips that show a high interannual variation in drip rate produce cone-shaped stalagmites. A change in the saturation level will also produce changes in form; this may be due to vegetation change on the surface above the cave due to fire or other vegetation change (Coleborn et al. 2018). Very high drip rates and waters with low saturation levels form flowstones, where calcite precipitation is controlled by degassing over a larger surface area, coupled with a turbulent flow regime that promotes more rapid gas exchange than the laminar flow on top of stalagmites.

Figure 6.4   Shawl formation 250 cm high in Junction Cave, Wombeyan, New South Wales, Australia.

Figure 6.5   Massive stalagmites in the Hall of the Thirteen, Gouffre Berger, France. Photo by Gareth Davies.

Helictites are elongated speleothems that may grow in any direction. They may be straight, smoothly curving, or spiral, and range in size from a few centimetres up to a metre in length. The most common minerals forming them are aragonite and calcite; the former tend to produce straight speleothems which may be called anthodites, while the calcite forms tend to be curved. All helictites have a central pore or tube of capillary dimensions fed by hydrostatic pressure. In carbonate helictites, the tip extends by the growth of calcite crystals around the pore as the moisture evaporates or degasses (Figure 6.6). The rate of flow is too low for a hanging drop to form at the tip; if the water flow rate increases then a straw may form. Moore (1954) explained helictite curvature by a combination of impurities, crystal axis rotation and stacking of wedge-shaped crystals.

Anthodites are clusters of slender, branching tubes which originate near a point and radiate from it (Figure 6.7). They are clearly closely related to helictites and a mix of aragonite and calcite, with transitional forms that are hard to classify. Frostwork consists of sprays of branching aragonite needles and may grow from thin films of carbonate hearing moisture over the crystal surfaces. It is usually found in caves below dolomitic bedrock and where there is strong evaporation, such as strong air drafts. If evaporation concentrates the

Figure 6.6 Helictite formation in Barellan Cave, Jenolan, New South Wales, Australia. Source: Photo by Andy Spate.

Figure 6.7 Quill anthodites composed of aragonite in Shishkabob Cave, Mole Creek, Tasmania, Australia.

solution sufficiently so that magnesium minerals precipitate, then blobs of moonmilk can form on the anthodite tips. Impressive examples of this speleothem are found in Lechuguilla Cave, New Mexico and in the caves of the Black Hills, South Dakota, USA.

6.2 Carbonates

Carbonates are the most common mineral deposits found in caves (Ford and Williams 2007; Hill and Forti 1997); the 10 secondary carbonate minerals listed by Hill and Forti (1997) are those associated with 'normal' karst waters (i.e. not derived from geothermal water or from ore bodies). Of these, calcite is by far the most important and, together with aragonite, constitutes perhaps 95% of all cave minerals. The incidence of other carbonates is occasional or rare and usually confined to particular chemical environments or environmental conditions (Table 6.2).

The main processes by which carbonate formations are deposited are by CO_2 loss or degassing, and by evaporation. In the case of CO_2 loss, percolation waters rich in dissolved carbon dioxide of soil origin enter the cave passage through fissures and pores (Figure 6.8). Once exposed to the cave atmosphere, which is relatively depleted in CO_2, the percolation waters will equilibrate with their new surrounds and CO_2 will be lost from solution.

Table 6.2 Carbonate minerals commonly found in caves.

Name	Formula	Crystal system	Distinctive properties
Aragonite	$CaCO_3$	Orthorhombic	Colourless or white
			Transparent to translucent
			Acicular habit: usually in the form of needles in radiating groups
			Common cave mineral
			Specific gravity determination:
			o Aragonite – 2.95
			o Bromoform – 2.85
			o Calcite – 2.75
			Calcite will float in bromoform while aragonite will sink
Calcite	$CaCO_3$	Trigonal	Usually colourless or white, but can be stained various shades of red, tan, or grey
			Transparent to translucent
			Hardness: 3
			Usually massive, but sometimes in the form of a scalenohedron or rhombohedron Extremely high birefringence
			The most common cave mineral Effervesces vigorously in cold dilute hydrochloric acid
Dolomite	$CaMg(CO_3)_2$	Trigonal	White, tan, or pink
			Transparent to translucent
			Effervesces slightly in cold dilute acid
			Rare in caves Must be distinguished from calcite by optical or X-ray diffraction techniques
Huntite	$CaMg_3(CO_3)_4$	Trigonal	White Dull, powdery, very fine-grained
			Rare in caves X-ray techniques required for positive identification
Hydromagnesite	$Mg_5(CO_3)_4$ $(OH)_2 \cdot 4H_2O$	Monoclinic (pseudo-orthorhombic)	White Vitreous to silky when wet, earthy and powdery when dry.
			Feels like cream cheese when rubbed between the fingers
			Extremely fine-grained Rare in most caves
			A common constituent of moonmilk Cannot be positively identified without an X-ray examination

(Continued)

Table 6.2 (Continued)

Name	Formula	Crystal system	Distinctive properties
Magnesite	$MgCO_3$	Trigonal	White Tasteless Powdery and fine-grained Rare in caves Should be identified by X-ray techniques
Monohydrocalcite	$CaCO_3 \cdot H_2O$	Hexagonal	Colourless to white Very rare; forms as a coldwater aerosol product in alpine caves Identification must be by X-ray diffraction
Nesquehonite	$MgCO_3 \cdot 3H_2O$	Monoclinic	White Massive, fine-grained Easily soluble in dilute acid Rare in caves Should be identified by X-ray techniques
Siderite	$FeCO_3$	Trigonal	Yellowish-brown to amber Isomorphous with calcite As cores in spar
Vaterite	$CaCO_3$	Hexagonal	White High-temperature polymorph of calcite Found in carbidimite formations

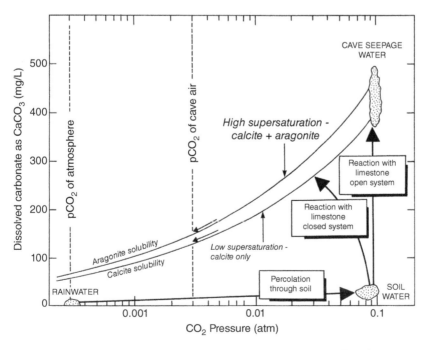

Figure 6.8 Effects of pCO_2 variation on the equilibrium solubility of calcite and aragonite in caves fed by percolation water. Closed systems tend to produce calcite deposition alone, while open systems produce supersaturated solutions depositing both calcite and aragonite. Source: From Gillieson (1996).

This raises saturation levels of the solution with respect to calcite (and, where applicable, other minerals in solution), creating the necessary preconditions for mineral precipitation. If suitable nuclei are available and kinetic barriers are overcome, calcite will inevitably precipitate.

Evaporation concentrates CO_2 and dissolved ions in the source waters; this also raises saturation levels and may trigger calcite precipitation. Degassing of CO_2 without the assistance of evaporation is considered to be the primary cause of carbonate deposition (White 1976). Calcite formed under evaporative conditions (usually near cave entrances, in draughty caves, and in caves in dry regions) is often microcrystalline and soft; by contrast, calcite formed by CO_2 degassing is usually hard and massively crystalline (Ford and Williams 2007), The most readily understood accounts of carbonate precipitation processes are provided by Dreybrodt (1980), Picknett et al. (1976), and White (1976).

The mechanisms by which calcite and other cave carbonates are formed has generated much interest over several centuries. Despite this, and notwithstanding the fact that the general processes of carbonate deposition are well understood, some variations on the above soil derived CO_2 degassing, and evaporation processes are worth noting. Such variations are important, for they may affect the mineralogy/polymorph or physical form of the deposit. For example, Dreybrodt (1982) considered the theoretical case of $CaCO_3$ deposition occurring in the absence of soil CO_2 so as to explain speleothem formation during glacial periods. Here, cold waters in equilibrium with atmospheric CO_2 percolate through bare or ice-covered karst and dissolve carbonate bedrock. The waters are warmed under closed-system conditions by their passage through the bedrock as well as upon entering the cave environment. Speleothems are thus deposited by CO_2 outgassing caused by temperature changes. This lack of biogenic input may explain the heavy $\delta^{13}C$ isotope signatures of some alpine cave deposits (Dreybrodt 1982; Harmon et al. 1983).

James (1977) showed that fluctuations in cave air CO_2 levels (rather than changes in the CO_2 of percolation waters per se) exert a major control on calcite deposition. Organic material washed into caves at Bungonia, New South Wales, Australia, during storm events and its subsequent decomposition by microorganisms contributed significantly to cave air CO_2 levels (over 5% in places at various times) and served to control carbonate precipitation/dissolution cycles in deeper cave passages.

Atkinson (1983) noted the following complex dissolution/mineral precipitation sequence from Castleguard Cave, Alberta, Canada:

1. Dissolution of calcite and dolomite bedrock by carbonic acid (derived from atmospheric CO_2)
2. Oxidation of pyrite to sulphuric acid and Fe hydroxides
3. Further dissolution of calcite and dolomite by the sulphuric acid evolved from 2
4. Precipitation of calcite following supersaturation owing to common-ion effects of 3.

An interesting version of the common-ion effect is invoked by Davis et al. (1990, 1992) in the development of subaqueous speleothems in Lechuguilla Cave, New Mexico, USA. Odd worm-like tubes or tendrils are found growing beneath the margins of gour pools; capillarity cannot be used to explain these eccentric formations akin to helictites. Instead, the common-ion effect has been invoked. Small trickles of water enter the pool after passing over or between gypsum blocks. The dissolution of the gypsum boosts the calcium content of the water and forces the less-soluble calcite to precipitate. The tubes or tendrils mark the entry points of the gypsum charged water into the ponds.

6.3 Controls over Carbonate Mineralogy

Carbonate mineralogy is controlled to a large extent by ion ratios and pCO_2, according to Ford and Williams (2007). Given that calcite and aragonite are the most common cave minerals, it is not surprising that much effort has been spent in trying to understand the conditions under which each is formed. The best reviews of these precipitation processes are those of Curl (1962) and Hill and Forti (1997).

In thermodynamic terms, aragonite is 11–16% more soluble than calcite (Harmon et al. 1983; Hill and Forti 1997). Thus, a solution which is just saturated with calcite is aggressive towards aragonite; under the temperature and pressure conditions likely to be encountered in caves, a given solution will preferentially precipitate calcite (Picknett et al. 1976). To this can be added the presence of a pre-existing calcite surface upon which further calcite growth is best suited compared with 'new' aragonite growth (Picknett et al. 1976).

Although its presence in speleothems was earlier thought to indicate warm conditions (e.g. Moore 1956), aragonite is found in a wide range of cave environments, including in high latitude alpine regions (Harmon et al. 1983). Thus, the idea that aragonite deposition is controlled by temperature is probably no longer tenable (Hill and Forti 1997); the principal reasons given for aragonite precipitation are the poisoning effect on calcite growth by foreign matter such as clays, and rapid CO_2 degassing at high supersaturation levels (Frisia et al. 2002).

Foreign matter, such as magnesium ions, is thought to poison calcite crystal growth (Bögli 1980; Bull 1983; White 1976). Inhibition of calcite precipitation moves supersaturation levels into the range of aragonite precipitation. Numerous examples exist which document aragonite precipitating in preference to calcite in the presence of high concentrations of Mg^{2+} ions (e.g. Cabrol and Coudray 1978; Gonzalez and Lohmann 1988; Roques 1964). Cabrol and Coudray (1978), for example, found aragonite speleothems in southern France confined to those portions of caves directly beneath dolomite bedrock. Zeller and Wray (1956) found that strontium and lead initiated aragonite, while Bischoff and Fyfe (1968) noted that sulphate can inhibit calcite growth and favour aragonite precipitation. Zinc leached from galvanized piping may also inhibit calcite deposition around the paths of tourist caves. Trace elements in speleothems are ably reviewed by Verheyden (2004).

Organic and/or clay particles have also been shown to be calcite inhibitors and thus indirect inducers of aragonite formation. In a petrographic study, Craig et al. (1984) observed that aragonite bands were initiated from clastic layers and solution boundaries in speleothems from Missouri, USA. In laboratory experiments, they demonstrated that the addition of successively larger quantities of clay minerals into supersaturated solutions inhibited calcite growth and favoured the precipitation of metastable polymorphs (in this case vaterite). White (1976) mentioned the importance of adsorption of organic substances (especially amino acids) onto the steps and kinks of calcite crystals in blocking further calcite growth and promoting aragonite crystallisation.

High levels of supersaturation and/or rapid degassing also induce aragonite precipitation. This is done most effectively via evaporation (Hill and Forti 1997), and in some ways evaporative and foreign ion effects are not mutually exclusive. In Carlsbad Caverns, New Mexico, Gonzalez and Lohmann (1988) found that evaporation played a significant role in aragonite precipitation. Aragonite was precipitated only after pore water Mg/Ca exceeded ~1.5; it was the dominant phase when the ratio extended beyond 2.5. High Mg/Ca contents were attributed to several factors. Initially, degassing caused precipitation of low magnesium calcite. This reduction in Ca^{2+} ions coupled with evaporation led to high Mg concentrations and the precipitation of aragonite followed by the precipitation of high magnesium calcite. Harmon et al. (1983) suggested that evaporation of Mg-enriched seepage waters triggered the precipitation of aragonite and other more soluble minerals in Castleguard Cave.

The incidence of the remaining carbonate minerals is occasional or rare. Hydromagnesite and other hydrated carbonates are microcrystalline aggregates formed under conditions of high supersaturation triggered by either evaporation and/or excessive degassing, and in most cases appear to be precipitated as residues (particularly as moonmilk) once thresholds of less soluble minerals have been breached (Hill and Forti 1997; White 1976; White 1982). Dolomite, which is relatively common as a secondary deposit in subaerial environments (e.g. dolocretes), is quite rare in caves. Thrailkill (1968) observed dolomite only as an alteration product of other metastable minerals in Carlsbad Caverns. However, Gonzalez and Lohmann (1988) found dolomite coating pool shelves and as layers within pool ledge deposits in the same environment. They ruled out an evaporative origin because other hydrated carbonates were absent; instead, they suggested that dolomite may precipitate under conditions where pool water is in equilibrium with cave pCO_2 which becomes undersaturated with respect to calcite and aragonite but retains supersaturation with respect to dolomite.

6.4 Other Cave Deposits Formed by Carbonate Minerals

Reviews of carbonate cave formations are found in White (1976), Bögli (1980), Hill and Forti (1997) and Ford and Williams (1989). Initiation, development, and the growth dynamics of basic soda straws, stalactites, and stalagmites are dealt with by Moore (1962), Curl (1972), Curl (1973), Cabrol and Coudray (1978), Franke (1975), and Gams (1981); these are well summarised by Hill and Forti (1997), including speleothem growth rates (Table 6.3).

There are some unusual speleothems formed from calcite. For instance:

- Moonmilk, a soft powdery calcite, has attracted a lot of attention primarily because of its unusual chemical composition and its use by humans. The precise reasons for its microcrystalline habit are unclear. Its formation is discussed by Bernasconi (1976, 1981) and in later work by Fischer (1988, 1992), Onac and Ghergari (1993), and Borsato et al. (2000).
- Some unusual subaqueous speleothems are noted by Davis (1989) and Davis et al. (1990), including helictites draped in calcite rafts.

Figure 6.9 Cave coralloid formations in Resurrection Cave, Mount Etna, Queensland, Australia.

- Rare, cubic cave pearls were reported from Castleguard Cave by Roberge and Caron (1983). Their flat facets are postulated to have formed by contact with neighbouring concretions.
- A model for the development of spathite (flared soda straws) was proposed by Hubbard et al. (1984). Spathite is comprised of a series of fan shaped cones that occur singly or in multiples, with the top of each new cone growing from the base of the one above.
- Observations on deflected soda straw stalactites in Wales, UK, by Sevenair (1985) suggested air movements were responsible for non-vertical growth with wind flow directions through the cave correlating well with the straw orientation. Davis (1986) suggested the possibility of differential seepage deposited minerals having the same effect.

Cave coral or 'popcorn' is a widespread formation (Figure 6.9) whose origin remains unclear. Strictly the formation should be termed 'coralloid' to describe the great variety of nodular, globular, botryoidal, and coral-like speleothems subsumed in a single term (Hill and Forti 1997). Clearly, some forms have developed under water, while others have formed at the interface of rock and air in both wet and dry environments. The following mechanisms have been suggested by Hill and Forti (1997):

- By bedrock seepage emerging and passing through the coralloid itself
- By thin films of water flowing over wall irregularities
- By splash from dripping water
- By the upwards capillary movement of water from pools onto walls
- By the condensation of water droplets and degassing.

Probably more than one mechanism can be operating at any one site and over time as the cave environment changes. It is still unclear whether organic substrates, such as fungal hyphae and plant roots, can act as condensation nuclei for coralloid speleothems, though numerous examples of this can be seen.

Pure calcite and aragonite speleothems are most likely to be translucent and colourless (Ford and Williams 2007). Clastic inclusions, trace elements, and organic compounds may impart coloration (Van Beynen et al. 2001). The colours most commonly seen range from white through yellow, orange and brown. These may be due to iron oxides and hydroxides or complex organic acids and phenolic compounds. White (1981) confirmed Gascoyne's (1977) conclusion that yellow-brown coloration may be caused primarily by high molecular weight fulvic acid and humic substances leached from overlying decomposing soils (Lauritzen et al. 1986). Latham (1981) discussed the effect of these organic substances on crystal growth.

6.5 Growth Rates of Speleothems

Rates of speleothem growth are highly variable depending on the mineralogy and context of individual growth. Straw stalactites generally grow fastest because they have the greatest extension per unit mass of crystal deposited. Beneath bridges straws can grow at very high rates (3000 mm/yr) where dissolved cement is re-precipitated in strongly evaporitic conditions (Ford and Williams 2007). Measured rates from show caves range between 0.2 and 2 mm yr^{-1} for straws, between <0.1 and 3 mm yr^{-1} for stalactites, and between <0.005 and 7 mm yr^{-1} for stalagmites.

Genty et al. (2001) summarised factors affecting speleothem growth rate from six cave sites in Europe. They demonstrated that the mean annual growth rate is primarily dependent on the dripwater calcium ion concentration. There is a correlation between measured growth rate and mean annual temperature due to the correlation between calcium ion concentration, soil pCO_2 and surface temperature. Measured growth rates from several sites are summarised in Table 6.3.

It has long been postulated that speleothems should show annual growth rings, similar to the varves of lake deposits. For this to happen, there must be an annual rhythm in the surface climate such as an annual monsoon, seasonal drought, or frozen soil. This rhythm must then be transferred to the speleothem via a change in the carbonate chemistry or included elements. Finally, the speleothem must be sensitive to the surface changes. If it is deep within a cave then the smoothing effect of stored groundwater may remove any annual signal. Thus, the expression of seasonal variation may be highly variable within a single cave. (Baker et al. 2008) have reviewed annual laminations in speleothems and they are indeed found in a variety of climatic settings. Three types of analysis are at present available for laminations:

- Visible laminations counted either manually or by semi-automated hyperspectral image analysis
- Calcite-aragonite couplets identified by microscopy or scanned using X-ray diffraction
- Trace elements showing annual variation being detected by fluorescence

The combination of annual lamination sequences up to 1000 years and stable isotope analysis provides a powerful proxy record for climatic variation on an annual or decadal scale (Fairchild and Baker 2012).

Table 6.3 Measured growth rates of speleothems.

Mineral/speleothem	Rate (mm yr^{-1})	Location
Calcite		
Stalagmite	0.05	Domica Cave, Slovakia
	0.17	Grotta Gigante, Italy
	0.25	Wyandotte Cave, Indiana, USA
	6.07	Clapham Cave, Yorkshire, UK
	7.66	Ingleborough Cave, Yorkshire, UK
Stalactite	0.06	Moaning Cave, California, USA
	0.03	Uamh an Tartair, Scotland, UK
	0.13–0.20	Postojna Cave, Slovenia
	0.15–0.20	Hand-Dug Cave, Kansas, USA
	0.17	Ingleborough Cave, Yorkshire, UK
	0.24–0.31	Brown's Folly Mine, England, UK
	0.32	Postojna Cave, Slovenia
	0.39–0.65	Grotte de Villars, France
	0.42	Grotte de Proumeyssac, France
	0.66–1.04	Grotte la Faurie, France
	1.00	Wyandotte Cave, Indiana, USA
	2.43	New Cave, Ireland
	3.20	Grotta Doria, Italy
Straw	0.21–0.45	Douchlata Cave, Bulgaria
	1.00	Rhine valley, Germany
	1.24	Han-sur-Lesse Cave, Belgium
	3.00–4.00	Slouperhöhle, Czech Republic
Crust	0.01	Postojna Cave, Slovenia
	0.17	Ochozerhöhle, Czech Republic
	0.25	Vypusterhöhle, Czech Republic
	0.36	Moaning Cave, California, USA
Gypsum		
Selenite crust	0.02	Mammoth Cave, Kentucky, USA
Selenite needles	0.07	Cove Knob Cave, West Virginia, USA

Note that these are maximum rates for fast-growing formations.
Source: Modified from Hill and Forti (1997), Genty et al. (2001).

6.6   Important Non-carbonate Minerals

There is a bewildering array of non-carbonate minerals found in caves. Ford and Williams (2007), quoting Hill and Forti, cite over 200. Many of these are unusual phosphatic minerals formed in the presence of bat guano. The most important are evaporites, phosphates and nitrates, oxides, and hydroxides.

6.6.1   Evaporites (Sulphates and Halides)

Evaporitic minerals have their surface analogues in playa/brine deposits of semi-arid to arid regions. Evaporite cave minerals are highly soluble in water and are thus commonly found in dry parts of caves or in caves located within regions receiving less than 250 mm annual average rainfall. They are not necessarily confined to warmer regions, however. Harmon et al. (1983) reported at least three sulphate minerals from Castleguard Cave in the Canadian Rockies (gypsum, mirabilite, and epsomite). Powdery and crystalline gypsum deposits are also found deep in caves like Selminum Tem and Atea Kananda in ever-wet New Guinea, where the annual average rainfall exceeds 6000 mm. In these situations, the source of the sulphate is pyritic impurities in the limestone or in interbeds, and vigorous air circulation in large passages aids evaporation (Figure 6.10).

Figure 6.10   Mixed calcite and gypsum stalagmite in Gua Tempurung, Malaysia.

Cave evaporites consist largely of sulphates and halides. Hill and Forti (1997) list 10 sulphate minerals derived from 'normal' karst processes; many others result from the interaction of karst waters with ore bodies. The most common evaporite mineral is gypsum, thought to be the third most common cave mineral overall after calcite and aragonite (Hill and Forti 1997). Gypsum is especially abundant in the caves of the American Southwest (USA). Of the halides, halite is the most abundant form. While the caves of the Nullarbor Plain, Australia, have some of the largest halite formations known, the caves of Mt. Sedom, Israel, are formed entirely in halite and have extensive halite speleothems.

Gypsum and other sulphate minerals require a source of sulphate ions. White (1988) lists four possible sources, which can be recategorized into two distinct source groups:

Carbonate bedrock sources:

- Dissolution of sulphate-bearing beds
- Oxidation of pyrite (FeS_2) occurring in the bedrock
- Weathering of anhydrite nodules

Deep groundwater sources:

- Movement of deep H_2S waters into contact with phreatic karst waters to form sulphuric acid, which may dissolve calcite and deposit gypsum.

To these, Hill and Forti (1997) add bat guano and basalts.

Together with other evaporites, gypsum deposition via evaporation is often regarded as a physical rather than a chemical process. However, while gypsum and related sulphate minerals are a result of evaporation, such precipitation is often a final stage of more complex chemical reactions. This is illustrated in two of the four mechanisms for gypsum deposition listed by Hill and Forti (1997).

The sensitivity of sulphates to cave environmental conditions is illustrated by Maltsev (1990). He found the growth of gypsum anemolites (wind-controlled speleothems) in a Ukraine cave varied with cave humidity fluctuations. In summer, when air flows into the cave and with humidity at 70–90%, accretion of up to 7 mm occurred. In winter, the air flow is reversed, humidity rises and dissolution of the fine gypsum structures takes place.

Halite and other halides are extremely soluble and are confined to relatively high temperatures and low humidities and, on present knowledge, are found in only a few caves situated in arid to semiarid regions (White 1976). Parent minerals are leached from the regolith by percolation waters and redeposited under evaporative conditions in the cave environment. Lowry (1967) discussed halite speleothems from the Nullarbor; Goede et al. (1992), in describing from the same region what may be the largest halite speleothems in the world, hypothesised their forming under conditions of low humidity and strong air currents.

Although evaporites form deposits similar to those formed by carbonates (especially stalactites and stalagmites), most halide and sulphate formations are characterised by unique morphologies (Figure 6.11). This is a reflection of both the crystallographic properties of the minerals involved and depositional mechanisms from which they are produced. As most occur from the evaporation of seepage waters, the common forms are crusts, incipient stalactites, and stalagmites (e.g. James 1991) and soil cements (e.g. Harmon et al. 1983). Selenite swords are perhaps a spectacular exception – they are thought to be sub-aqueous in origin. Hill and Forti (1997) and White (1976) provide the best descriptions of evaporite speleothems, although the latter is concerned largely with gypsum formations.

Figure 6.11 Gypsum encrustations in Ozernaja Cave, Ukraine. Source: Photo by John Gunn.

6.6.2 Phosphates and Nitrates

Phosphate and nitrate minerals are largely organic in origin, which has prompted Bögli (1980) to categorise them as 'organic' cave sediments. According to White (1976, p. 313), however, "the reaction products contain authigenic chemical species which must be regarded as cave minerals". The major source for the phosphates is the waste products and remains of cave-dwelling fauna, especially bats; nitrate sources may derive from within the cave as well as from overlying materials, especially soils (Ford and Williams 2007). Lists of major phosphatic minerals can he found in White (1976), Bögli (1980), and Hill and Forti (1997).

Phosphate and nitrate minerals are formed by the reaction of leached phosphate and nitrate compounds with limestone bedrock, secondary carbonate deposits, and/or unconsolidated cave sediments. The phosphate "solubilisation" process, as well as a description of a typical flow path from leachate to the precipitation of various phosphate species, is discussed in Hill and Forti (1997, Figure 85). Phosphate minerals rarely occur as pure or semi-pure forms produced by carbonate and evaporite minerals (although, see Dunkley and Wigley 1967 for a description of 300 mm biphosphammite stalactites). They occur primarily as crusts on breakdown material and bedrock walls, or as nodules and horizons within cave clastic sediments or guano piles (Hill and Forti 1997).

Cave nitrate minerals have generated considerable interest because of their use in the production of gunpowder. Hill (1981b) made the distinction between "saltpetre caves" (i.e. caves containing high nitrate concentrations in their inorganic, clastic sediments) and "guano caves" (i.e. caves containing nitrous organic bat guano but little clastic sediment). Saltpetre was mined from sediments and, strictly speaking, is a product derived from nitre (KNO_3) and not a primary cave mineral (Hill 1981b).

The formation of cave nitrate minerals is not entirely understood. Hill (1981a) and Hill and Forti (1997) list several nitrate sources: surface soils, volcano-sourced groundwaters, and cave guano piles. Since nitrate minerals are formed in the absence of guano piles, it is thought that their origin stems largely from the leaching of nitrates from overlying soils with the assistance from *Nitrobacter* (a nitrogen-fixing bacterium). This has been extensively studied by Hill (1981a). This view has been challenged in recent times by Lewis (1992), who ruled out a seepage water origin and instead argued that bacteria in cave sediments fix the nitrogen from the cave atmosphere. Barton (2006) and Barton and Northrup (2007) have reviewed these issues in the geomicrobiology of caves.

Because of their high solubility, nitrates are confined to very dry sections of caves and to caves in very dry regions. Hill (1981a) provides a phase diagram for the major nitrate minerals from US caves. They form mainly crust deposits on cave walls as well as coatings on guano piles and on and within cave soils (Hill and Forti 1997).

6.6.3 Oxides, Silicates, and Hydroxides

While many oxides, silicates, and hydroxides may be transported into the cave environment, some are chemically formed in situ. The cave environment is wet, mildly alkaline, and oxidising, conditions which favour the formation of certain Fe and Mn oxide minerals (White 1976). The source ions for these Fe and Mn deposits are weathered surficial deposits and decaying organic matter. White (1976) and Hill and Forti (1997) provide the most useful discussions.

Manganese deposits most commonly occur as dark coatings on stream bottoms and cave walls (Ford and Williams 2007), but may form unconsolidated fills (Hill and Forti 1997). Deposition is a result of mixing of Mn-bearing water with highly oxygenated cave water; bacteria may be involved in their precipitation (Peck 1986). Hill (1982) has documented these and other black deposits from US caves. Birnessite is thought to be the most common Mn oxide; others include pyrolusite and ranciéite (Laverty and Crabtree 1978), the former being common in surface soils and in calcretes.

Iron oxide minerals form from the oxidation of pyrite or other iron-bearing sources (White et al. 1985), (White 1988); reduction by bacteria may also be important (Davis et al. 1990; Hill and Forti 1997; Peck 1986). Haematite usually indicates thermal conditions (Hill and Forti 1997). Unlike Mn oxides, Fe minerals may form more extensive deposits, including stalactites and stalagmites (up to 3 m high – Szczerban and Urbani 1974).

The two most common forms of siliceous minerals are quartz and cristobolite (White 1976). Hill and Forti (1997) cite a range of silicate minerals (especially days) whose origin may be detrital rather than secondary; other silicate minerals, such as allophane (Webb and Finlayson 1984; Webb and Finlayson 1985), are secondary. Most silica is derived from the weathering of non-carbonate rocks, particularly igneous. Thus, cave groundwaters previously in contact with such sources may be rich in dissolved silica. Precipitation via evaporation and CO_2 degassing are the most common depositional processes (Hill and Forti 1997). White (1976) suggested that large euhedral quartz forms may indicate elevated temperatures of the depositing waters. Siliceous deposits occur in forms similar to those produced by carbonates (Hill and Forti 1997).

6.7 Ice in Caves

Ice speleothems are found in alpine and/or high latitude caves or in caves characterised by a particular set of air-flow conditions capable of maintaining low cave temperatures (White 1976). Ice can form either seasonally (usually as winter ice close to cave entrances) or permanent (Ford and Williams 1989). Caves supporting perennial ice are known as 'glacieres'. They often contain ice stalagmites (Figure 6.12) as well as deep layered floor deposits which may be of archaeological or paleoclimatic interest.

Cave ice forms from two mechanisms: the freezing of seepage water and the freezing of water vapour. According to Ford and Williams (1989), these produce at least seven types of cave ice: dripstones and flowstones; recrystallised 'old' ice; frozen ponds; hoarfrost (of which four forms are recognised); extrusion ice; intrusion ice; and as an ice matrix in clastic sediments. Detailed descriptions of ice formations are provided in Hill and Forti (1997)

Figure 6.12 Ice Stalagmites in Lofthellir, Iceland. Source: Photo by John Brush.

6.8   Other Minerals

Hill and Forti (1997) categorise other minerals under 'ore-related' and 'miscellaneous'. Ore-related minerals are especially interesting; they precipitate from groundwaters in contact with igneous bodies and may produce in any one cave a varied mineral assemblage. An excellent example is from Lilburn Cave, California, USA, (Rogers and Williams 1982), where petromorphic minerals, such as hornblende, occur in abundance.

6.9   Cave Deposits of the Nullarbor Plain, Australia

The Nullarbor Plain is Australia's largest karst (220 000 km²), with an arid to semi-arid climate: annual rainfall is between 150 and 250 mm while annual evaporation is 1250–2500 mm. The plain has no surface drainage but is punctuated by collapse dolines which are generally shallow. The caves are extensive, much modified by salt wedging and collapse processes, and commonly descend gently to near-static pools and lakes of brackish to saline water. The flooded tunnels of Cocklebiddy Cave are in excess of 6 km long – most of them explorable only by scuba diving. Withdrawal of hydrostatic support at times of lower sea level has led to collapse into these water table caves, allowing entry into large caverns such as Koonalda, Abrakurrie (Figures 6.13 and 6.14), and Weebubbie caves (Figure 6.15).

Figure 6.13   Entrance chamber of Koonalda Cave, Nullarbor, Australia. Black flint nodules are visible in the walls.

Figure 6.14 The main passage of Abrakurrie Cave, Nullarbor Plain, is typical of chambers that have been highly modified by exsudation. The floor area of the chamber is 9000 m².

A particular feature of the Nullarbor caves is the abundance of gypsum and halite, which produce both speleothems (Figure 6.16) and many weathering forms through wedging during recrystallisation. Gypsum speleothems are also abundant, with calcite speleothems less obvious largely because of destruction through salt wedging (Figure 6.17). Lowry and Jennings (1974) provide a good introduction to the geomorphology, while more recent detail is given in the review by Gillieson and Spate (1992).

The Nullarbor caves contain secondary minerals of great significance, a number of which have been identified as new minerals. More common minerals exist in previously unrecorded forms, and many of the mineral deposits are of great value as a source of palaeoenvironmental data for the Nullarbor Plain. Table 6.4 lists the secondary minerals that have been identified from the Nullarbor caves, their chemical formula, and location. Key references are Bridge (1973), Bridge and Clarke (1983), and Caldwell et al. (1982).

The calcite formations of the Nullarbor Caves are usually dark brown to black in colour. The black colour was shown by Caldwell et al. (1982) to be due to humic compounds and is also seen on the surface and down to the water table. Uranium series dating (Goede et al. 1990) indicated that its deposition ceased more than 350 000 years before present. Flow-stones of black calcite over a meter thick suggest that its deposition extended over a long period of time. More recent dating using the Uranium–Lead isotope method (Woodhead et al. 2006; Sniderman et al. 2016) provides dates well into the Pliocene, between 3.4 and 5.6 million years ago. The fossil pollen records preserved in speleothems from the Nullarbor

LONGITUDINAL SECTIONS

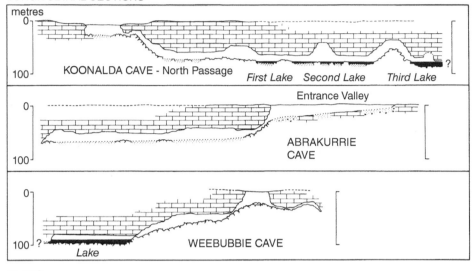

PLANS

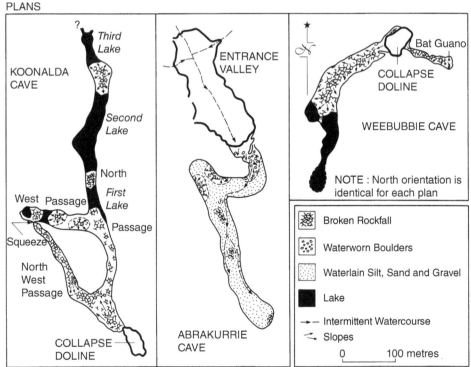

Figure 6.15 Longitudinal sections and plans of some deep Nullarbor Plain caves (Australia). Source: From Gillieson (1996).

Table 6.4 Secondary minerals in the Nullarbor caves (Australia).

Mineral	Composition	Location
Halides		
Halite	NaCl	Most caves
Sulphates		
Gypsum	$CaSO_4 \cdot 2H_2O$	Most caves
Selenite	$CaSO_4 \cdot 2H_2O$	Mullamullang N37, Thampanna N206
Mirabilite	$Na_2SO_4 \cdot 10H_2O$	Unknown cave
Taylorite	$(NH_4,K)SO_4$	Murra-el-Elevyn N47
Aphthitalite	$(K,Na)_3Na(SO_4)_2$	Petrogale N200, Murra-el-Elevyn N47
Syngenite	$K_2Ca(SO_4)_2 \cdot 2H_2O$	Murra-el-Elevyn N47
Carbonates		
Calcite	$CaCO_3$	Most caves
Aragonite	$CaCO_3$	Tommy Grahams N56
Phosphates		
Whitlockite	$Ca_8(Mg,Fe)H(PO_4)_7$	Petrogale N200, Murra-el-Elevyn N47
Stercorite	$H(NH_4)Na(PO_4) \cdot 4H_2O$	Petrogale N200
Newberyite	$MgHPO_4 \cdot 3H_2O$	Petrogale N200
Mundrabillaite	$(NH_4)_2Ca(HPO_4)_2 \cdot 2H_2O$	Petrogale N200
Monetite	$CaHPO_4$	Murra-el-Elevyn N47
Hannayite	$(NH_4)_2Ca(HPO_4)_2H_2O$	Murra-el-Elevyn N47
Carbonate-hydroxylapatite	$Ca_5(PO_4,CO_3)_3OH$	Murra-el-Elevyn N47
Brushite	$CaHPO_4 \cdot 2H_2O$	Murra-el-Elevyn N47
Biphosphammite	$(NH_4,K)H_2PO_4$	Petrogale N200, Murra-el-Elevyn N47
Archerite	$(K,NH_4)HPO_4$	Petrogale N200
Organics		
Guanine	$C_5H_3(NH_2)N_4O$	Murra-el-Elevyn N47
Oxammite	$(NH_4)_2C_2 \cdot 4H_2O$	Petrogale N200
Uricite	$C_5H_4N_4O_3$	Dingo Donga N160
Weddellite	$CaC_2O_4 \cdot 2H_2O$	Webbs N132, Petrogale N200
Iron and manganese oxides		
Goethite	$FeOOH$	Mullamullang N37
Haematite	Fe_2O_3	Mullamullang N37
Pyrolusite	MnO_2	Mullamullang N37
Clays		
Illite, Kaolin, Montmorillonite		From caves in the Mundrabilla area

Source: From Gillieson and Spate (1992).

Figure 6.16 Gypsum flowers in Easter Extension, Mullamullang Cave, Nullarbor, Australia. These are approximately 30 cm long.

reveal an abrupt onset of warm and wet climate early within the Pliocene, driving complete ecological change. Pliocene warmth thus clearly represents a discrete interval which reversed a long-term trend of late Tertiary cooling and drying.

The Nullarbor contains one speleothem form, closely allied to cave shields, which appears to be different from those reported elsewhere. The term 'stegamite' was coined by Webb (1991), who described it as a high ridge in a calcite floor, generally with a crack along its top, which separates the speleothem into two vertically standing shield-like plates (Figure 6.18). Stegamites are made up of finely banded black calcite from the medial crack in a similar fashion to that postulated for cave shields (Hill and Forti 1997), although there is some dispute as to their origin. They are sometimes found in clusters, and in a few cases, secondary stegamites intersect the primary form approximately at right angles.

The Nullarbor is unique among world karsts in demonstrating a very extensive modification of caves by the arid zone process of exudation (otherwise known as salt or crystal weathering). The process works by detachment of particles of various sizes from a rock surface (Figure 6.19) by the growth of crystals (mainly of gypsum, to a lesser extent halite) from percolating saline solutions. This process enlarges hollows in cave roofs to form domes, some of large size: in Mullamullang Cave there are several which are tens of metres in diameter, arching up from the general collapse surface of the original roof by several metres. This

Figure 6.17 Halite speleothems overgrowing gypsum, in turn overgrowing old calcite speleothems.

doming process may extend vertically upwards from cave voids below to create the vertical shafts which intersect the ground surface as blowholes. Much of the speleothem decoration (particularly the calcite) has been broken down by exudation.

Exsudation or salt weathering is responsible for the production of the famous 'coffee and cream' which is perhaps the most dramatic and unusual of the Nullarbor cave sediments. The best and most extensive deposits were to be found in Mullamullang Cave (Figure 6.20), but they are also known from a number of other sites. 'Coffee and cream' is an airfall deposit produced by crystal weathering. It is made up of fine crystalline powder and more granular material with small fragments of halite and gypsum cave flowers. Both the light- and the dark-coloured materials are made up of high magnesium calcite. The 'cream' is a cream, occasionally pink, colour containing about 2% of iron minerals. The strikingly contrasting 'coffee' is of similar chemical composition, but with about 8% iron compounds plus some manganese dioxide (Caldwell et al. 1982). The two forms appear to flow over one another. The whole forms an unusual entity of puzzling appearance, especially as the roofs above appear to be evenly coloured and textured, at least at a scale far smaller than the 'coffee and cream' facies. They probably deserve further investigation, but unfortunately the Mulla-mullang Cave deposits have been very much disturbed by visitors. It is believed that 'coffee and cream' is only found in caves beneath the Nullarbor Plain.

Figure 6.18 Stegamite in Gorringe Cave, Nullarbor, Australia.

Figure 6.19 Exsudation or salt wedging of cave ceiling, Webbs Cave, Nullarbor, Australia.

Figure 6.20 'Coffee and cream' speleothem banks in Mullamullang Cave, Nullarbor, Australia.

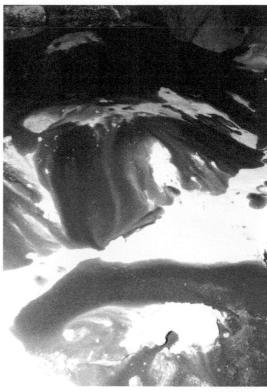

Figure 6.21 Spring Passage in Old Homestead Cave, Nullarbor, Australia. A typical phreatic passage modified by breakdown due to salt wedging.

The Nullarbor and its caves are currently in an arid climate, but recent dating of the dark brown calcite stalagmites (Blyth et al. 2010; Treble et al. 2017) has shown that most formed between three and four million years ago, during the mid-Pliocene warm period (Figure 6.21) when global surface temperatures are believed to have been several degrees higher than today, with relatively high atmospheric CO_2 concentrations. When the study of the age distribution, geochemistry, and palynology of the Nullarbor stalagmites is completed science will gain a wealth of information about the timing and effects of warmer periods in Earth history, which will provide analogues for the climates of the next 100 years.

Exsudation is also capable of shattering massive calcite speleothems. The crystal weathering splits massive speleothems as if they had been sawn down the middle, or smashes them up as if a determined vandal had taken to them with a sledge hammer. Outstanding examples occur in the central Nullarbor, particularly in Webbs Cave, Kelly Cave, Witches Cave, and Thampanna Cave.

References

Atkinson, T.C. (1983). Growth mechanisms of speleothems in Castleguard Cave, Columbia Icefields, Alberta, Canada. *Arctic and Alpine Research* 15: 523–536.

Baker, A., Smith, C.L., Jex, C. et al. (2008). Annually laminated speleothems: a review. *International Journal of Speleology* 37: 4.

Barton, H.A. (2006). An introduction to cave microbiology: a review for the non-specialist. *Journal of Cave and Karst Studies* 68: 43–54.

Barton, H.A. and Northrup, D.E. (2007). Geomicrobiology in cave environments: past, current and future perspectives. *Journal of Cave and Karst Studies* 69: 163–178.

Bernasconi, R. (1976). The physico-chemical evolution of moonmilk. *Cave Geology* 1: 63–88.

Bernasconi, R. (1981). Mondmilch (moonmilk): two questions of terminology. In: *Proceedings of the 8th International Congress of Speleology, 1981 Kentucky*, 113–116. Department of Geography and Geology, Western Kentucky University.

Bischoff, J.L. and Fyfe, W.S. (1968). Catalysis, inhibition, and the calcite-aragonite problem; [part] 1, the aragonite-calcite transformation. *American Journal of Science* 266: 65–79.

Blyth, A.J., Watson, J.S., Woodhead, J., and Hellstrom, J. (2010). Organic compounds preserved in a 2.9 million year old stalagmite from the Nullarbor Plain, Australia. *Chemical Geology* 279: 101–105.

Bögli, A. (1980). *Karst Hydrology and Physical Speleology*. Berlin: Springer.

Borsato, A., Frisia, S., Jones, B., and Van Der Borg, K. (2000). Calcite moonmilk: crystal morphology and environment of formation in caves in the Italian Alps. *Journal of Sedimentary Research* 70: 1179–1190.

Bridge, P.J. (1973). Guano minerals from Murra-el-Elevyn Cave, Western Australia. *Mineralogical Magazine* 39: 467–469.

Bridge, P.J. and Clarke, R.M. (1983). Mundrabillaite: a new cave mineral from Western Australia. *Mineralogical Magazine* 47: 80–81.

Bull, P.A. (1983). Chemical sedimentation in caves. In: *Chemical Sediments and Geomorphology: Precipitates and Residua in the Near-Surface Environment* (eds. A.S. Goudie and K. Pye), 301–319. London: Academic Press.

Cabrol, P. and Coudray, J. (1978). *Les phénoménes de diagenèse clans les concrétions carbonatees des grottes*, 82–85. Orsay, France.

Caldwell, J.R., Davey, A.G., Jennings, J.N., and Spate, A.P. (1982). Colour in some Nullarbor Plain speleothems. *Helictite* 20: 3–10.

Coleborn, K., Baker, A., Treble, P.C. et al. (2018). The impact of fire on the geochemistry of speleothem-forming drip water in a sub-alpine cave. *Science of the Total Environment* 642: 408–420.

Craig, K.D., Horton, P.D., and Reams, M.W. (1984). Clastic and solutional boundaries as nucleation surfaces for aragonite in speleothems. *National Speleological Society Bulletin* 46: 15–17.

Curl, R.L. (1962). The calcite-aragonite problem. *National Speleological Society Bulletin* 24: 57–73.

Curl, R.L. (1972). Minimum diameter stalactites. *National Speleological Society Bulletin* 34: 129–136.

Curl, R.L. (1973). Minimum diameter stalagmites. *National Speleological Society Bulletin* 35: 1–9.

Davis, D.G. (1986). Discussion: the deflected stalactites of Dan-Yr-Ogof. *National Speleological Society Bulletin* 48: 62–63.

Davis, D.G. (1989). Helictite bushes – a subaqueous speleothem. *National Speleological Society Bulletin* 17: 120–124.

Davis, D.G., Palmer, A.N., and Palmer, M.J. (1990). Extraordinary subaqueous speleothems in Lechuguilla Cave, New Mexico. *National Speleological Society Bulletin* 52: 70–86.

Davis, D.G., Palmer, A.N., and Palmer, M.V. (1992). Extraordinary subaqueous speleothems in Lechuguilla Cave, New Mexico: reply. *National Speleological Society Bulletin* 54: 34–35.

Dreybrodt, W. (1980). Deposition of calcite from thin films of natural calcareous solutions and the growth of speleothems. *Chemical Geology* 29: 89–105.

Dreybrodt, W. (1982). A possible mechanism for growth of calcite speleothems without participation of biogenic carbon dioxide. *Earth and Planetary Science Letters* 58: 293–299.

Dreybrodt, W. and Romanov, D. (2008). Regular stalagmites: the theory behind their shape. *Acta Carsologica* 37: 2–3.

Dunkley, J.R. and Wigley, T.M.L. (1967). *Caves of the Nullarbor*. Sydney: Speleological Research Council, University of Sydney.

Fairchild, I.J. and Baker, A. (2012). *Speleothem Science: From Process to Past Environments*. Blackwell.

Fairchild, I.J., Frisia, S., Borsato, A., and Tooth, A.F. (2007). Speleothems. In: *Geochemical Sediments and Landscapes* (eds. D.J. Nash and S.J. Mclaren), 200–245. Oxford: Blackwell.

Fischer, H. (1988). Etymology, terminology and an attempt of definition of Mondmilch. *National Speleological Society Bulletin* 50: 54–58.

Fischer, H. (1992). Type locality of Mondmilch. *Cave Science* 19: 59–60.

Ford, D.C. and Williams, P.W. (1989). *Karst Geomorphology and Hydrology*. London: Unwin Hyman.

Ford, D.C. and Williams, P.W. (2007). *Karst Hydrogeology and Geomorphology*. Chichester, England: Wiley.

Forti, P. (2009). Speleothems: part I. Teaching resources for speleology and karst, Società Speleologica Italiana. http://www.speleo.it/site/index.php/open-document/1018-teaching-resources-for-speleology-and-karst

Franke, H.W. (1975). Sub-minimum diameter stalagmites. *National Speleological Society Bulletin* 37: 17–18.

Frisia, S., Borsato, A., Fairchild, I.J., and Mcdermott, F. (2000). Calcite fabrics, growth mechanisms, and environments of formation in speleothems from the Italian Alps and southwestern Ireland. *Journal of Sedimentary Research* 70: 1183–1196.

Frisia, S., Borsato, A., Fairchild, I.J. et al. (2002). Aragonite-calcite relationships in speleothems (Grotte de Clamouse, France): environment, fabrics, and carbonate geochemistry. *Journal of Sedimentary Research* 72: 687–699.

Gams, I. (1981). Contribution to morphometrics of stalagmites. In: *Proceedings of the 8th International Congress of Speleology, 1981 Kentucky*, 276–278. Department of Geography and Geology, Western Kentucky University.

Gascoyne, M. (1977). Trace element geochemistry of speleothems. 205–208. Sheffield.

Genty, D., Baker, A., and Vokal, B. (2001). Intra-and inter-annual growth rate of modern stalagmites. *Chemical Geology* 176: 191–212.

Gillieson, D.S. and Spate, A.P. (1992). The Nullarbor karst. In: *Geology, Climate, Hydrology and Karst Formation: Field Symposium in Australia* (ed. D.S. Gillieson), 65–99. Canberra: Department of Geography and Oceanography, University College, Australian Defence Force Academy.

Goede, A., Harmon, R.S., Atkinson, T.C., and Rowe, P.J. (1990). Pleistocene climatic change in southern Australia and its effect on speleothem deposition in some Nullarbor caves. *Journal of Quaternary Science* 5: 29–38.

Goede, A., Atkinson, T.C., and Rowe, P.J. (1992). A giant Late Pleistocene halite speleothem from Webbs Cave, Nullarbor Plain, southeastern Western Australia. *Helictite* 30: 3–7.

Gonzalez, L.A. and Lohmann, K.C. (1988). Controls on mineralogy and composition of spelean carbonates: Carlsbad Caverns, New Mexico. In: *Paleokarst* (eds. N.P. James and P.W. Choquette), 81–101. New York: Springer New York.

Harmon, R.S., Atkinson, T.C., and Atkinson, J.L. (1983). The mineralogy of the Castleguard Cave, Columbia Icefields, Alberta, Canada. *Arctic and Alpine Research* 15: 503–516.

Hill, C.A. (1976). *Cave Minerals*. Huntsville, Alabama: National Speleological Society.

Hill, C.A. (1981a). Mineralogy of cave nitrates. *National Speleological Society Bulletin* 43: 127–132.

Hill, C.A. (1981b). Origin of cave saltpeter. *National Speleological Society Bulletin* 43: 110–126.

Hill, C.A. (1982). Origin of black deposits in caves. *National Speleological Society Bulletin* 44: 15–19.

Hill, C.A. and Forti, P. (1997). *Cave Minerals of the World*. Huntsville: National Speleological Society.

Hubbard, D.A., Herman, J.S., and Mitchell, R.S. (1984). A spathite occurrence in Virginia: observations and a hypothesis for genesis. *National Speleological Society Bulletin* 46: 10–14.

James, J.M. (1977). Carbon dioxide in the cave atmosphere. *Transactions of the British Cave Research Association* 4: 417–429.

James, J.M. (1991). The sulphate speleothems of Thampanna Cave, Nullarbor Plain. *Helictite* 30: 19–23.

Jennings, J.N. (1985). *Karst Geomorphology*. Oxford: Blackwell.

Latham, A.G. (1981) Muck spreading on speleothems. 356–357. Kentucky.

Lauritzen, S.E., Ford, D.C., and Schwarcz, H.P. (1986). Humic substances in speleothem matrix: Paleoclimatic significance. In: *Communications 9th International Congress of Speleolo 1986*, 77–79. Barcelona: Catalan Speleological Society.

Laverty, M. and Crabtree, S. (1978). Rancieite and mirabilite: some preliminary results on cave mineralogy. *Transactions of the British Cave Research Association* 5: 135–142.

Lewis, W.C. (1992). On the origin of cave saltpeter: a second opinion. *National Speleological Society Bulletin* 54: 28–30.

Lowry, D.C. (1967). Halite speleothems from the Nullarbor Plain, Western Australia. *Helictite* 6: 14–20.

Lowry, D.C. and Jennings, J.N. (1974). The Nullarbor karst, Australia. *Zeitschrift fur Geomorphologie* 18: 35–81.

Maltsev, V.A. (1990). The influence of seasonal changes of cave microclimate upon the genesis of gypsum formations in caves. *National Speleological Society Bulletin* 52: 99–103.

Moore, G.W. (1954). The origin of helictites. *National Speleological Society Bulletin*, Occasional Papers. 5–16.

Moore, G.W. (1956). Aragonite speleothems as indicators of palaeotemperature. *American Journal of Science* 254: 746–753.

Moore, G.W. (1962). The growth of stalactites. *National Speleological Society Bulletin* 24: 95–106.

Onac, B.P. (1997). Crystallography of speleothems. In: *Cave Minerals of the World*, 2e (eds. C.A. Hill and P. Forti). Huntsville, Alabama: National Speleological Society.

Onac, B.P. and Ghergari, L. (1993). Moonmilk mineralogy in some Romanian and Norwegian caves. *Cave Science* 20: 107–111.

Palmer, A.N. (2007). *Cave Geology*. Dayton, OH: Cave Books.

Peck, S.B. (1986). Bacterial deposition of iron and manganese oxides in North American caves. *National Speleological Society Bulletin* 48: 26–30.

Picknett, R.G., Bray, L.G., and Stenner, R.D. (1976). The chemistry of cave waters. In: *The Science of Speleology* (eds. T.D. Ford and C.H.D. Cullingford), 213–266. London: Academic Press.

Roberge, J. and Caron, D. (1983). The occurrence of an unusual type of Pisolite: the cubic cave pearls of Castleguard Cave, Columbia Icefields, Alberta, Canada. *Arctic and Alpine Research* 15: 517–522.

Rogers, B.W. and Williams, K.M. (1982). Mineralogy of Lilburn Cave, Kings Canyon National Park, California. *National Speleological Society Bulletin* 44: 23–31.

Roques, H. (1964). Contribution à l'étude statique et cinétique des systèmes gaz carbonique-eaucarbonate. *Annales De Spéléologie* 19: 255–484.

Sevenair, J.P. (1985). The deflected stalactites of Dan-Yr-Ogof. *National Speleological Society Bulletin* 47: 28–31.

Sniderman, J.M.K., Woodhead, J.D., Hellstrom, J. et al. (2016). Pliocene reversal of late Neogene aridification. *Proceedings of the National Academy of Sciences United States of America*: 1999–2004.

Szczerban, E. and Urbani, F. (1974). Carsos de Venezuela, 4: Formas carsicos en areniscas precambricas del Territorio Federal Amazonas y Estado Bolivar. *Boletin de la Sociedad Venezolana de Espeleologia* 5: 27–54.

Thrailkill, J.V. (1968). Dolomite cave deposits from Carlsbad Caverns. *Journal of Sedimentary Research* 38: 141.

Treble, P.C., Baker, A., Ayliffe, L.K. et al. (2017). Hydroclimate of the last glacial maximum and deglaciation in southern Australia's arid margin interpreted from speleothem records (23-15 ka). *Climate of the Past* 13: 667.

Van Beynen, P., Bourbonniere, R., Ford, D.C., and Schwarcz, H.D. (2001). Causes of colour and fluorescence in speleothems. *Chemical Geology* 175: 319–341.

Verheyden, S. (2004). Trace elements in speleothems: a short review of the state of the art. *International Journal of Speleology* 33: 95–101.

Webb, R. (1991). Stegamites – a form of cave shield? In: *Proceedings 18th Biennial Conference, Australian Speleological Federation, Perth*, 95–98. Western Australian Speleological Group Inc.

Webb, J.A. and Finlayson, B.L. (1984). Allophane and opal speleothems from granite caves in south-East Queensland. *Australian Journal of Earth Sciences* 31: 3419.

Webb, J.A. and Finlayson, B.L. (1985). Allophane flowstone from Newton Cave, western Washington state. *National Speleological Society Bulletin* 47: 45–48.

White, W.B. (1976). *Cave Minerals and Speleothems*. London: Academic Press.

White, W.B. (1981). Reflectance spectra and color in speleothems. *National Speleological Society Bulletin* 43: 20–26.

White, W.B. (1982). Mineralogy of the Butler Cave-sinking stream system. *National Speleological Society Bulletin* 44: 90–97.

White, W.B. (1988). *Geomorphology and Hydrology of Karst Terrains*. New York: Oxford University Press.

White, W.B., Scheetz, B.E., Atkinson, S.D. et al. (1985). Mineralogy of Rohrer's Cave, Lancaster County, Pennsylvania. *National Speleological Society Bulletin* 47: 17–27.

Woodhead, J., Hellstrom, J., Maas, R. et al. (2006). U–Pb geochronology of speleothems by MC-ICPMS. *Quaternary Geochronology* 1: 208–221.

Zeller, E.J. and Wray, J.L. (1956). Factors influencing the precipitation of calcium carbonate. *Bulletin of the American Association of Petroleum Geologists* 40: 140–152.

7

Cave Sediments

7.1 Introduction

Early in their underground career most cavers will become intimately acquainted with the unctuous cave muds that adhere to just about anything that contacts them. They will also be aware of the seemingly vast amount of haphazardly piled angular boulders that make up cave breakdown. The reward for negotiating these breakdown piles and mud wallows is the privilege of viewing pristine calcite formations. When most people think of cave sediments, they have in mind the clastic sediments made up of fine or coarse particles of mineral or organic matter. A great deal of research has been carried out on the material deposited along with these sediments (bones, pollen, artefacts) as a means of elucidating environmental or human histories. However, cave sediments may be intrinsically interesting, for a thorough understanding of their origins, processes of deposition, and chemical alteration can tell us a great deal about catchment hydrology and erosion, as well as climate changes.

Less research has been undertaken on the processes by which sediments are produced, transported, and deposited within the cave system. In part, this is because of the difficulty of observing and measuring these processes in flooding caves. As we have seen in Chapter 2, the bulk of material moving through a cave system is indeed clastic sediment. There is in fact very little difference in the nature of surface and underground clastic sediments, although we must devise more long-term models of sedimentation for cave sediments. In addition, there is an important interaction between cave sediments and cave morphology which produces a range of depositional structures and passage morphologies. Cave sediments are tricky to understand!

7.2 Clastic Sediment Types

Clastic sediments form from fragments of rocks (the regolith) that have been broken up by physical or chemical weathering processes. These fragments are further transformed by the winnowing effect of sediment transport and by chemical alteration (diagenesis) during long or short periods of repose in the surface or underground environment. Sediments are classified as either allogenic (origin outside the cave) or authigenic (origin within the cave), which is often reflected in whether their geologic origin differs from that of the cave itself. Cave sediment types (Table 7.1) are thus diverse and include organic debris and its

Caves: Processes, Development, and Management, Second Edition. David Shaw Gillieson.
© 2021 John Wiley & Sons Ltd. Published 2021 by John Wiley & Sons Ltd.

Table 7.1 Cave sediment types.

Type	Origin	Nature
Clastic	Allogenic or authigenic	Angular boulder debris; e'boulis; subangular to subrounded gravels and cobbles; sands and silts; finely laminated cave clays
Organic	Allogenic debris	Woody debris; humus and fine particulate organic matter; dung and spores; charcoal and ash; microartefacts (bone, shell, lithics)
	Authigenic deposits	Coprolites and bone; bat and bird guano; phosphatic mineral crusts and laminae; nitrate-rich deposits; subaerial stromatolites
Chemical (Chapter 6)	Allogenic	Tufa and travertine fragments; pisolites (iron nodules); caliche
	Authigenic	Calcite and gypsum speleothems and interlayered deposits; tufas and travertines
Ice	Authigenic cave drips, wall condensation, cave pools	Ice stalagmites and stalactites; pond ice; rime crystals

chemical derivatives (phosphate and nitrate minerals), inorganic chemical precipitates, and ice. In any single cave, the total assemblage will depend on both the past and present geologic and climatic environments. Cave sediments thus provide us with a potential library of environmental information – if we can read the language in which they are written. To do this we need to understand the nature of the materials and their processes of transport, deposition, and diagenesis.

7.3 Processes of Sedimentation

Cave sediments may be deposited by either gravity-fall or aqueous transport processes. The distinction between these becomes blurred when we consider such processes as turbidity currents sliding down steep sediment banks in a cave pool, or the injection of fluidised mudflows into tropical cave passages by mass movements (Gillieson 1986).

7.3.1 Gravity-Fall Processes

Gravity-fall processes generally refer to the slow or rapid movement of clastic sediments in air, either as a dry or as a saturated mass movement, whether or not this movement occurs on a solid surface. The deposition of air-fall tephra (volcanic ash) in caves is a special case of this type. The principal types of deposits moved by gravity-fall processes include cave breakdown, debris cones under blind shafts (Figure 7.1), cave wall fans below fissures, loess and related deposits, bushfire smoke and fine charcoal, mudflow deposits, and till. In addition, the deposition of lint and skin cells in tourist caves is a major process of sedimentation by gravity fall and may total several tonnes per year in heavily used caves.

Figure 7.1 Huge sand cone in Sand Cave, Naracoorte Caves, Australia. Source: Photo: Steven Bourne.

By far the most widespread form of gravity-fall sediment in caves is breakdown or incasion. The alteration of passage shape by breakdown seems to be the ultimate fate of most caves once the water which formed them is removed. As long as the passage is water-filled, stress lines in the rock are evenly distributed around the cavity (Figure 7.2); once the water is removed, local concentration of stress leads to failure of the arched section, usually along bedding planes. This process propagates upwards, leading to the development of breakdown domes (Figure 7.3) or blind shafts. In many cases this propagation intersects overlying non-limestone rocks, allowing this allogenic material to enter the cave.

Breakdown blocks vary in dimension from fist-sized rocks to boulders the size of houses. Their shape depends greatly on the thickness of the bedding. Thus, a thinly-bedded limestone will tend to produce platy fragments, and a massive limestone more cubic fragments. This purely stress-related breakdown may be modified by frost shattering near entrances to produce platy, angular debris of cobble to boulder size. Thus, in Greftsprekka Cave, Nordland, Norway, frost-shattered debris can be found up to 50 m below the surface (Figure 7.4), while more massive breakdown is found throughout the cave to a depth of 230 m.

The stages in the development of a passage by breakdown can be seen in Postojna Cave, Slovenia (Gospodaric 1976). Here, cave development by the underground Pivka River has persisted throughout several glacial-interglacial cycles (Figure 7.5). At the earliest stages, in the mid-Quaternary, the fluvial regime was predominantly erosional with little active sedimentation. Allogenic chert gravels and sands eroded the cave passage, and at a later stage quartz sands, chert sands, and limonite from bauxitic soils were deposited. An initial breakdown cone formed in the active stream passage; its development ponded water and led to the excavation of a lower passage. The first speleothems formed in the passage,

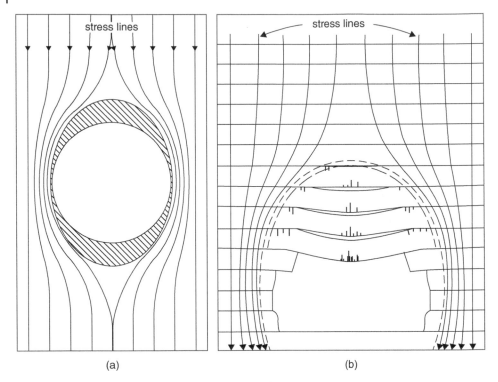

Figure 7.2 Distribution of stress lines around natural cavities in limestone: (a) distribution around water-filled void below the water table; (b) distribution around air-filled void after lowering of the water table. Note incipient collapses at areas of concentrated stress.

principally stalagmites and reddish flowstones. During the third stage, the Pivka eroded and transported rubble, gravel, and flood loam. The collapse of the speleothems occurred because of sapping of the sediments from below and removal of the collapsed material in the lower course of the Pivka River. In the most recent phases of development, the breakdown dome has developed upwards to intersect the surface and form the Stara Apnenica doline. There has thus been an alternation of fluvial and breakdown processes, with the breakdown sealing and, therefore, protecting the underlying waterlain sediments.

Extensive speleothem deposition has occurred in the dry galleries abandoned by the underground Pivka River, with the contemporary distribution of sediments in the cave reflecting the long history of deposition and erosion by a variety of processes. The elucidation of this sequence relied on careful inspection, analysis. and correlation of sediment sequences throughout the cave. The full sequence presented above may not be visible at any one point in the cave; it is an abstraction subject to revision in the light of future discoveries and refinements in technique. While initial age estimates based on radiocarbon suggested a minimum age of *c.*40 000 years ago for the flood loams in Postojna Jama, later electron spin resonance studies (Ikeya et al. 1983) indicated that these deposits may be older than 190 000 years ago. Initial underestimation of the age of cave sediments seems to be widespread, with examples coming from both Australia (Frank and Jennings 1978) and Yorkshire, England (Gascoyne et al. 1983).

Figure 7.3 An abandoned river canyon in Selminum Tem, Papua New Guinea, has breakdown piles and deep mudflows capped by speleothems dated by uranium series methods to 15 000 years ago.

7.3.2 Waterlain Clastic Sediments

Caves can be seen as underground gorges and floodplains in which clastic sedimentation proceeds in modes analogous to those of surface fluvial systems. This provides a conceptual scheme for sedimentation processes where water is the transporting agent. The major difference between the surface and underground fluvial systems is that, in the latter, the water and sediment are totally confined within a conduit. This has two main effects:

- Dramatic fluctuations in water level – caused by flood stage or to passage morphology – result in steep gradients in depositional energy along a cave passage. There is thus a greater diversity of sediment textures per unit length of channel than is found on the surface. This affects both estimation of past flow velocities from sediments and stratigraphic correlation.
- Subsequent flows of water down a particular cave passage may wholly or partially remove the sediment deposited by a prior event. The resistance of an individual 'parcel' of sediment to this process of reworking will depend on its texture and on the passage geometry at the site. The fate of a parcel of sediment entering a cave is to be shunted through the passage with successive removals of some constituents by sorting, and a gradual diminution of total volume (Figure 7.6). Thus, the life history of a parcel of cave sediment is

Figure 7.4 The 50 m entrance shaft of Greftsprekka Cave, Norway, has been heavily modified by frost shattering during colder climatic phases.

one of periodic reworking until its identity is lost, its volume becomes negligible, or it is deposited in a very low energy environment.

The degree of this reworking will depend largely on the texture of the sediment. Both the large boulder-size particles and the very fine cohesive clays will be resistant to reworking once emplaced in a cave. In Westmoreland Cave, Mole Creek, Tasmania, Australia, dolerite boulders of the Last Glacial age are wedged in the passages and appear little altered by weathering. In contrast, sand-sized sediments will be readily moved and reworked. This is a result of the velocity required for erosion being higher than the velocity needed to transport coarse and extremely fine particles. In Mammoth Cave, Kentucky, USA, backflooding deposits of silt can be moved to higher elevations by successive floods (Collier and Flint 1964). After every flood, a thin layer of clay is deposited over all submerged passages. This phenomenon is very common in epiphreatic caves and may act to retard solutional attack on cave walls in the tropics (Gillieson 1985).

Some parcels of sediment may be shunted into side passages during very high floods and will there remain unaltered beyond the reach of successive flow events. Other parcels may be sealed in by rapid flowstone growth or capped by very fine muds that are resistant to erosion. Thus, the Cricket Muds of Clearwater Cave, Sarawak, Malaysia have sat

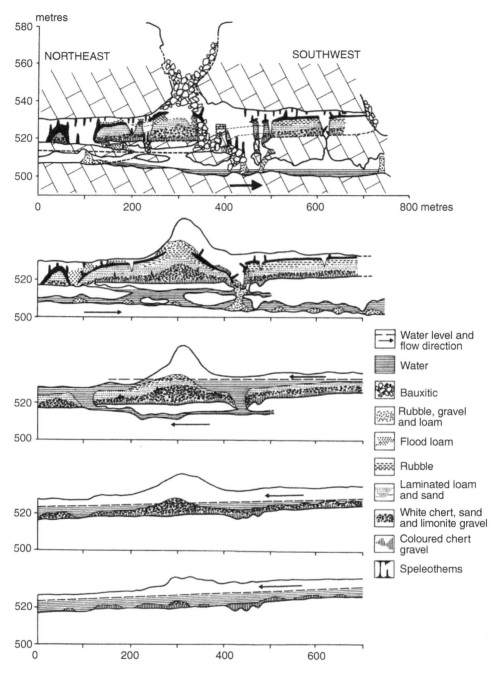

Figure 7.5 Developmental phases of the Otoska Jama, a part of the Postojna caves of Slovenia. The top diagram shows the present cave morphology in cross-section, dominated by collapse. The bottom diagram is the inferred original morphology of the phreatic passages excavated by the Pivka River. Progressive lowering of the level of the Pivka has led to several phases of cave roof collapse. Source: Modified from Gospodaric (1976).

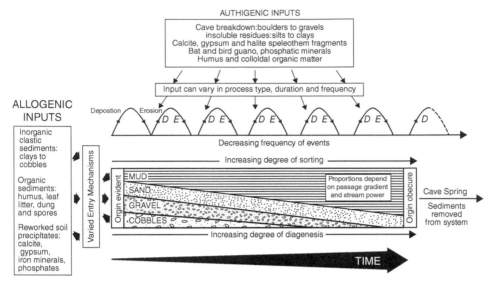

Figure 7.6 Processes affecting cave sediments through time. Cave sediments may be derived from allogenic or authigenic sources and may be highly variable in texture. They are sorted by phases of underground transport, which progressively removes some components and leads to an overall reduction in volume. The residuum may eventually emerge at cave springs.

undisturbed (apart from the burrowing of crickets) for the last three-quarters of a million years (Noel and Bull 1982). These very fine-textured muds often have a mean grain size smaller than 0.001 mm (10 phi). They are difficult to entrain, but once in suspension, they are easily transported, according to the Hjulström curve (Figure 7.7). On settling from suspension, they may accrete at steep angles, draped on underlying rock, flowstone, or sediment surfaces (Bull 1981). Quite commonly minor slumping occurs, producing flame structures analogous to those in turbidites, of which they are really a specialised case. Bull (1981) has described surge marks in these fine muds produced by the swash and backwash of pulsed floods in cave passages. These fine muds are useful in that they may provide a palaeomagnetic record and may also preserve pollen.

In Castleguard Cave, Alberta, Canada, there are three phases of fine silt filling, which have been interpreted by Schroeder and Ford (1983) as varved sequences deposited under full glacial conditions. These three silts persist for several kilometres in the cave and were clearly deposited under passage-full conditions. The youngest silt is older than flowstone dated at *c.*140 000 years ago, but is younger than a stalagmite which is >720 000 years ago. Schroeder and Ford (1983) believe the other two silts to be younger than this stalagmite, but Gale (personal communication) has isolated pollen from the silts that suggests lowland warm-temperate rainforest of probable Late Miocene age. However, the pollen in this material may be reworked from older terrestrial deposits. Thus, the precise age of the extensive silt fill of Castleguard Cave is open to conjecture.

It is clear from studies in tropical caves (Smart et al. 1986; Williams et al. 1986) that these fine cave fills may be very old and may predate the existing terrain by several million years. Laureano et al. (2016) have reconstructed a record of flooding and fluvial aggradation in Lapa Doce and Torrinha caves, NE Brazil, over all of the Pleistocene. The stratigraphy

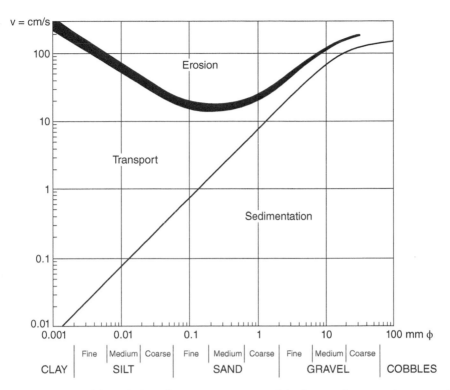

Figure 7.7 Empirical relationships between mean particle diameter in mm, and stream velocity in cm s^{-1}, according to Hjulström (1935). Source: After Bögli (1980).

suggests periodic flooding with phases of standing water during which paragenetic processes formed new passages.

Extensive gravel trains in cave passages may relate to the winnowing of fines by successive flows of water that are unable to move the gravels by themselves (though some slight movement by bedload traction may occur). The shapes of the gravels are held to be different in caves: Siffre and Siffre (1961) suggested that they were flatter because of their transport through inverted syphons, though Bull (1978) thought that flattening was more likely related to thin bedding. Studies by Kranjc (1981) have shown a marked decline in gravel roundness with distance of within-cave transport. Other downstream trends are indicated by Valen and Lauritzen (1990) for erratic boulders in Sirijorda Cave, Norway. There is a consistent decrease in mean size away from the points of injection of the boulders, presumably transported in subglacial meltwaters. The boulders are up to 1 m in diameter, and extremely large discharges must have been involved.

In Postojna Cave, Slovenia, examination of a sediment section in the underground River Pivka allows reconstruction of a partial history of the cave passage. A false floor of flowstone indicates that the sediment fill once nearly sealed the passage (Figure 7.8). While it was intact a flood loam was deposited on top. Since that time, incision has exposed the underlying sediments – interlayered gravels and loams. It is clear that the flow that eroded the loam (sample 404) deposited the gravel (sample 405), as these interfinger in both cross-section and long-section. The whole deposit has a downflow dip, which suggests that it infilled a pool, much like a surface point bar deposit. Consideration of the cumulative particle size curves (Figure 7.8) reveals that the lowest gravels are poorly sorted (wide range of sizes and large dispersion about the mean grain size) and that sorting improves up the section, although mean particle size increases. This suggests increased fluvial energy in the cave prior to the deposition of the loam. An increase in depositional energy also occurred after the loam was laid down.

Clearly flow in the passage declined or stopped so that the speleothems could form. The mineralogy of the gravels and sands suggests that each size component of the sediment has a distinctive source in the catchment, with the coarser gravels being dominantly limestone while the sands are quartz-rich. By dating the speleothems, it would be possible to gain a minimum age for the gravels and a maximum age for the flood loams; this would

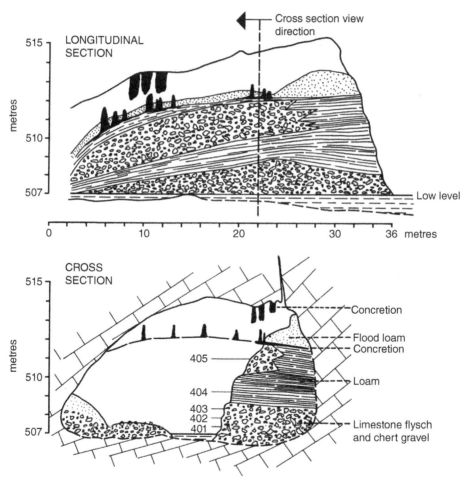

Figure 7.8 Stratigraphy and particle size characteristics of cave sediments from Postojna Cave, Slovenia. These sediments have been transported by the underground Pivka River and display many stratigraphic characteristics of fluvial sediments, sealed by a flowstone layer. The lower diagram provides details of the particle-size characteristics of sediment samples whose locations are given in the middle cross-section. Source: After Gospodaric (1976).

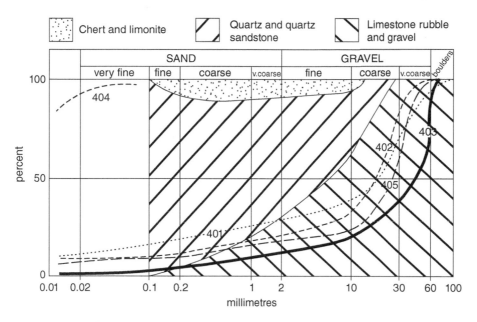

Figure 7.8 (*Continued*)

allow inferences about climatic control on the sedimentation to be tested. So, a great deal of information about the sequence of flows through a cave passage, and the likely source of the sediments in the catchment feeding the cave can be gained from detailed particle size analysis coupled with stratigraphic analysis.

Careful analysis of individual sediment sections (Figure 7.9) permits estimation of both the regime and energy of past water flows through a cave passage. Such estimation is based on hydraulic and sediment transport theory developed for open channels on the surface (Allen 1985), so results must be interpreted with caution. Nevertheless, some useful reconstructions have been made. In the Planinska Jama (Slovenia), Gospodaric (1976) has used the particle-size characteristics of fluvial sediments deposited from the River Rak to reconstruct water velocities (Figure 7.10). In the Great Hall of this cave, an eroded fill of laminated loam is overlain by gravels, the whole section being partially buried by breakdown blocks. The velocity of deposition can be estimated from hydraulic theory and was initially low, then increased during the phase of gravel deposition. There is a trend to diminishing velocity with the younger sediments, expectable with declining flows through a passage. The sediment textures themselves show a fining upward trend from gravels through coarse and medium sands. While most samples are poorly sorted (low gradient of the cumulative curves in Figure 7.10), sample 4 is well sorted, and this may reflect either a change in provenance or unusual persistence in river flow. Some information may thus be gained about both magnitude and persistence of river flow.

In the Cheat River canyon of West Virginia, US, Springer and Kite (1997) investigated the relationship between slackwater deposits preserved in caves and those found on the surface at tributary junctions. Following a disastrous major flood in 1985, sediments, woody debris and urban debris were deposited along the Cheat River and in its caves. Polystyrene balls adhering to the cave walls indicated a peak flood stage of 15 m, while slackwater

Figure 7.9 A flowstone dated by uranium series methods at c. 50 000 years ago caps deep fluvial gravels in Eagles' Nest Cave, Yarrangobilly, New South Wales, Australia.

sediments lay within 1 m of this stage. A sequence of older slackwater sediments, representing five palaeofloods, in one cave were inferred as being >400 000 years ago from valley incision rates. In contrast, surface slackwater sediments at tributary junctions were not closely related to peak flood stage indicators (woody debris and other urban debris), and the flood stage was only 6–11 m. Thus, analysis of the surface deposits underestimated peak flows by 50%. Alternative approaches to this problem involve scanning electron microscopy of included quartz grains (Bull et al. 1989) or estimation of flow velocity from erosional scallops on cave walls (Curl 1974), but these may require good sampling strategies to be representative.

Another interesting study from Postojna Cave, Slovenia, sheds light on both regional tectonics and cave development (Sasowsky et al. 2003). During excavation of a tunnel linking two sections of the cave, a fluvial sediment section 1.5 m thick was exposed. The laminated sediments showed clear evidence of slickensides, micro-faults indicative of tectonic movement in the rock mass. The sediments themselves indicated water transport in a cave conduit along or near a major fault system at Postojna. Faults can act to channel water flow in karst as inception horizons or alternatively act as aquicludes. Palaeomagnetic analysis of the sediments revealed that the sediments had been emplaced between 770 000 and

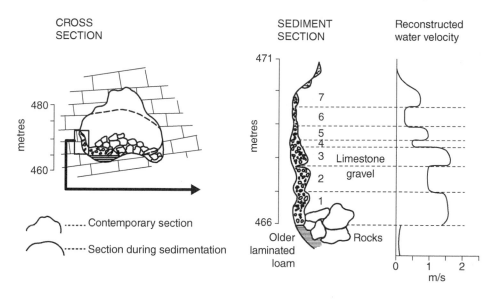

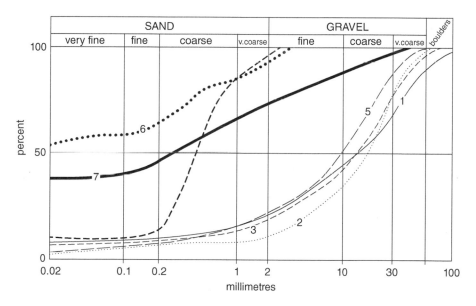

Figure 7.10 Pluvial sediments of the underground River Rak in Planinska Jama, Slovenia. A sediment section in a former phreatic passage has been analysed in detail; from these particle-size data, the former water velocities in the passage can be reconstructed. Source: After Gospodaric (1976).

990 000 years ago and were magnetically reversed (Jaramillo event) and that there had been regional tectonic activity since that time. The conduit itself was likely to have been active for 100 000 years in an environment of active valley downcutting, with cave deposits at different levels providing a prospect of estimating rates of that downcutting.

7.3.3 Cave and Rockshelter Entrance Deposits

Both gravity fall and waterlain sediments are often found associated with the entrance deposits of caves (Woodward and Goldberg 2001). The entrance zone, extending inwards to the dark zone, can be seen as transitional between surface environments and the more stable interior of the cave. The external facies of these sediment deposits are derived from surface wash, gravity fall, and in situ deposition, and may be subject to a range of geomorphic processes including slumping, erosion by water at the dripline, and collapse (Table 7.2). These result in complex depositional patterns and sediment stratigraphy that is subject to greater fluctuations in moisture and temperature. In contrast, the internal facies of cave sediments is in an environment where geomorphic processes are less active and may be confined to episodic floods and some collapse. All of these disturbance processes operate at varying rates over time and at different intensities as a result of the climatic and tectonic regime. Thus, in very high energy tropical environments, such as the montane karsts of Southeast Asia and Melanesia, the shelf life of cave entrance deposits may only be a few thousand years before they erode away.

A good example of these processes is provided by the work of Dykes (2007) at Niah Cave in Sarawak, Borneo. The cave contains very large deposits of guano from cave swiftlets. These are several metres thick in places and in the West Mouth of the cave a mass-movement deposit originated as a guano mudflow up the North Passage about 40 000 years ago. Sediment analyses showed that failure of the slope would have required huge inputs of water to the slope at that time. The texture of the guano is very powdery, making it prone to collapse when saturated. The most likely explanation for this collapse is that a wetter climatic phase or intense rainstorms coupled with seismic events triggered several shallow mudslides in the cave guano deposits.

Table 7.2 Important syndepositional and postdepositional processes and agents acting in prehistoric caves.

Physical	*Pene-contemporaneous reworking by wind, water, and humans*
	Slumping and faulting
	Trampling by humans and other animals
	Burrowing and digging by humans and other animals
	Vertical translocation of particles by percolating water
	Dumping of material, midden formation
	Hearth cleaning and rake-out
Chemical	Calcification (speleothems, tabular travertine, tufa, cemented 'cave breccia')
	Decalcification
	Phosphate mineralisation and associated bone dissolution
	Formation of other authigenic minerals: gypsum, quartz, opal, haematite

Source: From Goldberg and Sherwood (2006).

7.4 Sediment Transport and Particle Size

The approximate velocity of water flow at the moment of deposition can be estimated from the mean or modal particle size of the sediment, allowing estimation of past stream velocities in a cave passage. Waterlain clastic sediments in caves range from boulders to fine clays. The particle-size distribution is often skewed to the fines because of the winnowing or filtering effects of transport though cave conduits. They are generally well sorted and often show a considerable degree of particle flattening (especially the gravels). Thus, they do not necessarily behave as the spherical particles implicit in sediment transport theory. Nevertheless, this theory can be applied and provides useful results, if only in a relative sense.

The theory of sediment transport is complex (Richards 1982) and relies on fluid dynamics theory. Stream power at the sediment bed is a function of several factors:

$$\Omega = \rho g Q q \tag{7.1}$$

where Ω is the gross stream power, ρ the fluid density, g the gravitational acceleration, Q is the velocity in m.s^{-1} and q the channel slope (Bagnold 1966). This is related to shear stress by the equation

$$w = \frac{\Omega}{W} = t_0 v \tag{7.2}$$

where w is specific stream power, W is stream width, t_0 is mean shear stress (kg m^{-2}) and v is stream velocity (m s^{-1}). The shear stress is a measure of the force of moving water against the bed of the channel.

Sediment is entrained when the shear stress at the bed exceeds the gravitational and cohesive forces holding the individual particles in place. For spherical particles, the critical shear stress is

$$t_{crit} = 0.06(\rho_s - \rho)gD \tag{7.3}$$

where D is the particle diameter in mm and ρ_s is the particle density.

The maximum size of sediment particle that can be moved in a given channel is given by

$$D_{max} = 65t_0^{0.54} \tag{7.4}$$

For deposition, the settling velocity v_t is given by Stokes's law:

$$v_t = \frac{(\rho_s - \rho)}{18\mu}gD^2 \tag{7.5}$$

where μ is the dynamic viscosity. Thus, for a given settling velocity, the mean size of particles that will fall from suspension can be found. Once entrained, several modes of transport are possible:

- rolling of grains on a stationary bed,
- saltation of grains above the bed,
- sliding bed with varying depths of particle movement, and
- suspension load.

Whether sediment is transported by each or all of these modes will depend on mean grain size, the persistence of flow velocity, and the roughness of the bed. Each will tend to produce distinctive sedimentary structures that can be used to infer their environments of deposition.

An alternative approach to these theoretical equations is to use the empirical relationships between velocities of flow, transport, and sedimentation developed by Hjulström (1935). In Figure 7.7, these relationships are displayed according to mean grain size of the particles. The velocity of transport is always less than the velocity of erosion, as more hydraulic force is needed to tear the particles away from the cohesive bed of sediment than is needed to keep them in suspension or saltation. The critical velocities of erosion for fine particles, such as silts and clays, are as great as those required for coarse sands and gravels. This is because fine particles adhere to each other by electrostatic forces and the smooth surface of the sediment bed reduces turbulence. Thus, it is very hard to erode fine clays, but once eroded they can be kept in suspension very easily. In contrast, sands are very easily eroded and transported, and their velocities of sedimentation are not much less than their critical erosion velocities. Thus, sands tend to move through a cave system in a series of hops from temporary storage to temporary storage and may move right out of the cave altogether, whereas slits and clays can remain in place for a very long time. To some extent, this explains the markedly bimodal nature of cave sediments in which gravels and muds predominate.

If we consider the particle size distributions of selected cave sediments, then it is apparent that many lack a single central peak and are thus poorly sorted. Cave sediments tend to be skewed, bimodal, or polymodal in their particle-size distributions. Using a measure of central tendency such as the mean grain size is perhaps misleading, and it is probably better to employ one or more of the modes. Clayey gravels display bimodal distributions and act like 'plum puddings' in that the surface roughness caused by protruding gravel particles enhances the erosion of the fine clays by locally increasing flow velocity. Thus, there are important factors in sediment erosion and transport that are not accounted for by sediment transport theory.

For this reason, we must look at sediment structures if we wish to understand a cave history, fully. The best way to do this is to cut a vertical face on a sediment bank and examine the layering carefully. The thickness and inclination of sediment layers will tell much about the duration and energy of flow in a cave passage, while the disruption of those layers by slumping or by erosion will provide information on fluctuations in flow that may correspond to seasonal or longer-term perturbations in the karst system. The most difficult thing to determine is whether a thick bed of sediment is the result of one or many flow events in the cave. The only way to resolve this, apart from a very careful examination of the sediment structures, is by high resolution dating of a number of layers in the bed.

Relationships between water flow and depositional energy, as expressed in clastic sediment structures (Figure 7.11), can be organised along gradients from low to high energy, and from pools of deeper water to shallower cave streams (unconfined to confined channels). Within deep, virtually static water bodies (after floods or in the phreatic zone) sedimentation can result on very steep slopes of up to 70° by the slow settling of colloidal clay particles

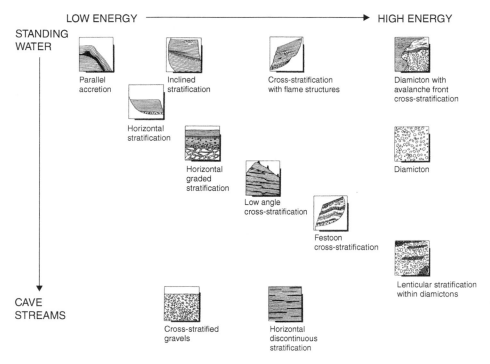

Figure 7.11 Scheme of fluvial cave sediment structures in relation to depositional energy and passage morphology. Source: From Gillieson (1986).

(Figure 7.12). These sediments accrete parallel to underlying rock or sediment surfaces (Bull 1981) and are often known as cap muds. Because of their mode of deposition, they have the potential to align with the prevailing geomagnetic field, and can thus provide a palaeomagnetic record (Noel and Bull 1982). Higher energy flows can dump of pulses of sediment into cave pools, producing inclined beds of sediments analogous to prograding deltas. At very high levels – those associated with flash flooding – the extreme turbulence connected with pulsed flow produces curled flame structures in the finer sediments. Rapid slumping of sediments on the deltaic slope may also occur in ways analogous to deep-sea turbidity currents.

Where the flow of water is constrained, as in cave streams, then there is a sequence of sedimentary structures ranging from horizontally stratified muds and sands to increasing degrees of cross-stratification. With growing energy levels, upward grading of beds from coarse to fine becomes common. At high, turbulent flow conditions, scouring of the bed associated with standing ripples and waves produces lensed cross-stratification. These features are very obvious in cross-section (Figure 7.11) and allow interpretation of the hydraulic regime under which deposition occurred as well as the degree of confinement of water flow. In combination with sediment transport theory, these allow the processes leading to the deposition of cave sediment to be reconstructed.

Figure 7.12 Alternate bands of oxidised and reduced fine clay sediments fill an old pool deposit in Selminum Tem cave, Papua New Guinea.

7.5 Diagenesis of Cave Sediments

Cave sediments tend to be much less subject to physical and chemical alteration than the surrounding rock. Physical alteration can include deformation due to loading and drying, leading to polygonal desiccation cracks and shrinking away from the walls of the cave. Some of the desiccation cracks can be up to 1 m deep, allowing superimposed sediments to intrude into lower layers.

Bioturbation by burrowing animals can be a significant disruptive process, especially in cave entrance and rockshelter deposits. Further into caves, the activities of cave invertebrates may be noticeable only in their effect on laminated clays, degrading their palaeomagnetic properties (Chapter 8).

Once emplaced into a cave, any allogenic sediment is subject to total darkness, near-constant high humidity, and near-constant temperature. This reduces the amount of chemical alteration that can occur. However, with time some migration of solutes into and out of the sediment will occur. This may be as the result of floods causing wetting and drying. In this context the porosity and mean grain size of the sediments has a great influence on the degree of diagenesis. Blue and red-banded clay sediments in Selminum Tem, New Guinea (Figure 7.12), owe their banding to alternate layers of very fine clay and slightly

Figure 7.13 Gypsum surface layers formed near a constriction in the passage, Lagangs Cave, Mulu, Sarawak, Malaysia.

coarser silt. Reducing conditions are maintained in the fine clays, with a dominance of ferrous iron salts, while increased porosity in the silts allows oxidation and dominance of ferric iron. Sediment banks are commonly cemented by iron oxide bands or by calcite from drips. Truly ancient sediments have reaction rims, often of calcite or phosphatic minerals, that may penetrate into the adjoining rock surfaces. Drying and capillary rise through the upper layers can cause other chemical precipitates may form at the surface of sediment banks. Gypsum veneers are the most common form of this, and may often form where constrictions in a passage accelerate airflow (Figure 7.13). These veneers may break up clay sediments to produce a sandy surface deposit of coated clay grains.

7.6   Stratigraphy and its Interpretation

Depositional energy affects cave sediment structures and can be used to reconstruct past discharges if interactions between cave sedimentation and passage hydrology are taken into account. Sediment surface textures gained from scanning electron microscopy can also be used to reconstruct cave hydrology (Gillieson 1983).

When we come to the task of interpreting the flow conditions under which a sediment was deposited, we can approach the problem in several ways: by using sediment transport theory, which relates the size of individual sediment particles to velocities of erosion, transport and deposition; or by using either empirical or theoretical relationships between sedimentary structures and the hydrologic regime. The former method produces more quantified results, but cannot account for the effects of bed roughness at the sediment-fluid interface. The latter method is not so amenable to quantification, but provides more information about the turbulence of flow, the frequency of deposition, and reworking of the sediments. Both approaches have their adherents, and probably the best solution is to use elements of each one to achieve the most comprehensive understanding possible.

One tool that has proven extremely useful is that of sediment micromorphology (Karkanas and Goldberg 2013). This is done by collecting small monoliths of in situ sediment and cutting thin sections after resin impregnation of the block. Several lines of evidence can be obtained from the thin sections. First, depositional structures such as fine waterlain laminations, slump structures, burrowing, and bioturbation can be detected. Secondly, chemical alteration of the sediment by weathering, fire, or decomposition of included organic remains can be detected. Thirdly, included microartefacts (flaked stone fragments), bone, coprolites, and charcoal can be readily identified. All of these provide greatly enhanced information about the mode of emplacement of the sediment and its subsequent history.

7.7 Provenance Studies

Cave sediments have the potential to provide us with a record of the geomorphic history of their catchments. When the catchment has distinctly varied lithology, analysis of the mineralogy of the gravel and sand fractions of cave sediments permits identification of sediment source areas. Thus, in Wales, Bull (1978) related the lithology of stream gravels in Agen Allwedd to autogenic sources (primarily limestone breakdown) or allogenic sources at the cave extremities. These mixed gravels showed significant downstream changes in lithological composition, and these changes could be related to the geomorphic evolution of the cave system. Magnetic minerals could also be used as tracers in this way.

This has been eloquently demonstrated using environmental radionuclides ^{137}Cs, ^{226}Ra, and ^{232}Th to provide a signature for sediments entering the Jenolan Cave system, New South Wales, Australia (Stanton et al. 1992). Recent build-up of sediments within the tourist section of this long cave system has led to backflooding of walkways and deposition of fines on flowstones and stalagmites. Concentrations of the fallout radionuclide ^{137}Caesium in these recent sediments indicated that most of the deposition had occurred since the mid-1950s. The ratio of ^{226}Radium to ^{232}Thorium in the recent cave sediments was compared to values for soils in different sub-catchments upstream of the caves. The results indicated that the recent cave sediments were derived from several catchments in which forestry activities had commenced during the early 1950s. The most likely source was erosion from unsealed forestry access roads in steep terrain.

Cave sediment studies, therefore, have the potential to assist land managers in identifying sources of accelerated erosion in karst terrains, and in reconstructing the long-term hydrology of karst catchments. However, there is still poor understanding of the mechanisms of transport, deposition, and diagenesis of cave sediments, especially during flood events. Scanning electron microscopy may offer a very good chance of relating sediment features to depositional energy, but this will require the use of recirculating flume experiments as well as within cave studies.

7.8 Cave Sediments and Environmental History at Zhoukoudian, China

Southwest of Beijing lies the World Heritage site of Zhoukoudian, often called the Peking Man site. A number of limestone caves, some unroofed, lie around Dragon Bone Hill. Here a deep sequence of sediments (Figure 7.14), up to 35 m thick, contains skeletal material, bone, and artefacts associated with the hominid *Homo erectus pekinensis*. These have been dated by a variety of techniques to between 770 000 and 300 000 years ago (Gao et al. 2017).

Skulls of the fossil hominids were excavated by D. Black, Teilhard de Chardin, and C.C. Young in 1929 and reported shortly afterwards (Black et al. 1933). Later excavations by Chinese scientists (Pei and Zhang 1985; Pei 1934) have expanded the areas of excavation,

Figure 7.14 Main fossil site (Western section) viewed from the entrance of the Pigeon Hall Cave, Zhoukoudian, China.

recovered a great deal of cultural material and refined the chronology; excavation at the site is ongoing. The skulls were transported for safekeeping in the United States of America at the onset of World War II, but were mysteriously lost in transit. Fortunately, casts had been made and exist in museum collections. Peking Man used small quartz flakes to slice and scrape muscle from bone and hefty chopping tools made of sandstone for bigger jobs, such as cutting through the rib cage of a large carcass. De Chardin used stratigraphic analysis to suggest that the hominids had used fire and were hunters, but alternative explanations have suggested their remains (or indeed the living hominids) may have been eaten by cave hyenas.

The cave formed as an enlargement of a vertical cleft or fault in which silty and angular rockfall accumulated over a very long time (Figure 7.15). There is also evidence of fluvial activity at the site with associated sediment transport. Over 90 years ago, the original investigators noted 'the evidently burnt condition of many of the bones, antlers, horn cores and pieces of wood found in the cultural layers, [and] a direct and careful chemical test of several specimens has established the presence of free carbon in the blackened fossils and earth. The vivid yellow and red hues of the banded clays constantly associated with the black layers are also due to heating or baking of the cave's sediments' (Pei and Zhang 1985).

The lowest archaeological unit is Layer 10, which is about 50–65 cm thick and can be divided into two units. The upper part is a compacted silty clay, pink to reddish-yellow in colour. The lower part is a yellowish-red and reddish-brown silt, which becomes more laminated with depth. This sediment was regarded by earlier workers as clear evidence of the use of fire on the site. Weiner et al. (1998) collected samples for micromorphological and infrared spectral analysis with bone material for chemical analysis.

In prehistoric deposits, fire ash should contain fine-grained calcite or carbonated apatite, with a small amount of silica aggregates. The sediments from the upper part of Layer 10 are comprised of quartz (45%), carbonated apatite (40%), and clay (15%). From the micromorphological analysis, the sediments are more likely to be due to diagenesis with secondary silica. Weiner et al. (1998) inferred that the carbonated apatite was not derived from ash and that there is no evidence for the use of fire in Layer 10. They also concluded that the much younger deposits of Layer 4 were not the result of fire in hearths, but more likely burnt bone and artefacts transported into the cave by running water. Further micromorphological studies by Goldberg et al. (2001) confirmed that the bulk of the sediment had been derived from loess and local hillwash, and was augmented by organic material from human occupation.

According to Goldberg et al. (2001), the deep accumulation of sediments at Zhoukoudian reveal evidence of three main depositional processes at the site: initial fluvial deposition associated with the lowest layers (basal sediments and unit 10), inwashing of material from the surrounding hillslopes (units 6–10), and windblown loess deposition in the upper layer 4. The basal laminated sediments indicate a fluvial origin, possibly as slackwater deposits or infilling depressions. The hillslope deposits contain kaolinitic clays derived from soil horizons, mixed with some loess material from further afield. The Pleistocene loess is widespread in the region and blankets hillslopes around the site. The bulk of the sediments are fault breccias and collapse material from the cave itself, suggesting intermittent tectonic activity coupled with normal cave collapse. Biogenic deposits include seeds of hackberry, which grows around the site (Figure 7.16), plant phytoliths, burnt bone,

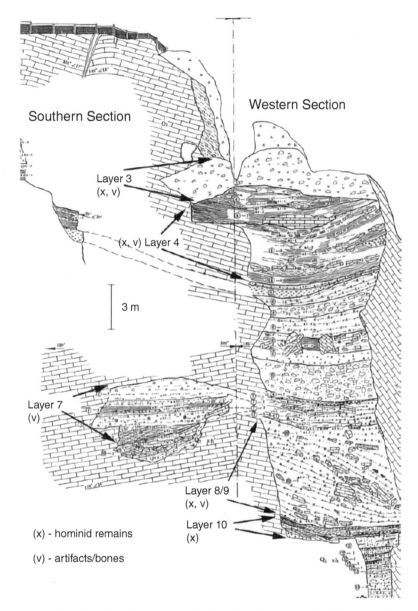

Figure 7.15 Stratigraphy of archaeological deposits at Location 1, Zhoukoudian. Earlier interpretations of the site associated burning with hominid remains and artefacts in Layers 10 and 4. Source: From Goldberg et al. (2001).

and stone tools. It is likely that the Zhoukoudian site shows evidence of regional climate fluctuations throughout much of the Pleistocene, including both moister, monsoonal conditions during interglacials (similar to today), and drier, windier glacials, during which the loess would have been deposited.

More recent research at Zhoukoudian (Gao et al. 2017) has re-opened the debate over the use of fire by hominids. The Chinese scientists have had greater access to finds from

Figure 7.16 Exterior of the Pigeon Hall Cave, Zhoukoudian, China.

different areas of the site and have carried out more excavations and associated geoarchaeological analyses.

Several criteria can be used for evaluating the presence of in-situ burning at archaeological sites (Goldberg, Miller, and Mentzer 2017, Weiner et al. 2000: p. 221). These are:

- The presence of constructed fireplaces and hearths in association with burned bones and stone tools,
- Clear presence of wood ash in a cave at a site where trees would have been absent, and
- Burned bones associated with stone tools in a stratigraphic unit.

At Zhoukoudian, in Layer 4, there is a clear association between burned bones and tools in stone outlined features. The sediments show high redness and magnetic susceptibility values characteristic of burning (Gao et al. 2017). Layers 8 and 9 have yielded more evidence of human fire use including burned objects and fire ash, along with three crania of *Homo erectus*. The upper unit of Layer 10 could be partially related to human activity due to the presence of blackened bones and stone tools.

This debate is a normal part of any scientific enquiry and serves to illustrate that we should have open minds about research findings and be prepared to accept alternative views that are soundly based on evidence. Further research at Zhoukoudian will help to refine the long history of human occupation at this iconic site (Figure 7.17). Analysis of the cave sediments and their included artefacts will resolve some issues and hopefully raise new questions as the excavations there move into their second century.

Figure 7.17 Main Western section at Location 1, Zhoukoudian, China. Stratigraphic units 5 to 8–9 are clearly visible.

References

Allen, J.R.L. (1985). *Physical Processes of Sedimentation*. London: Allen & Unwin.

Bagnold, R.A. (1966). An approach to the sediment transport problem from general physics. *US Geological Survey Professional Paper*. pp. 422–421.

Black, D., de Chardin, P.T., Young, C.C. et al. (1933). Fossil man in China: the Choukoutien cave deposits with a synopsis of our present knowledge of the late Cenozoic in China. *Memoirs of the Geological Survey of China. Series A* 11: 1.

Bögli, A. (1980). *Karst Hydrology and Physical Speleology*. Berlin: Springer.

Bull, P. (1978). A study of stream gravels from a cave: Agen Allwedd, South Wales. *Zeitschrift für Geomorphologie* 22: 275–296.

Bull, P.A. (1981). Some fine-grained sedimentation phenomena in caves. *Earth Surface Processes and Landforms* 6: 11–22.

Bull, P.A., Yuan, D., and Hu, M. (1989). Cave sediments from Chuan Shan tower karst, Guilin, China. *Cave Science* 16: 51–56.

Collier, C.R. and Flint, R.F. (1964). Fluvial sedimentation in Mammoth Cave. Kentucky. *US Geological Survey Professional Paper*. 475D: pp. D141–D143.

Curl, R.L. (1974). Deducing flow velocity in cave conduits from scallops. *Bulletin of the National Speleological Society of America* 36: 1–5.

Dykes, A.P. (2007). Mass movements in cave sediments: investigation of a ~40,000-year-old guano mudflow inside the entrance of the great cave of Niah, Sarawak, Borneo. *Landslides* 4: 279–290.

Frank, R.M. and Jennings, J.N. (1978). Development of a subterranean meander cutoff: the Abercrombie caves, New South Wales. *Helictite* 16: 71–85.

Gao, X., Zhang, S., Zhang, Y., and chen, F. (2017). Evidence of Hominin use and maintenance of fire at Zhoukoudian. *Current Anthropology* 58: S267–S277.

Gascoyne, M., Ford, D.C., and Schwarcz, H.P. (1983). Rates of cave and landform development in the Yorkshire Dales from speleothem age data. *Earth Surface Processes and Landforms* 8: 557–568.

Gillieson, D.S. (1983). Geoarchaeological applications of scanning electron microscopy. *The Artefact* 8: 43–54.

Gillieson, D.S. (1985). Geomorphic development of limestone caves in the Highlands of Papua New Guinea. *Zeitschrift für Geomorphologie* 29: 51–70.

Gillieson, D.S. (1986). Cave sedimentation in the New Guinea Highlands. *Earth Surface Processes and Landforms* 11: 533–543.

Goldberg, P. and Sherwood, S.C. (2006). Deciphering human prehistory through the geoarcheological study of cave sediments. *Evolutionary Anthropology: Issues, News, and Reviews* 15: 20–36.

Goldberg, P., Weiner, S., Bar-Yosef, O. et al. (2001). Site formation processes at Zhoukoudian, China. *Journal of Human Evolution* 41: 483–530.

Goldberg, P., Miller, C.E., and Mentzer, S.M. (2017). Recognizing fire in the Paleolithic archaeological record. *Current Anthropology* 58: S175–S190.

Gospodaric, R. (1976). The Quaternary caves development between the Pivka basin and Polje of Planina. *Acta Carsologica* 7: 5–135.

Hjulström, F. (1935). Studies of the morphological activities of rivers as illustrated by the River Fyris. *Bulletin of the Geological Institution of the University of Upsala* 25: 221–527.

Ikeya, M., Toshikatsu, M., and Gospodaric, R. (1983). ESR dating of Postojna cave stalactite. *Acta Carsologica* 11: 117–130.

Karkanas, P. and Goldberg, P. (2013). Micromorphology of cave sediments. In: *Treatise on Geomorphology* (eds. J. Schroeder and A. Frumkin). San Diego, United States: Academic Press.

Kranjc, A. (1981). Sediments from Babja Jama near Most na Soci. *Acta Carsologica* 10: 201–211.

Laureano, F.V., Karmann, I., Granger, D.E. et al. (2016). Two million years of river and cave aggradation in NE Brazil: implications for speleogenesis and landscape evolution. *Geomorphology* 273: 63–77.

Noel, M. and Bull, P.A. (1982). The palaeomagnetism of sediments from Clearwater Cave, Mulu, Sarawak. *Cave Science* 9: 134–141.

Pei, W.C. (1934). A preliminary report on the late-Palæolithic cave of Choukoutien 1. *Bulletin of the Geological Society of China* 13: 327–358.

Pei, W. and Zhang, S. (1985). A study of the lithic artifacts of Sinanthropus. *Palaeontologica Sinica* 12.

Richards, K. (1982). *Rivers: Form and Process in Alluvial Channels*. London, Methuen and Co. Ltd.

Sasowsky, I.D., Šebela, S., and Harbert, W. (2003). Concurrent tectonism and aquifer evolution >100,000 years recorded in cave sediments, Dinaric karst, Slovenia. *Environmental Geology* 44: 8–13.

Schroeder, J. and Ford, D.C. (1983). Clastic sediments in Castleguard Cave, Columbia Icefields, Alberta, Canada. *Arctic and Alpine Research* 15: 451–461.

Siffre, A. and Siffre, M. (1961). Le façonnement des alluvions karstique. *Annales De Spéléologie* 16: 73–80.

Smart, P., Waltham, T., Yang, M., and Zhang, Y. (1986). Karst geomorphology of western Guizhou, China. *Transactions of the British Cave Research Association* 13: 89–103.

Springer, G.S. and Kite, J.S. (1997). River-derived slackwater sediments in caves along Cheat River, West Virginia. *Geomorphology* 18: 91–100.

Stanton, R.K., Murray, A.S., and Olley, J.M. (1992). Tracing recent sediment using environmental radionuclides and mineral magnetics in the karst of Jenolan caves, Australia. In: *Erosion and Sediment Transport Monitoring Programmes in River Basins* Proceedings of a symposium held at Oslo, August 1992. IAHS Publ. no. 210 (eds. J. Bogen, D.E. Walling and T.J. Day). Wallingford, UK: IAHS Press.

Valen, V. and Lauritzen, S.E. (1990). The sedimentology of Sirijorda Cave, Norland, northern Norway. In: *Proceedings of the 10th International Congress of Speleology*, 125–126. Budapest: Hungarian Speleological Society.

Weiner, S., Xu, Q., Goldberg, P. et al. (1998). Evidence for the use of fire at Zhoukoudian, China. *Science* 281: 251.

Weiner, S., Bar-Yosef, O., Goldberg, P. et al. (2000). Evidence for the use of fire at Zhoukoudian, China. *Acta Anthropologica Sinica* 19 (Suppl): Proceedings of 1999 Beijing International Symposium on Paleoanthropology, 218–223.

Williams, P.W., Lyons, R.G., Wang, X. et al. (1986). Interpretation of the palaeomagnetism of cave sediments from a karst tower at Guilin. *Carsologica Sinica* 6: 119–125.

Woodward, J.C. and Goldberg, P. (2001). The sedimentary records in Mediterranean rockshelters and caves: archives of environmental change. *Geoarchaeology* 16: 327–354.

8

Dating Cave Deposits

8.1 The Importance of Dating Cave Deposits

Most of the conceptual advances in the scientific study of caves have derived from the application of new dating techniques to cave deposits, such as calcite and gypsum speleothems, and to layered sediments. This development has occurred at a rapid pace over the last two decades, and new analytical methods are steadily appearing as a result of cooperation between geomorphologists, geochemists, and physicists. We now realise that many caves are time-transgressive structures, with long histories extending well into the Tertiary, and older in some cases. This has far-reaching consequences for our perception of rates of geomorphic process operation, and the construction of schemes of regional denudation chronology.

The dating of cave deposits has wider import than merely providing a chronology for the clastic sediments and their included flora and fauna, long the preserve of archaeologists and palaeontologists. While cave sediment chronology remains an important area of speleology, there is keen interest in finding the minimum ages of cave passages leading to a better understanding of the rate of valley deepening. Usually the dating of speleothems or sediments will only give us an age before which a passage must have been formed. In some cases, new passages will have formed after a deposit was laid down; this provides a maximum age for the passage. If the oldest deposits in a vertical sequence of caves also form a time sequence (Figure 8.1), then the time of formation of individual passages can be more precisely determined. If those passages are graded to a present or former river valley or erosion surface, then we can infer rates of landscape evolution from the sequence of cave passages and their deposits. This is perhaps the most powerful application of cave chronology, as here the caves are preserving evidence of past landscape development which has elsewhere been obliterated by surface processes. Such applications have radically revised thinking about the age and rate of evolution of landscapes as diverse as the Swiss Alps, NE Brazil, the Yorkshire Dales of England, and Fiordland in New Zealand.

Caves: Processes, Development, and Management, Second Edition. David Shaw Gillieson.
© 2021 John Wiley & Sons Ltd. Published 2021 by John Wiley & Sons Ltd.

Figure 8.1 Daxiao Shan, a perched phreatic tunnel north of Longgong, Guizhou, China, is testimony to rapid uplift and valley deepening on the margin of the Himalayas. The passage is 50 m high and wide and 800 m long.

8.2   Dating Techniques and the Quaternary Timescale

At an early stage in any speleological investigation, decisions must be made about the dating techniques to be employed. Those decisions will rest on the perception of the likely age of the cave passages and their deposits, on the availability of funds or access to a dating laboratory, and the questions being asked. Thus, a geomorphologist interested in the long-term evolution of a landscape will likely use a method with a potential time range of several million years, while an archaeologist concerned with patterns of human occupation of a site since the Last Glacial will probably opt for the radiocarbon technique. The time range of several widely used techniques is given in Figure 8.2. One obvious feature is the big gap in available techniques in the time range from *c*.400 000 years to 1 million years and older.

Unfortunately, we are today finding many sites whose antiquity falls in or beyond that gap. New techniques involving Uranium and Lead isotopes have been developed to deal with these timescales, but the techniques, although powerful, are still in the developmental stage. The challenge for the future is to devise, test, and apply new dating methods that can cope with the time range from 10^4 to 10^6 years. We can divide the currently available techniques into three groups:

- *Comparative,* where a preserved physical or chemical parameter of the sample is compared with values of the same parameter in a dated sequence – for example, palaeomagnetism;

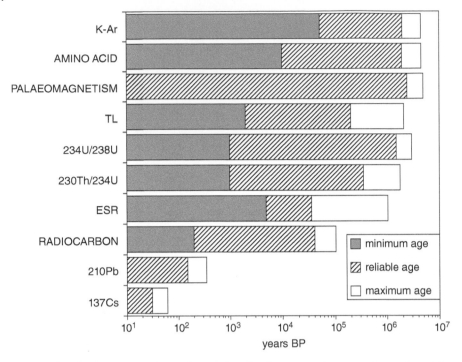

Figure 8.2 The Quaternary timescale and the effective range of comparative, radiometric, and radiogenic dating methods. Source: From Gillieson (1996).

- *Radiometric,* where the statistically random but predictable decay of a radioisotope incorporated into the sample is measured – for example, uranium-series and radiocarbon methods; and
- *Radiogenic,* where the crystal defects created in a sample by the radioactive decay of an isotope are measured – for example, electron spin resonance (ESR) and thermoluminescence (TL) dating. These techniques were investigated in the 1990s, and their low precision has not led to their adoption in speleothem science. They are, however, widely used in archaeological dating.

8.3 Palaeomagnetism

One comparative technique which does cover the time range from tens of thousands to a few million years is the comparative method of palaeomagnetic analysis. Unfortunately, it is a low precision method, in which individual samples can only be placed into broad time classes in which the magnetic polarity of the material is either normal or reversed. This magnetic timescale (Figure 8.3) extends back in time for the last 3 million years, i.e. the whole Quaternary, although deep-sea drilling extends the method almost yearly.

So far, most success has been achieved by analysing the remnant magnetic field preserved in fine-grained clastic sediments (pond deposits) within caves. These settle slowly in accordance with the prevailing geomagnetic field, although the dip of individual particles is some

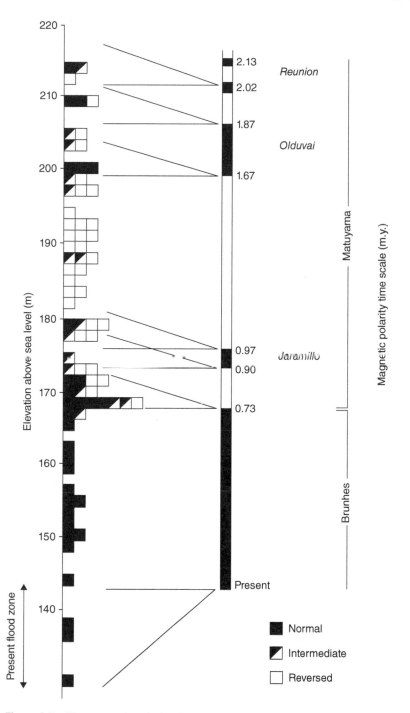

Figure 8.3 The magnetic polarity timescale and the record of normal and reversed magnetic declinations obtained from clays samples of sands, silts, and clays throughout the Mammoth-Flint Ridge-Rappel-Proctor Cave system of Kentucky, USA. Source: From Schmidt (1982).

30° shallower than the earth's field on account of final random motions on irregular sediment banks. In the cave environment, these pond deposits are little affected by chemical change, although some oxidation may occur, and calcite deposition may take place under drips. These secondary effects have their own magnetic field, which can usually be 'cleaned' from the specimen during analysis in the laboratory.

In most cases, the results are compared with the geomagnetic record of reversals, such as in the Mammoth-Flint Ridge-Roppel-Proctor Cave system of Kentucky (USA) (Schmidt 1982). In Figure 8.3, the record of normal and reversed magnetic declinations of fine-grained sediments from this vast cave system are related to the magnetic polarity timescale. The lowest and youngest passages have normal polarity, relating to the progressive downcutting of the Green River. Samples above 180 m elevation in the cave (above Gothic and Grand avenues) have reversed polarity and probably predate the Brunhes–Matuyama transition at 730 000 years. There is a suggestion that the highest passages (above 200 m elevation, the level of Collins Avenue) may date to the normal polarity Olduvai epoch at 1.6 million years, or slightly older. A problem with this approach is the patchy nature of the cave magnetic record; it is punctuated by large gaps, unlike deep-sea cores. Thus, a reversal may be a result of deposition in any of several epochs. Only where a large number of samples has been taken – in the above case, 500 oriented samples of clays, sands, and silts – can any confidence be attached to age estimates.

Magnetostratigraphy can also be used to make inferences about regional denudation chronology, as cave systems may encapsulate elements of the geomorphic history of the overlying landscape. Pease et al. (1994) have studied the polarity of sediment samples from the multi-level Wyandotte Cave, southern Indiana, USA. This system lies to the north of Mammoth Cave and, like that cave system, its base level has also been a tributary of the Ohio River. The sediments in the uppermost levels of Wyandotte Cave have a normal polarity, while those lower in the sequence have reversed polarity and correlate with old tube passages such as Upper Salts Avenue, Grand Avenue, and Cleaveland Avenue. These passages were thus active not less than 780 000 years BP (Brunhes-Matuyama transition). The uppermost passages are infilled by sediments not less than 2.6 million years old (Matuyama–Gauss transition), and these correlate with the residual sediments of the Upper Mitchell Plain of Indiana and the Pennyroyal Plateau of Kentucky. This provides a minimum age for this widespread land surface that has controlled karst development.

In truly ancient materials, the preserved magnetic field may be compared to the known position of the geomagnetic pole (the Apparent Polar Wandering Path) over geological time, and an age estimate obtained. If a thick sequence of sediments is present, then samples taken at close depth intervals may be used to construct a record of magnetic secular variation. This can be matched with long records of secular variation gained from analysis of lake sediments or deep-sea sediment cores. In Peak Cavern, Derbyshire, England, Noel (1986) has matched a one-metre sequence of laminated clays with lake magnetostratigraphy, implying an early Holocene age. Palaeocurrent directions in the cave passage were also obtained.

Calcite speleothems acquire some chemical remnant magnetisation during the crystallisation process and subsequent alterations. Detrital grains from floodwater may also be incorporated into the speleothem and may provide a stronger magnetic signal. In either case the strength is low, requiring use of very sensitive magnetometers. Palaeomagnetic dating can thus be used to supplement absolute dating methods: if the sample is older than 350 000 years, then normal or reversed polarity will indicate if it is younger or older than the Brunhes–Matuyama reversal at 730 000 years BP – or another geomagnetic reversal. Although results (including secular variation) have been obtained for speleothems from North America (Latham et al. 1986), problems with the identity of the speleothem magnetisation remain unsolved (Latham and Schwarcz 1990). Recent developments in Uranium series dating using very small sample sizes (typically 1 mg) may make magnetostratigraphy of speleothems a viable prospect for older samples and may be used to detect tectonic movements at individual cave sites.

8.4 Uranium Series; Uranium-Thorium, Uranium-Lead

The radiometric method known as uranium series disequilibrium is the most widely used and most successful for the dating of calcite speleothems. The first successful uranium series dates on speleothems were obtained by Duplessy et al. (1970) on material from the Aven D'Orgnac, France. The early work from the late 1960s to the 1980s used alpha-spectrometry, with large sample sizes and slow counting times limiting the number of analyses that could be performed. By the 1990s, thermal ionisation mass spectrometry (TIMS) revolutionised both sample size and counting time, and improved analytical precision allowed reliable dating at both young and old extremes of the U-Th dating range (Cheng et al. 2013). Both of these techniques have also led to the ability to date samples with very low uranium content. Uranium series dating has an effective range from a few thousand to 600 000 years BP for Uranium/Thorium and ages reported to >400 million years for Uranium/Lead (Richards and Dorale 2003). Precision for U-Th is around 0.1% and for U-Pb in the range 1–5%. In terms of age range, U-Th dating is limited by the half-life of ^{230}Th which is 75 584 years, so instead of accumulating indefinitely (as is the case for the U-Pb system) it approaches secular equilibrium with its parent isotope ^{234}U, whereby it decays as fast as it is produced and gives the U-Th dating system an effective upper age limit of somewhat over 500 000 years. Advances in instrumentation such as multi-collector inductively coupled plasma mass spectrometry (MC–ICPMS; Hellstrom 2003), further decreasing the sample size needed to 1 mg or less, mean that material as old as 600–650 000 years old can now be dated, with the other end of the datable age spectrum as recent as a few decades (Dorale et al. 2007). Advances in coring techniques (Spotl and Mattey 2012) and the use of broken speleothems have meant that sampling is now far more conservative than before.

The primary source of uranium isotopes is the weathering of igneous rocks and the incorporation of the weathering products in the hydrological cycle (Figure 8.4).

Two natural parent isotopes, ^{238}U and ^{235}U, exist. These have exceptionally long half-lives, and decay by emission of α and β particles to stable lead isotopes ^{206}Pb and ^{207}Pb. Intermediate daughter products such as ^{234}U and radon (^{226}Ra) are also suitable for dating. Upon weathering, more daughter ^{234}U is released than the parent ^{238}U and ^{235}U. All three species become oxidised and are readily transported in solution as complex ions (Figure 8.4).

They may be precipitated in calcite. The daughter products protactinium (^{231}Pa) and thorium (^{230}Th) are insoluble and may bond to the charged surface of clay particles or

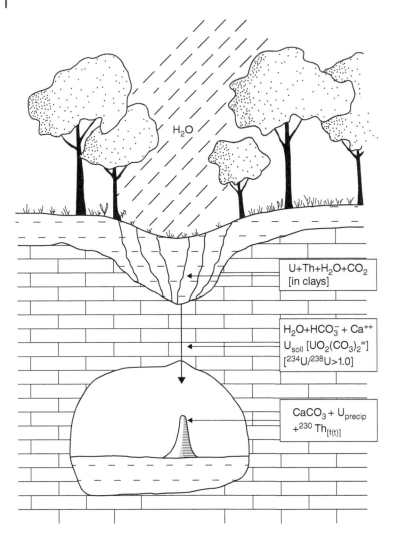

Figure 8.4 Schema of the geochemical pathways of uranium and thorium into caves and speleothems. Source: From (Gillieson 1996).

organic free radicals. They will be present in the calcite only as detrital layers or impurities. Thus, in pure calcite, any ^{230}Th will be the result of the decay of ^{234}U over time since crystallisation. This forms the basis of the uranium series method as applied to speleothems (Figure 8.5). A complex equation (Ford and Williams 2007) governs the decay of ^{234}U to ^{230}Th. This is summarised in Figure 8.6. The initial ratio of ^{234}U to ^{238}U is normally always greater than 1.0 and the ratio of ^{230}Th to ^{234}U is greater than 0. Through time ratios evolve to the right, to a point where the steepening isochrons cannot be resolved. This is at an age around 350 000 years.

Several criteria must be met for reliable uranium series dating of speleothems:

- There must be sufficient uranium in the calcite. The absolute minimum is 0.01 ppm, while 90% of samples have more than this concentration, up to a recorded maximum of

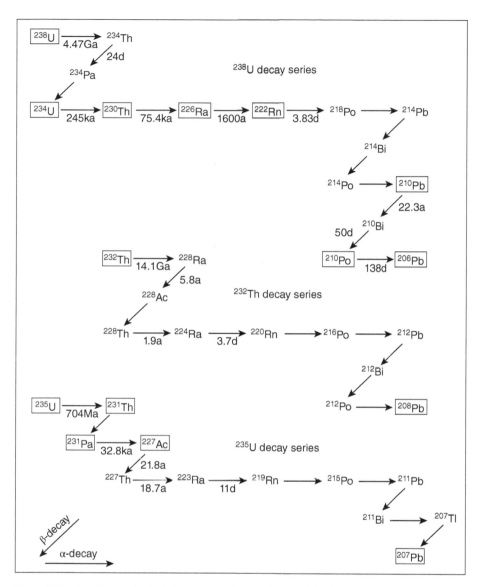

Figure 8.5 The three principal decay series for uranium and thorium nuclides including those measured in ^{234}U-^{230}Th, ^{235}U-^{231}Pa, ^{210}Pb, ^{226}Ra, ^{238}U-^{206}Pb, and ^{235}U-^{207}Pb dating, Half-lives are shown except where very short, and the more long-lived isotopes are enclosed in a box. Source: From Fairchild and Baker (2012).

120 ppm. At some sites where speleothem growth is dependent on direct rainwater input only, uranium concentrations may be too low.

- The calcite must be closed to exchange after the co-precipitation of calcite and uranium salts. This assumption is frequently violated, and careful examination of individual samples for areas of porous calcite, of resolution, or of reprecipitation is necessary. For this reason, porous tufas and stalactites are avoided. This entails careful sampling

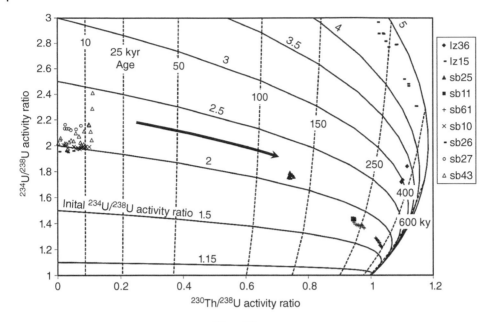

Figure 8.6 Graphical illustration of the ^{230}Th/^{234}U dating method. Speleothem calcite starts with a Uranium composition on the y-axis and its subsequent evolution follows the arrow shown, parallel to lines defining the ratio of radioactivity of ^{234}U to ^{238}U at the time of formation (this ratio falls over time because of faster decay of ^{234}U). The ultimate end-point is secular equilibrium, when the ratio of radioactivities of the isotopes in the decay chain ^{238}U $\rightarrow$ ^{234}U $\rightarrow$ ^{230}Th are one. Source: From Fairchild and Baker (2012).

and sectioning, with the real possibility that a sample from a geomorphologically meaningful context will have to be rejected. There is also a question of conservation involved, as any sampling is destructive to the cave environment. Time spent on personal reconnaissance is seldom wasted.

- No ^{230}Th or ^{231}Pa must be deposited in the calcite. In reality, most calcites contain some of these insoluble isotopes as contaminants. Usually, the concentration of ^{232}Th is measured in the sample, an initial ratio of (^{230}Th)/(^{232}Th) assumed, and a correction applied for allogenic ^{230}Th. Newer, more sophisticated methods use partial leaching and assay techniques (Latham and Schwarcz 1990). If the ratio of (^{230}Th)/(^{232}Th) is greater than 20, radiogenic thorium is assumed to dominate, and contamination is minimal. Highly contaminated specimens are either avoided or subjected to repeat determinations. The latter is a costly process. The major sources of detrital ^{230}Th are clays, deposited from floodwaters, organic colloids in soil, or the fine carbon particles of bushfire smoke. At Yarrangobilly, in the Snowy Mountains of Australia, thin black layers in flowstones are made of carbon dust which contains abundant thorium liberated from the soil during the intense bushfires which burn the area once a decade or so.

Recent developments in uranium series dating using isotope dilution mass spectrometry (Li et al. 1989) offer the prospect of reliably dating speleothems with U contents as low as 0.08 ppm to 400 000 years. The new method offers high precision with counting errors reduced significantly. For material with higher U content (>3 ppm), it should be

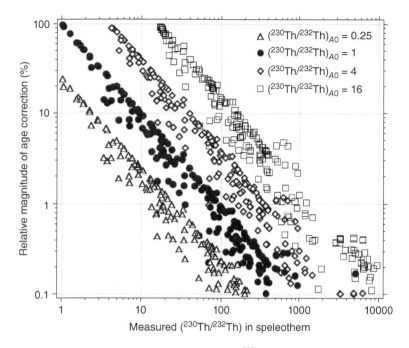

Figure 8.7 The effect of correction for initial ^{230}Th illustrated using a large, diverse body of speleothem U-Th analyses previously undertaken at the University of Melbourne (Australia). The data are corrected assuming a range of initial $(^{230}$Th$)/(^{232}$Th$)$, plotted as the percentage by which this correction changed the calculated age vs. measured $(^{230}$Th$)/(^{232}$Th$)$. Source: From Hellstrom (2006) © 2006 Elsevier.

possible to date material as old as 600 000 years. The problem of detrital thorium remains, however.

This increased precision in U-Th determinations has focussed attention on the significant uncertainty introduced into U-Th ages by the presence of detrital Th transported by groundwater. This thorium is bonded to fine organic matter, colloidal material and fine sediments and can cause a substantial overestimate of the U-Th age. The range of values of detrital radiogenic to stable Th $(^{230}$Th$)/(^{232}$Th$)$ varies with rock and soil types and typically ranges from 0.2 to 18 (Hellstrom 2006). Figure 8.7 shows the effect of variation in $(^{230}$Th$)/(^{232}$Th$)$ on U-Th ages (Hellstrom 2006), and clearly demonstrates either the need for accurate determinations of the detrital thorium component or on use samples where the ratio of $(^{230}$Th$)/(^{232}$Th$)$ in the speleothem is greater than 300. The proportion of detrital $(^{230}$Th$)/(^{232}$Th$)$ may vary over time; it is usually assumed to be constant and given an arbitrary range of values ($\sim$0.9 $\pm$ 0.45) but this may significantly underestimate the actual range (Hellstrom 2006). Using interval counting growth layers on more recent stalagmites coupled with high-resolution U-Th determinations with interval counting would allow an assessment of the variability of $(^{230}$Th$)/(^{232}$Th$)$ over time and thus reduce the age uncertainty (Domínguez-Villar et al. 2009).

Within the U-series family another technique – U-Pb – is capable of dating ancient speleothems, based on the two radioactive decay series from ^{238}U and ^{235}U to stable ^{206}Pb and ^{207}Pb, respectively (Richards et al. 1998). This is not a routine technique because it is very time-consuming and expensive and requires samples of exceptionally well-defined chemical composition (Woodhead et al. 2006). Speleothem samples must be rich in U (ideally at least 1 ppm) and contain very little initial (so-called common) Pb from the local environment. Under ideal circumstances, a sample for dating would contain negligible Pb at the time of formation, and thus any Pb measured subsequently could be confidently assumed to be the product of radioactive decay. If initial Pb was present at the time of speleothem deposition, significant corrections have to be applied, increasing the age uncertainty. This is particularly the case if the speleothem is less than a few million years old. Research by Woodhead et al. (2010) based on 650 speleothem samples suggests that cleaned samples typically have 1–30 ppb Pb, substantially lower than suggested in the literature. Thus, more samples may be amenable to this method than previously thought.

In the U-Pb dating scheme, radiogenic (produced in situ) ^{206}Pb can be completely obscured by common (initial) ^{206}Pb, meaning that age determinations are usually undertaken by multiple sample analysis and isochron construction (Woodhead et al. 2006). As with U-Th, an important consideration in U-Pb dating of speleothem material is the initial $(^{234}U)/(^{238}U)$ ratio, which is measured directly where possible, but must usually be estimated for samples more than about 2 million years old (Pickering and Kramers 2010; Woodhead et al. 2006). Without taking the initial excess of ^{234}U into account, ages can be greatly over-estimated, and where initial $(^{234}U)/(^{238}U)$ cannot be directly determined, it can lead to additional uncertainties of hundreds of thousands of years.

U-Pb dating of speleothems requires very accurate measurement of the ratios $(^{206}Pb)/(^{238}U)$, $(^{207}Pb)/(^{206}Pb)$, and $(^{234}U)/(^{238}U)$ and is analytically more difficult than U-Th. It is therefore important to screen samples for detrital Th and Pb, often carried out as a precursor to U-Th dating (Woodhead et al. 2010). Some laboratories still use TIMS for the Pb measurements, although MC-ICP-MS allows much greater sample throughput; in either case, sample size requirements are up to 100 times greater than for U-Th.

The usual approach to viewing U-Pb data for speleothem samples with variable common Pb content is to plot sample data in the so-called Tera-Wasserburg isochron construction plot of $(^{238}U)/(^{206}Pb)$ vs $(^{207}Pb)/(^{206}Pb)$ (Figure 8.8; Woodhead et al. 2006). In this diagram analyses are plotted corrected for analytical blank, but are not required to be corrected for common Pb content, which is useful in this context as most traditional methods for correction of common Pb involve measurement of isotopes which can be either in very low abundance in speleothems (^{207}Pb, ^{208}Pb) and/or are compromised by mass spectrometer interferences (^{204}Pb). The diagram in Figure 8.8 has two important features. The first is the concordia curve, which represents the line along which undisturbed radiogenic Pb (i.e. those producing an identical age from the $(^{235}U)/(^{207}Pb)$ and $(^{238}U)/(^{206}Pb)$ systems) should lie. Tick marks along this line represent ages that would be derived from entirely radiogenic Pb. In the conventional approach to U-Pb geochronology of speleothems, several individual sub-samples (ideally all from the same growth layer) are analysed and will often comprise variable mixtures of these radiogenic and common Pb components. If the sources of common Pb within a given growth layer are unique, and only the amounts vary, the data will define a line in the diagram from which the age of the sample can be determined via its intersection with the concordia curve. In relatively low Pb samples, analyses are likely to plot near to the radiogenic component. In other more Pb-rich samples, where

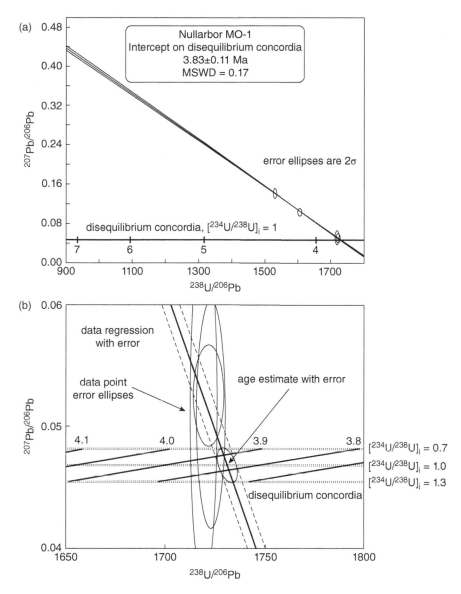

Figure 8.8 (a) U-Pb data for Nullarbor sample LBCM01 ('M0-1') plotted using the Tera-Wasserburg construction. The quasi-horizontal line in this plot represents a disequilibrium concordia plotted for $(^{230}Th)/(^{238}U)i = 0$, $(^{234}U/^{238}U)i = 1$. Tick marks with numbers on this curve are ages in millions of years. A linear regression with its associated 2σ uncertainty envelope is also shown passing through the individual blank- corrected U-Pb analyses. The calculated age is derived simply from the intersection of the two. (b) Detail of the area of intersection in the lower RHS of (a) showing a family of possible disequilibrium concordia representing sample evolution under different $(^{234}U/^{238}U)i$ conditions and chosen, in this case, to show the best estimate of the true value ±30%. Parallel lines in bold running across these concordia represent disequilibrium isochrons with ages marked in Ma. Source: From Woodhead et al. (2006) © 2006 Elsevier.

much of this Pb is likely to be common Pb, individual aliquots may plot much closer to the common Pb end member at the top of the y-axis in Figure 8.8.

In Figure 8.8b the enlarged area of intersection of the disequilibrium concordia with the analytical data incorporates the analytical uncertainties and the variance in $(^{234}U)/(^{238}U)$ ratios. The range of likely ages can be estimated from the ellipse of overlap.

At higher latitudes, the cessation of speleothem deposition during glacial periods has been well documented (Baker et al. 1993; Lauritzen 1991; Lauritzen 1995; Richards and Dorale 2003). The record from Hulu Cave, China (Figure 8.9; Wang et al. 2001), reveals precise changes of monsoonal activity related to the last ice-age termination, and to all the stadials and interstadials of the last glacial cycle. This record, based on multiple stalagmites became the stalagmite record to which later speleothem-based studies of the last glacial period compared their records. The data from Sofular Cave, in northern Turkey, produced from two stalagmites, yielded a largely continuous 50 000-year record through

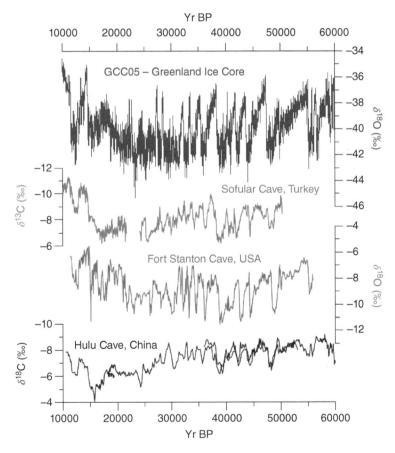

Figure 8.9 Stalagmites throughout the Northern Hemisphere experienced changes in climate synchronous with the changes recorded in the Greenland ice cores. Records in this figure were retrieved from stalagmites in North America (stalagmite FS2, Fort Stanton Cave, southwestern United States), Europe (stalagmites So-1 and So-2, Sofular Cave, northern Turkey), and Asia (stalagmites H82, MSD, MSL, PD, and YT, Hulu Cave, eastern China), and are compared to the Greenland ice core record GCC05. Source: From Polyak and Denniston (2012) © 2012 Elsevier.

the last glacial. Both the carbon and oxygen values were characterised by isotope variations that matched the Greenland ice core stadials and interstadials.

In Switzerland, there is strong evidence for glacier advances during the last 400 000 years from the St. Beatus Caves (Häuselmann et al. 2008). The sediment sequences in the caves are capped by speleothems which show a non-uniform sequence of deposition. Glacial advances occurred within the following time windows: >350, 235–180, 160–135, 114–99, 76–54, and 30–16 thousand years ago. The research also describes the following dated periods of glacial valley deepening: 805–760 m asl (>350 000 years), 760–700 m asl (235–180 000 years), 700–660 m asl (160–135 000 years), and 660–560 m asl (30–16 000 years).

8.5 Radiocarbon

This is an absolute radiometric method relying on the steady decay of naturally occurring radiocarbon or carbon-14. This radioisotope is produced by cosmic ray bombardment of nitrogen molecules in the upper atmosphere. Roughly 1 in 10 000 atoms of carbon in the atmosphere is radiocarbon. Radioactive isotopes are unstable and decay spontaneously, emitting radioactivity. This decay occurs at a set rate for each radioisotope and is unaffected by external influences. The rate is defined by its half-life $\left(T\frac{1}{2} \right)$ the time taken for half of the atoms to decay following the exponential law

$$N = N_0 e^{-\lambda t} \tag{8.1}$$

where N is the number of atoms present at time t, N_0 was the number present at the start of decay, and λ is the half-life or decay constant. For radiocarbon (^{14}C) the half-life is 5715 years; this gives an effective upper limit of six or seven half-lives, or around 40 000 years.

When calcite is precipitated in the cave, we assume that the system is closed, that is carbon locked into the crystal lattice cannot subsequently exchange with any later HCO_3 ions passing over the calcite. The decay of radiocarbon thus measures the time since the calcite was precipitated. The carbon in calcium carbonate may come from two sources. Half should come from the bedrock calcium carbonate, where ^{14}C is absent due to its great age. This is usually termed 'dead' carbon. The other 'live' half should come from the air or from humic compounds in the soil organic matter. In reality, the proportions are quite different. A comparison with U-Th analyses demonstrated that the dead carbon in speleothems is in the range 5–40% (Genty et al. 1999; Genty et al. 2001; Genty and Massault 1999), with typical values 12–20% (Genty et al. 2001). The dead carbon percentage can vary over time with variations in the range 5–10% over time periods of 10^2–10^3 year. This really limits the utility of radiocarbon dating of speleothems unless one can measure the isotopic composition of cave drip waters over several seasons, calculate a correction factor, and assume that this also applied in the past. For this reason, and the short time range afforded by the technique, radiocarbon age determinations on speleothems have been very limited in scope. The method has, however, been extensively used to date organic matter in cave entrance deposits (Figure 8.10), primarily archaeological sites (Gillieson and Mountain 1983). Radiocarbon has been successfully used to provide a late Pleistocene and Holocene

Figure 8.10 Excavation of organic-rich entrance facies dating to the late Pleistocene, Nombe Cave, New Guinea Highlands.

chronology for stalagmites in Drotsky's Cave, Botswana (Burney et al. 1994). A good correlation between uranium series and radiocarbon ages was obtained from a 60 cm. core taken from two coalesced formations, and extraction of the pollen preserved in the growth rings has provided a post-glacial vegetation history for this site on the edge of the Kalahari Desert. However, Holmgren et al. (1994) have also carried out a similar analysis on a stalagmite from Lobatse Cave in Botswana; they note a good correlation between uranium series and radiocarbon ages younger than 20 000 years, but before that the relationship breaks down, with a discrepancy of 20 000 years for older material with a uranium series age of 50 000 years.

A significant development has been the determination of ^{14}C in speleothems that post-date the atmospheric nuclear bomb testing era in the 1960s (Hua 2009). These tests created a spike in atmospheric ^{14}C, which is only just returning to pre-bomb levels (Figure 8.11). This atmospheric ^{14}C signal is transformed through storage in both the vegetation and soil, and subsequent mixing of groundwater of different ages, leading to a damped and lagged increase in 14C in speleothems. Genty and Massault (1999) and Genty et al. (1999) demonstrated the use of multiple ^{14}C determinations in modern stalagmites to not only confirm modern deposition through the presence of elevated ^{14}C (often greater than 100% modern carbon), but to determine the rate of transfer of carbon from the surface to the stalagmite.

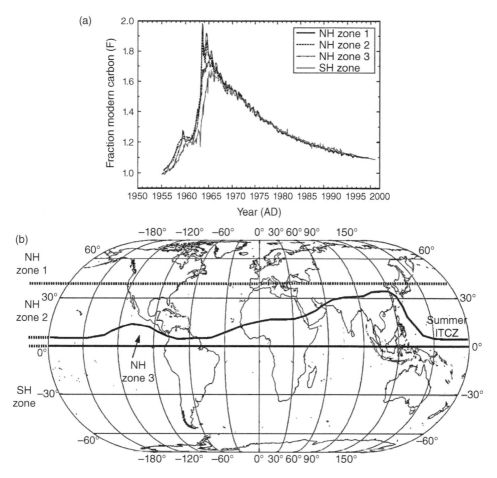

Figure 8.11 (a) Regional tropospheric ^{14}C curves for the period ad 1955–2001 for four different zones (Northern Hemispheric zones 1-3 and Southern Hemispheric zone). (b) The four zones into which the tropospheric ^{14}C data have been grouped. Source: From Fairchild and Baker (2012).

8.6 Other Dating Methods: Cosmogenic Radionuclides, and Tephrochronology

As well as creating carbon isotopes in the upper atmosphere, cosmic rays strike exposed rock surfaces and create isotopes at the surface of the earth. The isotopes that have been most widely used for dating in karst systems are ^{10}Be and ^{26}Al. These are found in quartz which is common in the sand and pebbles making up cave sediments and have a fixed ratio of $1:6.8$ in the quartz. ^{10}Be has a half-life of 1.38 million years while ^{26}Al has a half-life of 0.72 million years. Quartz exposed at the ground surface will be exposed to cosmic rays and once the material is transported underground the fixed ratio will generate a steady decay curve as the ratio is lowered for the two isotopes. The effective dating range of the method is from 100 000 to ~5 million years (Granger and Muzikar 2001). The $^{10}Be/^{26}Al$ age for a given sediment in a cave is a maximum residence time since it was last exposed at the earth's surface. There have been some useful applications of the method. Granger et al. (1997) have dated river gorge deepening in Virginia which confirms that the age of Mammoth Cave and other caves of the Cumberland Plateau, Kentucky may be as much as 5 million years ago (Anthony and Granger 2004, 2007), while the extensive multilevel cave system Siebenhengste-Barenschach in Switzerland may be up to 4.4 million years old (Häuselmann et al. 2007; Häuselmann and Granger 2016).

Tephrochronology is a technique which has been widely applied to archaeological sediments in caves and rockshelters and has potential at other sites (Lowe and Alloway 2015). Stock et al. (2006) have used both cosmogenic dating and tephrochronology to estimate incision rates for the Bighorn River in Wyoming, USA. Spence Cave is a phreatic passage formed along an anticline and subsequently truncated and exposed by 120 m of river incision (Figure 8.12). A windblown sand deposit just inside the entrance yielded a $^{26}Al/^{10}Be$ burial age of 0.31 ± 0.19 million years. This represents a minimum age for the development of Spence Cave and provides a maximum incision rate for the Bighorn River of 0.38 ± 0.19 mm yr^{-1}.

Horsethief Cave is a complex phreatic cave system located 43 km north of Spence Cave on a plateau surface about 340 m above the Bighorn River. Electron microprobe analyses

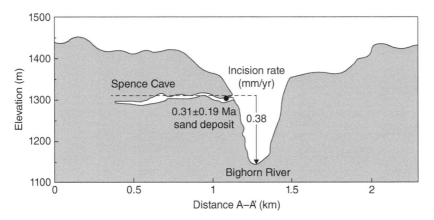

Figure 8.12 Topographic cross section showing position of Spence Cave in relation to Bighorn River where it incises the Sheep Mountain anticline, Wyoming, USA. Dashed line indicates approximate water table position during cave development. Minimum cave age of 0.31 ± 0.19 million years suggests a maximum Bighorn River incision rate of 0.38 ± 0.19 mm yr^{-1}. Source: From Stock et al. (2006).

Table 8.1 Bighorn River (Wyoming, USA) incision rates based on cave sediment ages.

Cave	Height above bighorn river (m)	Adjusted height (m)	Age (millions of years)	Incision rate (mm yr^{-1})	Adjusted incision rate (mm yr^{-1})
Spence	119 ± 2	119 ± 2	0.31 ± 0.20	0.38 ± 0.19	0.38 ± 0.19
Horsethief	343 ± 5	221 ± 6	00.639 ± 0.002	0.54 ± 0.01	0.35 ± 0.19

Adjusted heights and incision rates account for paleo-hydraulic gradient indicated by the dip of cave passages.
Source: From Stock et al. (2006).

of white, fine-grained sediment in the Powder Mountain section of Horsethief Cave confirm that this deposit is Lava Creek volcanic ash, erupted from the Yellowstone Plateau volcanic field about 0.64 million years ago. Using this as a minimum age for Horsethief Cave, extrapolation of the cave profile gradient westward to the Bighorn River gorge suggests a maximum incision rate of 0.35 ± 0.19 mm yr^{-1}. This matches the estimated rate from Spence Cave (Table 8.1) and suggests significant incision of the river in the Pleistocene. These may well be minimum ages for the caves as they would substantially postdate the phreatic development.

In New Zealand, the rugged glaciated karst mountains of northwest Nelson have yielded some of the deepest caves in the southern hemisphere. One of the deepest and longest, Bulmer Cavern, has been the subject of cosmogenic dating by Holden (2018). Rapid uplift in the area from the Eocene to the Miocene seems to have been followed by a slowing of uplift, despite rapid uplift along the Alpine Fault to the south. Until this study, there were no data on uplift rates for the Pliocene and Pleistocene.

Bulmer Cavern is New Zealand's longest cave at 72 km and third deepest at 755 m. The cave runs northwards from multiple entrances on Mount Owen. The cave is developed on six levels with some phreatic tube remnants and canyons, with shafts connecting various levels (Figure 8.13).

Cave sediments range from well-sorted sands and silts to poorly sorted conglomerates, in places covered by thin flowstones or other speleothems. This suggests minimal disturbance since emplacement. Samples were collected from all levels and where possible duplicates were collected from the same passage. Burial ages from the various cave passages are given in Table 8.2. These were then combined with passage elevations (from aneroid barometer and cave surveys) to yield uplift rates by least squares regression. The results suggest an uplift rate of 0.13 mm yr^{-1} from the mid-Pliocene (2.76 million years ago) to the start of the Pleistocene, and a decreased rate of 0.067 mm yr^{-1} since the start of the Pleistocene.

8.7 Timing Glacial and Interglacial Events in New Zealand

In the far southwest of the South Island of New Zealand, there are deep glacial valleys with evidence of multiple glaciations, forming the Fiordland region. Thrust faults and shear zones associated with the main Alpine Fault dissect slivers of marble into bands very reminiscent of the stripe karst of northern Norway (Lauritzen 2001). Access to this area is difficult, but karst features and some caves have been found. One of the best known is the Aurora-Te Ana-au Cave on the western flanks of the Lake Te Anau glacial trough (Figure 8.14). Cave deposits can reveal the glacial history of the area because caves beneath the walls of glacial troughs are like dip-sticks that reveal the occurrence and depth of glacial ice (Williams 2017).

Steeply dipping limestones extend from near the tree-line at 1100 m to below lake level. The Tunnel Burn drains from Lake Orbell in the Murchison Mountains into Aurora Cave and resurges at Te Ana-au Cave (a tourist cave) beside the shore of Lake Te Anau at 203 m (Figure 8.15). Dry upper-level passages mark earlier stream routes. Aurora-Te Ana-au Cave descends 267 m down the valley side and has about 8 km of passages (Figure 8.16). The cave

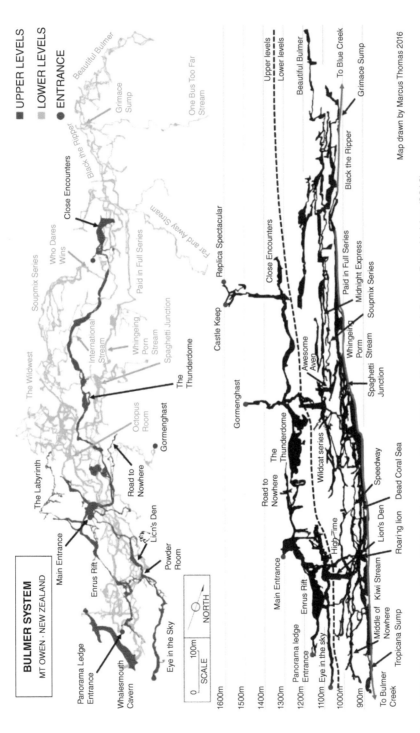

Figure 8.13 Plan and long section of Bulmer Caverns, northwest Nelson, New Zealand. Source: From Holden (2018).

Table 8.2 Cosmogenic nuclide concentrations, burial ages and erosion rates from Bulmer Caverns, New Zealand.

Sample	(26Al) (x10³ atoms g⁻¹)	(10Be) (x10³ atoms g⁻¹)	(26Al/10Be)	Burial age (million years)	Erosion rate (mm yr⁻¹)
Bear Pit	57.3 ± 17.8	10.14 ± 0.96	5.65 ± 1.83	0.37 ± 0.70	
Castle Keep	9.74 ± 9.81	2.35 ± 0.86	4.15 ± 4.44	1.01 ± 2.27	
Cobble Passage	247 ± 25	74.2 ± 1.5	3.33 ± 0.34	1.47 ± 0.33	
Eurus Rift		60.5 ± 1.9			
Eye in the Sky	59.1 ± 6.2	33.3 ± 1.8	1.78 ± 0.21	2.76 ± 0.42	
Mt Owen 1 (surface)	1120 ± 142	180 ± 5	6.66 ± 0.80	N/A	0.06 ± 0.01
Octopus Room	146 ± 25	26.0 ± 1.3	5.63 ± 0.98	0.38 ± 0.41	
Road to Nowhere	13.5 ± 19.0	4.56 ± 0.54	2.96 ± 4.17	1.71 ± 3.00	
Soupmix 2	109 ± 26	31.9 ± 1.1	3.40 ± 0.82	1.42 ± 0.60	
Wind in the Willows	92.7 ± 15.2	25.6 ± 0.9	3.61 ± 0.61	1.30 ± 0.44	
Yelsgup	43.3 ± 11.1	9.40 ± 0.83	4.61 ± 1.24	0.80 ± 0.62	

Erosion rates calculated using the online calculators, formerly known as CRONUS. Analytical uncertainties are reported at 1σ precision.
Source: From Holden (2018).

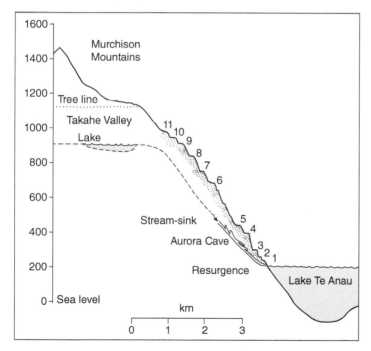

Figure 8.14 Location of Aurora-Te Ana-au Cave beneath the side of Lake Te Anau glacial trough, Fiordland (New Zealand). Eleven lateral moraines deposited by past glaciers are schematically shown extending up the trough to about 1000 m. The cave underlies the lower part of the trough with its resurgence close to modern lake level. Source: From Williams (2017) © 2017 Elsevier.

Figure 8.15 Lake Te Anau (New Zealand) with an incised valley of Tunnel Burn on right side of image. Source: Photo by Neil Collinson.

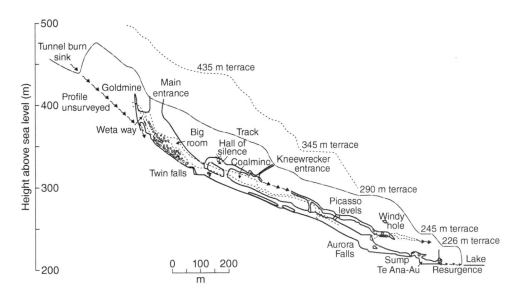

Figure 8.16 Projected long profile of Aurora-Te Ana-au Cave (New Zealand) from about 450 m to lake level. Superimposed over the cave is the overlying surface profile of lateral moraine terrace steps. Source: From Williams (2017) © 2017 Elsevier.

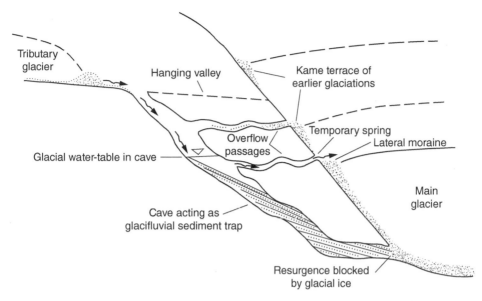

Figure 8.17 Model of Aurora Cave (New Zealand) as a glaci-fluvial sediment trap. Source: From Williams (2017) © 2017 Elsevier.

contains speleothems dated to at least 270 000 years ago (Williams 1996), which may coincide with the penultimate interglacial.

The large Te Anau Glacier filled the valley during Pleistocene glacial phases, and the cave was overrun by ice several times. The cave passages acted as a glaciofluvial sediment trap (Figure 8.17). After the ice retreated the Tunnel Burn re-entered the cave and flushed out most of the glaciofluvial sediments (Figure 8.18). During the warmer interglacials and interstadials water circulation allowed speleothems to grow again. This created a sequence of dateable speleothems between glaciofluvial gravels in the cave that has provided a detailed chronology of glacial advances. This has added greater detail to understanding of New Zealand's changing climates over at least the last 130 000 years, with less detail prior to that (Williams et al. 2015).

During glacial periods the cave outlet was blocked by ice, with the cave being flooded by meltwater and blocked by fluvial sediment. The water and sediment were derived from both the tributary Takahe valley glacier and the main valley glacier in the Te Anau trough. High-level passages are still choked with glaciofluvial sediments, suggesting that the cave was completely blocked by sediment. During smaller glacial advances only the lower part of the cave was blocked, and the water found higher level outlets.

The cross-section in Figure 8.19 can be seen as a response to three advances of the Takahe valley glacier, the earliest commencing after 60 000 years ago. Erosion of the sediment fill was followed by deposition of the lower flowstone. A second advance terminated with the erosion of the fill and deposition of another flowstone. A third and final advance terminated and the upper flowstone was deposited at 16 000 years ago.

The formation of speleothems in the Hall of Silence at 40 000, 48 000, and 50 000 years ago provides evidence of non-glacial conditions with a glacial advance evidenced by sediments bracketed by the 40 000 and 48 000 dates. Cave sediments graded to a surface terrace at

Figure 8.18 Twin Falls in Aurora Cave, Fiordland, New Zealand. Source: Photo by John Brush.

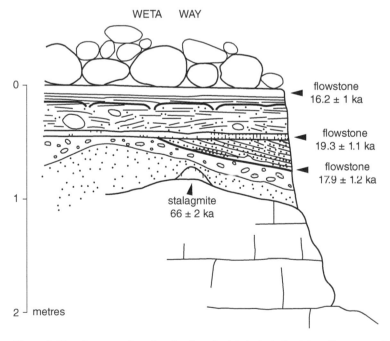

Figure 8.19 Cross-section showing interbedded glaciofluvial sediments and speleothems in Weta Way passage (New Zealand). Source: From Williams (1996).

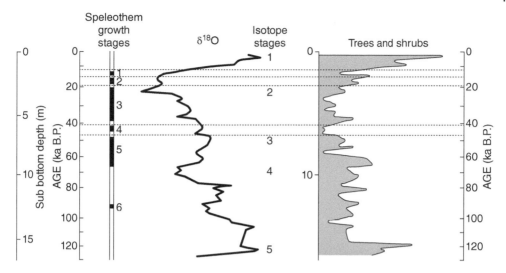

Figure 8.20 Comparison of Aurora Cave (New Zealand) speleothem growth stages with oxygen isotope record from deep-sea cores and pollen record from DSDP site 594, Chatham Rise. Source: From Williams (1996).

330 m above sea level bury abraded flowstone dated to 40 000 years ago and supports the existence of a glacial event at that time.

The evidence from Aurora Cave suggests that the last glaciation consisted of several short glacial advances with intervening interstadials that were also brief. This supports ice-core evidence of rapid climate changes during the Last Glacial Maximum, and that the coldest period of the glaciation (shown by the deepest oxygen isotope trough in Figure 8.20) may have been too dry for significant glacial advances. Maximum glacier extent may have been later when the climate started to ameliorate and become wetter (Williams 1996).

References

Anthony, D.M. and Granger, D.E. (2004). A Late Tertiary origin for multilevel caves along the western escarpment of the Cumberland Plateau, Tennessee and Kentucky, established by cosmogenic 26Al and 10Be. *Journal of Cave and Karst Studies* 66: 46–55.

Anthony, D.M. and Granger, D.E. (2007). A new chronology for the age of Appalachian erosional surfaces determined by cosmogenic nuclides in cave sediments. *Earth Surface Processes and Landforms* 32: 874–887.

Baker, A., Smart, P.L., Edwards, R.L., and Richards, D.A. (1993). Annual growth banding in a cave stalagmite. *Nature* 364: 518–520.

Burney, D.A., Brook, G.A., and Cowart, J.B. (1994). A Holocene pollen record for the Kalahari Desert of Botswana from a U-series dated speleothem. *The Holocene* 4: 225–232.

Cheng, H., Edwards, R.L., Shen, C.C. et al. (2013). Improvements in 230Th dating, 230Th and 234U half-life values, and U–Th isotopic measurements by multi-collector inductively coupled plasma mass spectrometry. *Earth and Planetary Science Letters* 371: 82–91.

Domínguez-Villar, D., Fairchild, I.J., Baker, A. et al. (2009). Oxygen isotope precipitation anomaly in the North Atlantic region during the 8.2 ka event. *Geology* 37: 1095–1098.

Dorale, J.A., Edwards, R.L., Alexander, E.C. et al. (2007). Uranium-series dating of speleothems: current techniques, limits, and applications. In: *Studies of Cave Sediments: Physical and Chemical Records of Paleoclimate*. Revised ed. (eds. I.D. Sasowsky and J.E. Mylroie), 177–197. Dordrecht: Springer.

Duplessy, J.C., Labeyrie, C.L., and Nguyen, H.V. (1970). Continental climatic variations between 130000 and 90000 years BP. *Nature* 226: 631–632.

Fairchild, I.J. and Baker, A. (2012). *Speleothem Science: From Process to Past Environments*. Chichester: Wiley.

Ford, D.C. and Williams, P.W. (2007). Chapter 5, Karst hydrogeology. In: *Karst Hydrogeology and Geomorphology*, 103–144. Chichester: Wiley.

Genty, D. and Massault, M. (1999). Carbon transfer dynamics from bomb 14C and d13C times series of a laminated stalagmite from SW France: modelling and comparison with other stalagmites. *Geochimica et Cosmochimica Acta* 63: 1537–1548.

Genty, D., Massault, M., Gilmour, M. et al. (1999). Calculation of past dead carbon proportion and variability by the comparison of AMS 14 C and TIMS U/Th ages on two Holocene stalagmites. *Radiocarbon* 41: 251–270.

Genty, D., Baker, A., Massault, M. et al. (2001). Dead carbon in stalagmites: carbonate bedrock paleodissolution vs. aging of soil organic matter. Implications for 13C variations in speleothems. *Geochimica et Cosmochimica Acta* 65: 3443–3457.

Gillieson, D.S. (1996). *Caves: Processes, Development and Management*. Oxford: Blackwell.

Gillieson, D.S. and Mountain, M.J. (1983). The environmental history of Nombe Rockshelter, Papua New Guinea highlands. *Archaeology in Oceania* 18: 53–62.

Granger, D.E. and Muzikar, P.F. (2001). Dating sediment burial with in situ-produced cosmogenic nuclides: theory, techniques, and limitations. *Earth and Planetary Science Letters* 188: 269–281.

Granger, D.E., Kirchner, J.W., and Finkel, R.C. (1997). Quaternary downcutting rate of the New River, Virginia, measured from differential decay of cosmogenic 26Al and 10Be in cave-deposited alluvium. *Geology* 25: 107–110.

Häuselmann, P., Granger, D.E., Jeannin, P.-Y., and Lauritzen, S.-E. (2007). Abrupt Glacial Valley incision at 0.8 ma dated from cave deposits in Switzerland. *Geology* 35: 143–146.

Häuselmann, P. and Granger, D.E. (2016). Dating of caves by cosmogenic nuclides: method, possibilities, and the Siebenhengste example. *Acta Carsologica* 34: 43–50.

Häuselmann, P., Lauritzen, S.E., Jeannin, P.Y., and Monbaron, M. (2008). Glacier advances during the last 400 ka as evidenced in St. Beatus Caves (BE, Switzerland). *Quaternary International* 189: 173–189.

Hellstrom, J. (2003). Rapid and accurate U/Th dating using parallel ion-counting multi-collector ICP-MS. *Journal of Analytical Atomic Spectrometry* 18: 1346–1351.

Hellstrom, J. (2006). Uranium-thorium dating of speleothems with high initial thorium using stratigraphical constraint. *Quaternary Geochronology* 1: 289–295.

Holden, G. (2018). Cosmogenic nuclide dating of the sediments of Bulmer Cavern: implications for the uplift history of southern Northwest Nelson, South Island New Zealand. A thesis submitted to Victoria University of Wellington in partial fulfilment of the requirements for the degree of Master of Science in Geology. University of Wellington.

Holmgren, K., Lauritzen, S.-E., and Possnert, G. (1994). 230Th/234U and 14C dating of a late Pleistocene stalagmite in Lobatse II Cave, Botswana. *Quaternary Science Reviews* 13: 111–119.

Hua, Q. (2009). Radiocarbon: a chronological tool for the recent past. *Quaternary Geochronology* 4: 378–390.

Latham, A.G. and Schwarcz, H.D. (1990). Magnetization of speleothems: detrital or chemical? In: *Proceedings of the 10th International Congress of Speleology, 1990 Budapest*, 82–84.

Latham, A.G., Schwarcz, H.P., and Ford, D.C. (1986). The paleomagnetism and U-Th dating of Mexican stalagmite, DAS2. *Earth and Planetary Science Letters* 79: 195–207.

Lauritzen, S.E. (1991). Uranium series dating of speleothems: a glacial chronology for Nordland, Norway. In: *Proceedings of the Late Quaternary Stratigraphy in the Nordic Countries 150,000–15,000 BP: The XXIV Uppsala Symposium in Quaternary Geology*, 127. Societas Upsaliensis Pro Geologia Quaternaria.

Lauritzen, S.E. (1995). High-resolution paleotemperature proxy record for the last interglaciation based on Norwegian speleothems. *Quaternary Research* 43: 133–146.

Lauritzen, S.E. (2001). Marble stripe karst of the Scandinavian Caledonides: an end-member in the contact karst spectrum. *Acta Carsologica* 30: 47–79.

Li, W.X., Lundberg, J., Dickin, A.P. et al. (1989). High-precision mass-spectrometric uranium-series dating of cave deposits and implications for palaeoclimate studies. *Nature* 339: 534–536.

Lowe, D.J. and Alloway, B.V. (2015). Tephrochronology. In: *Encyclopedia of Scientific Dating Methods* (eds. W.J. Rink and J.W. Thompson). Dordrecht: Springer.

Noel, M. (1986). The palaeomagnetism and magnetic fabric of sediments from Peak Cavern, Derbyshire. *Geophysical Journal of the Royal Astronomical Society* 84: 445–454.

Pease, P.P., Gomez, B., and Schmidt, V.A. (1994). Magnetostratigraphy of cave sediments, Wyandotte Ridge, Crawford County, Indiana: towards a regional correlation. *Geomorphology* 11: 75–81.

Pickering, R. and Kramers, J.D. (2010). Re-appraisal of the stratigraphy and determination of new U-Pb dates for the Sterkfontein hominin site, South Africa. *Journal of Human Evolution* 59: 70–86.

Polyak, V. and Denniston, R. (2012). Palaeoclimate records from caves. In: *Encyclopedia of Caves* (eds. W.B. White and D.C. Culver). Amsterdam, The Netherlands: Elsevier/Academic Press.

Richards, D.A. and Dorale, J.A. (2003). Uranium-series chronology and environmental applications of speleothems. In: *Uranium-Series Geochemistry, Reviews in Mineralogy and Geochemistry* (eds. B. Bourdon, G.M. Henderson, C.C. Lundstrom and S.P. Turner). Washington, D.C.: Mineralogical Society of America.

Richards, D.A., Bottrell, S.H., Cliff, R.A. et al. (1998). U-Pb dating of a speleothem of Quaternary age. *Geochimica et Cosmochimica Acta* 62: 3683–3688.

Schmidt, V.A. (1982). Magnetostratigraphy of sediments in Mammoth Cave, Kentucky. *Science* 217: 827.

Spotl, C. and Mattey, D. (2012). Scientific drilling of speleothems – a technical note. *International Journal of Speleology* 41: 29–34.

Stock, G.M., Riihimaki, C.A., and Anderson, R.S. (2006). Age constraints on cave development and landscape evolution in the Bighorn Basin of Wyoming, Usa. *Journal of Cave and Karst Studies* 68: 74–81.

Wang, Y.J., Cheng, H., Edwards, R.L. et al. (2001). A high-resolution absolute-dated late Pleistocene monsoon record from Hulu Cave, China. *Science* 294: 2345–2348.

Williams, P.W. (1996). A 230 ka record of glacial and interglacial events from Aorora cave, Fiordland, New Zealand. *New Zealand Journal of Geology and Geophysics* 39: 225–241.

Williams, P.W. (2017). *New Zealand Landscape: Behind the Scene*. Amsterdam: Elsevier Inc.

Williams, P.W., Mcglone, M., Neil, H., and Zhao, J.-X. (2015). A review of New Zealand palaeoclimate from the last interglacial to the global last glacial maximum. *Quaternary Science Reviews* 110: 92–106.

Woodhead, J., Hellstrom, J., Maas, R. et al. (2006). U–Pb geochronology of speleothems by MC-ICPMS. *Quaternary Geochronology* 1: 208–221.

Woodhead, J., Reisz, R., Fox, D. et al. (2010). Speleothem climate records from deep time? Exploring the potential with an example from the Permian. *Geology* 38: 455–459.

9

Cave Deposits and Past Climates

9.1 Introduction

Estimates of past climatic conditions have long held interest for the natural sciences, in that they provide some explanation for the changing patterns of distribution of plant and animal species on the planet, including our own species. Recently there has been renewed interest in past climate reconstruction as a means of providing analogues for the atmosphere likely to result from global warming as a result of the Greenhouse Effect. Most attention in this context has focused on the mid-Holocene (4000–7000 years BP) and the Last Interglacial (c.120 000–130 000 years BP). Climatic shifts can be inferred from the pollen record by analogy with present plant communities of known climatic tolerances. This indirect method makes many assumptions, not the least of which is that past tolerances reflect the present; on the long timescale, this may be questionable. More direct methods have been pioneered using the record preserved in deep-sea cores and ice cores, through stable oxygen isotope analysis.

Over the last 20 years, speleothem science has developed to provide long archives of paleoclimate (Fairchild et al. 2006). In Chapter 6, we see that stalagmites are built up layer by layer, often on an annual basis, and thus a longitudinal section through such a stalagmite provides a micro-stratigraphy that can span thousands to tens of thousands of years. Uranium-series dating provides an absolute chronology than can extend back to the mid-Pleistocene or much older (Woodhead et al. 2006).

Profiles can be made of oxygen isotope ratios, carbon isotope ratios, strontium isotope ratios, and any other trace elements that might have a climate signature. Similarly, the colour banding, luminescence banding, and crystal growth textures can also be measured. The feed water for the speleothems starts as rainwater, percolates through the soil and the epikarst, and eventually is incorporated into the speleothem calcite. It should, therefore, carry a signature of the surface climate at that time, though it may combine varying time intervals. Much of the current research is focused on understanding the pathways leading to speleothem signatures and how to reliably interpret those signatures. The removal of stalagmites from caves and their subsequent analysis is time-consuming and expensive. Paleoclimate studies to date have tended to focus on understanding the magnitude of temperature and other changes during glacial-interglacial cycles and during the Holocene.

Future research will attempt to gain much longer, continuous climate records at a timescale of millions of years, and to understand the variability of speleothem feedwater signatures within individual cave systems.

9.2 Oxygen Isotope Analysis

The element oxygen has three stable isotopes: ^{16}O, ^{17}O and ^{18}O. Oxygen-16 is about 500 times more abundant than oxygen-18, but the ratio of these two isotopes varies a little in nature and can be used to estimate past temperatures. The ratio of ^{18}O to ^{16}O is usually measured on gas in a stable isotope mass spectrometer. The value is generally measured as the difference between the sample ratio and that of a standard: for water, this is standard mean oceanic water (SMOW), while for calcite the standard is a fossil belemnite, Pee Dee Belemnite (PDB). This method eliminates systematic errors inherent in measuring small concentrations. It is calculated as follows:

$$\delta^{18}O\text{‰} = \frac{^{18}O/^{16}O\ sample - {}^{18}O/^{16}O \times standard \times 1000}{^{18}O/^{16}O\ standard} \tag{9.1}$$

This variation occurs as a result of the process of fractionation. In this process, one isotope is preferentially transferred when there is a change of state – for instance, from liquid to vapour. For example, when water evaporates from the surface of the ocean, the vapour contains a higher proportion of the light atom ^{16}O than was present in the ocean water. The vapour is thus depleted in ^{18}O, while the ocean water is enriched in ^{18}O. This enrichment (of the order of 1.1‰ during full glacial conditions) may be preserved in oceanic organisms (such as foraminifera), which use the water for their metabolism, and eventually die and become part of the deep-sea sediment. In contrast, ice formed from the vapour is depleted in ^{18}O. These two related processes are known as the ice volume effect, affecting stable isotope ratios at high latitudes or preserved isotopic ratios in fossils or crystals formed during the Last Glacial.

Fractionation also occurs during the process of crystallisation. If a molecule containing oxygen crystallises from a saturated solution, or if calcite or aragonite are extracted biochemically from water, then the calcite or aragonite will be slightly enriched with ^{18}O relative to the water. This enrichment is temperature-dependent: it is known as the crystallisation effect, and its magnitude is -0.24‰ $°C^{-1}$.

Finally, and most usefully, fractionation depends on temperature during condensation and freezing (Table 9.1). Evaporation and condensation of water at lower ambient temperatures produce lower values of $\delta^{18}O$ in precipitation; thus summer precipitation is

Table 9.1 Temperature dependence of oxygen isotope enrichment in calcite.

Temperature (°C)	$^{18}O/^{16}O$ ratio in solution	$^{18}O/^{16}O$ ratio in calcite
0	1:500	1.026:500
25	1:500	1.022:500

Source: From Bowen (1978) © 1978 Elsevier.

isotopically heavier than winter precipitation, and has values of $\delta^{18}O$ that vary from +0.17 to +0.39‰ $°C-1$ greater (Harmon et al. 1978). This is known as the precipitation effect. The temperature relationship is as follows:

$$T(°C) = 16.9 - 4.38(\delta^{18}O_c - \delta^{18}O_w) + 0.10(\delta^{18}O_c - \delta^{18}O_w)^2 \qquad (9.2)$$

where O_c = oxygen in calcite and O_w = oxygen in water. The $\delta^{18}O$ may be reduced by the ice volume effect in high latitudes or during glacial times and may be increased by the crystallisation effect. These two are usually allowed for before final calculations.

From Eq. (9.2), to determine absolute temperature it is necessary to know the $\delta^{18}O$ of the formation water. This is achieved by measuring the $2H:1H$, or D/H ratio (Deuterium/Hydrogen), on small inclusions of water trapped in the calcite and sampled by crushing the calcite in a vacuum vessel (Harmon et al. 1979; Yonge 1981). It has been repeatedly shown that there is an excellent correlation between the D/H ratio and $\delta^{18}O$ in water. This is known as the meteoric water line:

$$\delta D = (8.17 \pm 0.08)\delta^{18}O + (10.56 \pm 0.64)‰$$

$$(r = 0.997, \; standard\, error = \pm3.3\%) \qquad (9.3)$$

If the D/H ratio cannot be determined; then the relative temperature can be assessed from graphs of $\delta^{18}O$ change with time. To date, most studies have presented relative temperatures on account of expense and difficulty in measuring D/H ratios of fluid inclusions.

For reliable, stable isotope analysis of cave calcite, the mineral must have been deposited in isotopic equilibrium – that is, no evaporation must interfere with the fractionation. This assumption can be tested by taking multiple samples along a single growth layer (Figure 9.1) in the stalagmite. These samples should yield similar values, indicating that slow loss of CO_2 is the dominant process, rather than evaporation, which would tend to produce different

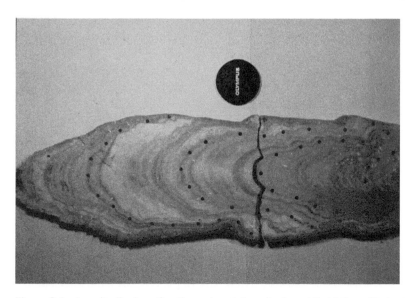

Figure 9.1 Longitudinal section through a stalagmite from Mimbi Cave, Kimberleys, Western Australia, showing growth layers and locations of samples for isotope analysis.

values on the outside of the stalagmite. In addition, there should be no correlation between the $^{18}O/^{16}O$ ratio and the $^{13}C/^{12}C$ ratio along a single growth layer. Such a correlation would again suggest a rapid loss of CO_2 by aeration rather than slow random processes. If these requirements can be met, then samples taken along the growth axis of a stalagmite dated by other means can provide a relative or absolute temperate curve for the time covered by the speleothem growth (Gascoyne 1992).

9.3 The Last Glacial-Interglacial Temperature Record

In 1992, Winograd and co-workers published a 500 000 year $\delta^{18}O$ record from a subaqueous speleothem (Winograd et al. 1992). Devils Hole (Nevada, USA) is a fault-controlled fissure which intersects the regional groundwater at approximately 15 m depth. The groundwater temperature has remained constant over recent decades (between 33 °C and 34 °C), reflecting the large size of the aquifer, which is recharged by the Spring Mountains (80 km northeast) and the Nevada high ranges (400 km northeast). The residence time of the groundwater is estimated at 2000 years. Most of the Devils Hole is lined with thick mammillary calcite which has been deposited underwater, and this has been analysed using thermal ionisation mass spectrometry (Figure 9.2). Resampling of the deposit has allowed extension of the record to the Holocene at 4500 year before present (Winograd et al. 2006).

The Devils Hole record generated significant debate within the palaeoclimate community because the $\delta^{18}O$ clearly showed a glacial-interglacial signal, but with the timing of glacial terminations earlier than that expected. Coplen et al. (1994) later presented a $\delta^{13}C$ record for the period 566 000–60 000 years, demonstrating that this was inversely correlated with $\delta^{18}O$ ($r = -0.75$) with a lead of up to 7000 years over other glacial records: the $\delta^{13}C$ record was interpreted as a record of changes in vegetation extent or density.

A long and accurately dated speleothem record has been produced from the Eastern Mediterranean region by Bar-Matthews et al. (2003). The study presented accurately dated 250-kyr $\delta^{18}O$ and $\delta^{13}C$ records determined from speleothems of the Peqi'in and Soreq Caves in Israel (Figure 9.3). There was a strong regional climatic signal in both cave records. Low $\delta^{18}O$ minima in the Peqi'in profile for the 250 000 to 185 000 year period (interglacial marine isotope stage 7) were indicative of high rainfall in the region at these times and matched increased organic marine deposition (sapropel – S1–S6) events. Maximum rainfall and lowest temperature conditions occurred at the beginning of the sapropel events and were followed by a decrease in rainfall and an increase in temperatures, leading to arid conditions. The record for the last 7000 years showed a trend towards increasing aridity and agreed well with climatic and archaeological data from North Africa and the Middle East.

Variations in the oxygen isotope composition of speleothems from Chinese caves have received worldwide attention from the palaeoclimate and geosciences communities. These have been interpreted as a record of the East Asian "summer monsoon intensity", and thence correlated with the Indian monsoon and northern hemisphere summer insolation. A number of oxygen isotope records from Chinese caves have provided long-term data on changes in both the strength of the East Asian monsoon and global climate generally. The extended Chinese record covers the past 640 000 years from several sites (Figure 9.4; Cheng et al. 2009, 2016). The length and precision of the record have permitted testing of the

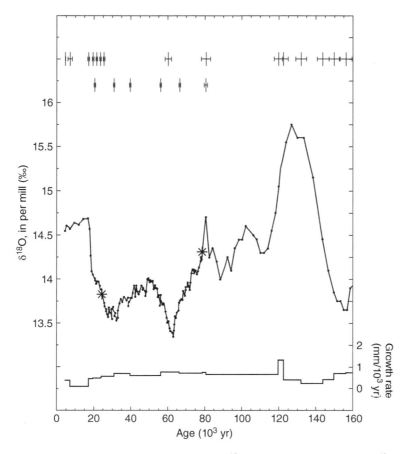

Figure 9.2 The Devils Hole (Nevada, USA) $\delta^{18}O$ series from 160 to 4.5 ka. $\delta^{18}O$ values are expressed relative to VSMOW. Asterisks mark the tie points between the three subaqueous speleothems. Location and uncertainty associated with U-Th analyses are shown above the $\delta^{18}O$ record and the mean deposition rates. Source: From Winograd et al. (2006).

idea that insolation changes caused by orbital changes of the Earth drove the terminations of each of the last seven ice ages, as well as the long periods of reduced monsoon rainfall associated with each of the terminations. In particular, they conclude that there has been a decline in rainfall since the mid-Holocene and that there was reduced monsoonal rainfall during glacial stages. These conclusions have been questioned by Maher and Thompson (2012). The speleothem $\delta^{18}O$ variations conflict with independent palaeoclimate proxies (cave $\delta^{13}C$, loess palaeosol magnetic properties, and $\delta^{13}C$ alkanes), which indicate no systematic decline in rainfall from the mid-Holocene and no glacial rainfall maxima. Using mass balance calculations – which incorporate seasonality effects in both $\delta^{18}O$ concentration and the amount of precipitation – they demonstrated that the cave $\delta^{18}O$ variations are better explained by variations in the rainfall source area. Indian monsoon-sourced $\delta^{18}O$ may have dominated at times of high boreal summer insolation, and more local Pacific-sourced moisture (from typhoons) at low insolation. The reduction of summer monsoonal rainfall during glacial stages may rather reflect reduced sea and land surface temperatures and the

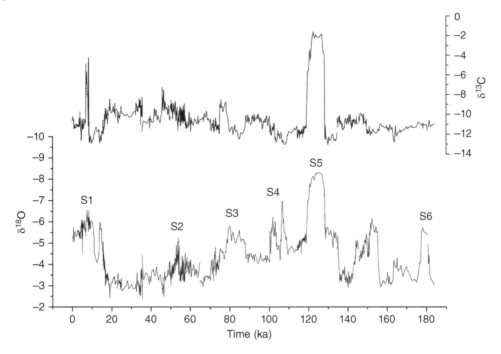

Figure 9.3 The Soreq Cave (Israel) $\delta^{18}O$ and $\delta^{13}C$ composite record, as archived at the World Data Centre for Paleoclimatology. The archived composite is made up of 21 speleothem samples constrained by 95 TIMS U-Th analyses and over 2000 stable isotope analyses. Source: From Fairchild and Baker (2012).

impacts of increased regional dust veils due to loess. Despite these differences in interpretation, the Chinese speleothem records remain the current longest continuous terrestrial records of palaeoclimate.

One of the longest continuous records has come from the Flint Ridge-Mammoth Cave system of Kentucky, USA, covering the period from 230 to 100 ka – that is, the penultimate glacial and the last two interglacials (Figure 9.5; Harmon et al. 1978). A section of a long columnar stalagmite was sampled in Davis Hall, a high-level passage with abundant decoration. For this stalagmite, three uranium-series dates were obtained, and a continuous timescale was derived assuming constant growth rate between these dated sections. The response of $\delta^{18}O$ to temperature change was determined from fluid inclusion isotope measurements. Major cold periods occur from 215 to 195 ka and from 160 to 130 ka, while warm periods occur from 180 to 165 ka and from 125 to 105 ka. These correspond with the marine isotope record but are better dated. Cave temperatures in this region may have differed by as much as 8 °C between glacial and interglacial times.

A very good relationship between solar insolation and monsoon variability from South America was reported by Cruz et al. (2006). Two stalagmite time series from Botuverá Cave, Brazil cover the past 116 000 years and are characterised by oxygen isotopic ratios that track the sinusoid of southern summer solar insolation. Within the Greenland ice cores, the last glacial is marked by rapid ameliorations in climate (Interstadials), classified as Dansgaard-Oeschger cycles. These are represented by sawtooth-shaped variations in $\delta^{18}O$ of 3–4‰ amplitude. In Europe, Genty et al. (2003) demonstrated that in southwestern

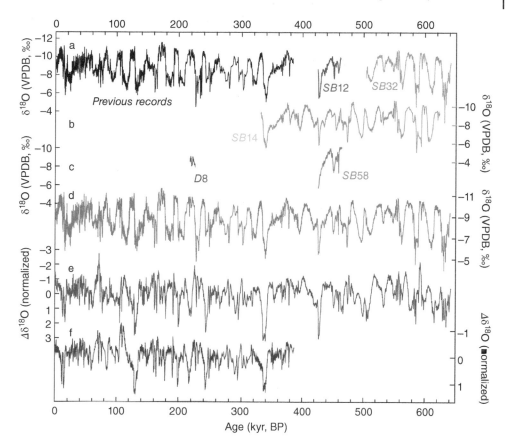

Figure 9.4 Chinese speleothem $\delta^{18}O$ records over the past 640 000 years, from Cheng et al. (2016). a: Previously published $\delta^{18}O$ records over the past 384 000 yr. (black) from Hulu, Dongge, and Sanbao caves (China), and new SB-12 (blue) and SB-32 (olive) $\delta^{18}O$ records; b: New SB-14 $\delta^{18}O$ record; c: New SB-58 (purple) and D-8 (pink) $\delta^{18}O$ records; d: Composite Asian Monsoon $\delta^{18}O$ record over the past 640 000 years. The record is constructed from previous data (<384 000 years) and new data from four stalagmites from Sanbao Cave (≥384 000 years) e: Detrended Asian Monsoon record ($\Delta\delta^{18}O$) is obtained by using the z-standard method to remove the insolation component (21 July insolation at 65° N). f: Detrended Asian Monsoon results from Barker et al. (2011). Source: From Cheng et al. (2016).

France, the $\delta^{13}C$ proxy responded more strongly to these Greenland Interstadials than did $\delta^{18}O$.

The cooling of cave air during a glacial will result in calcite which is richer in $\delta^{18}O$ than the dripwater. This water feeding the stalagmite is ultimately derived from the oceans, which will be poorer in $\delta^{18}O$ during a glacial because of the ice volume effect. These effects, therefore, work in opposition and may result in underestimation of temperature lowering. Other unknown processes of fractionation may occur between the ocean and the cave drip. It is, therefore, necessary to look at the isotopic properties of karst water in any study region before undertaking detailed research of stalagmite calcite and fluid inclusions.

The relationship between the isotopic composition of fluid inclusions and climate variables can be approached in several ways (Lachniet 2009). It should be said at the outset

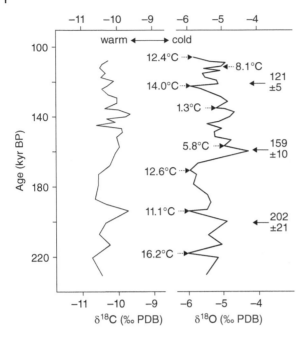

Figure 9.5 $\delta^{18}O_c$ and $\delta^{13}C_c$ variations along the growth axis of a stalagmite from Mammoth Cave, Kentucky, USA. Derived isotopic temperatures and dated points are also shown. Source: Modified from Harmon et al. (1978).

that speleothem palaeoclimatology is many years behind comparable disciplines such as dendroclimatology and the use of documentary climate sources, but significant progress has been made. Where annually laminated stalagmites are available, annual layer thickness records can be obtained using trace element variation. This permits correlation with the instrumental record and has yielded either a precipitation- or temperature-dominated signal, or one which is a mixed precipitation and temperature proxy (Fairchild and Baker 2012). For example, a stalagmite MA1 from Madagascar (Figure 9.6; Brook et al. 1999) comes from a seasonally dry climate region with a soil moisture deficit where speleothem dripwaters are likely to be supply-limited. Figure 9.6 shows the published correlation between MA1 and SOI.

An alternative calibration approach has been used by Lauritzen and Lundberg (1999) and Mangini et al. (2005), which does not involve direct regression with an instrumental series. This has been used where there is no tight chronology and slow growth rates, precluding the use of instrumental calibration. The authors developed a transfer function between stalagmite $\delta^{18}O$ and temperature, using a four- to five-point regression over a much longer time period (Figure 9.7). Problems in using this approach are both practical and conceptual. Dating uncertainties generate large error bands in the x-axis (temperature). The most positive $\delta^{18}O$ calibration point is based on a temperature of $-0.2\,°C$, presuming a temperature correlation with $\delta^{18}O$, and that temperature is around freezing, neither of which is actually known. Conceptually, the approach forces a calibration to temperature, rather than any other parameter, and presumes a stationary relationship between $\delta^{18}O$ and temperature over time (Fairchild and Baker 2012).

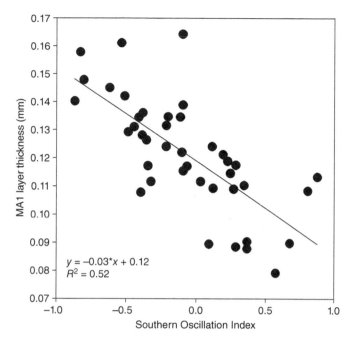

Figure 9.6 Example of calibration of a speleothem proxy against climate; annual growth rate series MA1 versus the Southern Oscillation Index (SOI). Source: From Brook et al. (1999).

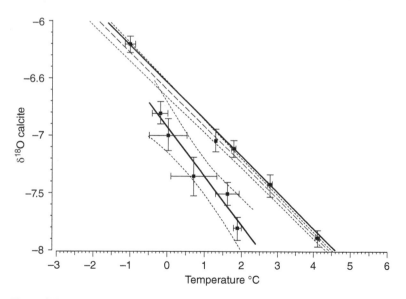

Figure 9.7 Transfer functions developed for stalagmites from northern Norway (upper points: Lauritzen and Lundberg 1999) and the Austrian Alps (lower points: Mangini et al. 2005). Source: From Lauritzen and Lundberg (1999).

9.4 Carbon Isotopes and Environmental Changes

Although the $^{13}C/^{12}C$ ratio has been widely used as an index of evaporation in stable oxygen isotope studies and in the correction of radiocarbon dates, the $\delta^{13}C$ values have not been widely used for their own palaeoenvironmental content. A measure of relative evaporation can be gained by considering the correlation of $\delta^{13}C$ and $\delta^{18}O$ along the stalagmite growth axis: sections with correlation indicate evaporation, uncorrelated sections indicate more mesic conditions. There are significant differences between the $\delta^{13}C$ ranges of C3 plants (trees and temperate grasses) and C4 plants (tropical grasses and some weeds). These ranges are shown in Figure 9.8, along with those of other materials likely to contribute carbon to speleothems. Since soil organic matter is derived from plant matter, the $\delta^{13}C$ values in feedwater will be derived from the surface vegetation above the cave as well as the passage of water through the bedrock. The major exception to this is where hydrothermal leaching of limestone enriches calcite in $\delta^{13}C$ (Bakalowicz et al. 1987). According to Ingraham et al. (1990), more water is lost by leakage from cave pools than from evaporation, and the rate of carbon isotope fractionation in cave pools is limited by the exchange of ^{13}C between water and vapour. The carbon isotope composition of cave deposits has been modelled by Dulinski and Rozanski (2016) with a good degree of success. Thus, the system is relatively well understood, and a significant change in $\delta^{13}C$ values of cave calcite over time may reflect a major change in vegetation. An enrichment of calcite $\delta^{13}C$ from values around −24‰ to −15‰ would imply a change from C3 to C4 plants, possibly because of warming. Studies of this kind have been carried out on soil profiles and on tufas (Marčenko et al. 2016), but there is considerable potential to apply this analysis to speleothems. A modelling approach has been carried out by Dulinski and Rozanski (2016) to better explain pathways leading to $\delta^{13}C$ values of cave calcite.

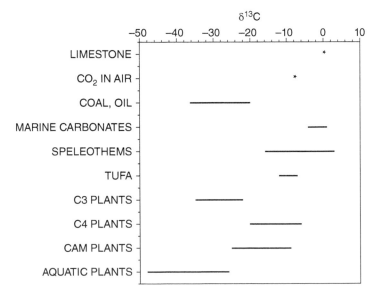

Figure 9.8 Range of $\delta^{13}C$ values for different materials found in the natural. Source: From Gillieson (1996).

In the USA, stalagmite samples from Mystery Cave, Spring Valley Cavern, and Coldwater Cave have been analysed by Dorale et al. (1992) and Denniston et al. (1999) and cover a wide region of the central north of the continent. They observed that the $\delta^{13}C$ records in the samples are quite consistent across the region and correlate with surface vegetation records. This is held to be due to a C3-C4 vegetation change dominating the proxy archives. The more positive $\delta^{18}O$ in some samples is explained as being caused by evaporative enrichment of $\delta^{18}O$ in the soil, which was more likely to occur with changing land cover to prairie vegetation, where soil water $\delta^{18}O$ would be more susceptible to evapotranspiration. Denniston et al. (1999) further demonstrated that at Coldwater Cave there was a relationship between surface topography (north versus south-facing slopes) and $\delta^{18}O$ enrichment.

Other organic compounds, such as humic and fulvic acids; and terpenes, have been isolated from calcite speleothems (Caldwell et al. 1982; Lauritzen et al. 1986). Much of the pigmenting material in stalagmite and flowstone is organic material, though there are dear cases where the colour is owing to inorganic trace elements (White 1981; White and Brennan 1990). Spate and Ward (1980) have shown that the black speleothems at Yarrangobilly, New South Wales, Australia, are because of finely divided organic carbon, not manganese as hitherto thought. Similarly, the black to dark brown calcite formations of the Nullarbor Plain may be on account of organic matter, though here the colour is probably owing to iron, manganese, and organic matter complexing with clays. Many calcite speleothems exhibit luminescence in the wavelength range 400–500 nm. The brightest luminescence is produced by fulvic acid, with several well-defined emission spectral peaks. Changes in the emission spectrum reflect the continuum of molecular structures between humic and fulvic acid. It should be possible to examine the emission spectra along a growth axis and gain information about the mix of humic material that has been captured by speleothem calcite over time. This could be calibrated against the humic acid emission spectra of contemporary vegetation types, and thus provide a long-term record of environmental change. This will be one of the challenges for karst research in the twenty-first century.

9.5 Cyclone History in the Indo-Pacific Region

Climate change scenarios all predict an increased risk from tropical cyclones in the Indo-Pacific region, consistent with increased risk to life and infrastructure from typhoons and hurricanes elsewhere on the planet. Until recently, our basis for assessing risk was the instrumental record over the last century. This record varies in accuracy and geographical distribution, making modelling difficult. Longer historical records, such as the Chinese records of typhoons (Liu et al. 2001) and their effects on trade in southeast Asia, only provide snapshots of cyclonic activity and their severity.

A new technique uses the oxygen isotope signal in speleothems to provide a high-resolution record of tropical cyclones in Australia. This method is based on the known depletion of $\delta^{18}O$ in rainfall derived from tropical cyclones (Frappier et al. 2007). Tropical cyclonic rain is strongly depleted in $\delta^{18}O$ because of widespread fractionation during the condensation of uplifted air and the continuous and high levels of rainfall (amount effect). Tropical cyclonic rain typically contains $\delta^{18}O$ levels between −5‰ and −15‰ (vSMOW). An isotope gradient also occurs across the cyclone, with the eyewall

Figure 9.9 An active stalagmite from Fern Cave, Chillagoe, Queenland, Australia, was sampled for the oxygen isotope study.

region showing lowest levels of $\delta^{18}O$. Low levels also occur within the zones of uplifted air around the cyclone known as spiral bands. Once this depleted water has become incorporated into speleothems, isotope values have been measured between −5‰ and −10‰ (Vienna Pee Dee Belemnite, vPDB).

A constant diameter stalagmite from Fern Cave (Figure 9.9), Chillagoe, tropical north Queensland (Australia) was first analysed using this technique (Haig et al. 2014). Chillagoe is approximately 100 km inland from Cairns, a regional city on the Great Barrier Reef. A comparison of speleothem layer thickness with annual rainfall (Figure 9.10) showed excellent agreement between wet and dry season layer component thickness and the respective seasonal rainfall totals for the period from 1902 to 2003 CE.

The speleothem layering showed ochre-coloured laminae followed by pure white laminae, corresponding to seasonal flushes of organic acids and colloidal clay though the epikarst; 777 couplets were sampled for $\delta^{18}O$ (Figure 9.11) and the more recent data compared with the historical cyclone record.

Each of the 27 peaks in the $\delta^{18}O$ depletion curve (Figure 9.12) corresponds to the passage of a cyclone within 400 km of Chillagoe. Twenty of the cyclones passed within 200 km of Chillagoe, 22 within 230 km, 23 within 270 km, and the other four within 400 km. The speleothem record accounts for 63% of all cyclones that passed within 200 km of Chillagoe since 1907 CE.

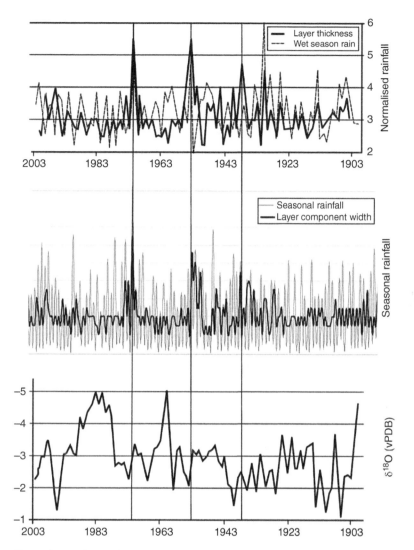

Figure 9.10 Comparisons of stalagmite layer thickness with wet season rainfall, layer component width (ochre and white components which together form one layer or couplet) and seasonal (wet and dry) rainfall, and $\delta^{18}O$‰ (vPDB) level. Source: From Nott et al. (2007) © 2007 Elsevier.

The size of each of the depletion peaks is a function of the $\delta^{18}O$ in cyclone rain and the level of $\delta^{18}O$ in percolating soil water from the previous wet season mixing with it. This lag effect means that a more appropriate measure of the level of isotope depletion for an individual cyclone is the difference in $\delta^{18}O$ between the depletion crest and the preceding trough in the curve (Figure 9.13).

The most intense cyclone in the instrumented record occurred in 1911 and by today's standard would be a Category 5 event. Using the isotope difference value of -2.5‰ for that cyclone as a reference, it is clear that the period between 1600 and 1800 CE had many more intense or hazardous cyclones impacting the site than the post 1800 CE period. This

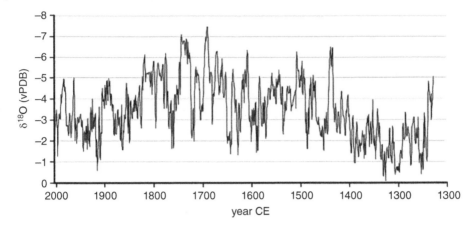

Figure 9.11 Annual $\delta^{18}O‰$ (vPDB) 1226–2003 CE for a speleothem from Chillagoe (Queensland, Australia). Source: From Nott et al. (2007) © 2007 Elsevier.

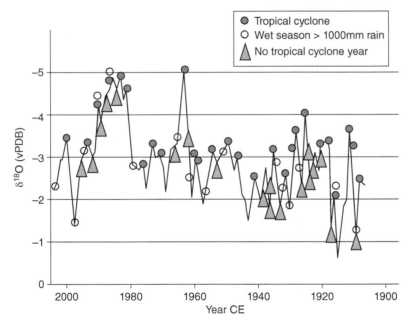

Figure 9.12 Timing of tropical cyclone occurrences, high wet season rainfalls and no cyclone years relative to $\delta^{18}O‰$ (vPDB) record for period 1907–2003 CE. Source: From Nott et al. (2007) © 2007 Elsevier.

is the period during which European explorers were mapping the east coast of Australia. The period from 1400–1500 CE had two events that had considerably lower isotope difference values and hence were presumably more hazardous than the 1911 CE event (Haig et al. 2014).

In comparison, the luminescence line record within corals of the central Great Barrier Reef shows a much higher frequency of large magnitude terrestrial floods and freshwater run-off to the sea between 1600 CE and approximately 1750 CE, which coincides with

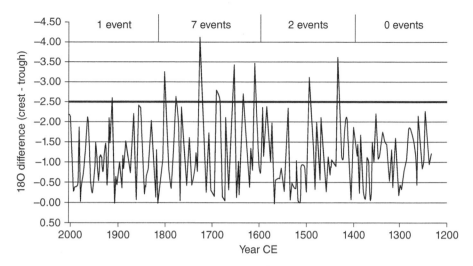

Figure 9.13 Detrended $\delta^{18}O‰$ (vPDB) values to account for $\delta^{18}O$ soil water dilution for landfalling tropical cyclones from 1228 to 2003 CE. The critical value for moderate to severe hazard impact at-a-station is −2.5‰. Note no events above this value between 1200–1400 CE and 1500–1600 CE, two events between 1400 and 1500 CE, seven events between 1600 and 1801 CE, and one event (just above this value) since 1801 CE. Source: From Nott et al. (2007) © 2007 Elsevier.

the speleothem record of the highest frequency of intense landfalling tropical cyclones. The speleothem record shows that the variability in intense landfalling cyclones is greatest at a centennial-scale compared to seasonal and decadal fluctuations. The changes between regimes seem to have been quite rapid. This study emphasises the importance of obtaining high-resolution long-term records of tropical cyclone activity so as to reliably assess risk for coastal populations in the tropics.

9.6 Other Proxy Records (Trace Elements, Annual Laminae, Pollen, Lipid Biomarkers)

Variations in the flux of water and dissolved gases through the epikarst should result in changes in the thickness of annual calcite laminae in stalagmites. The first step is to compare the thickness of the most recent laminae with a nearby instrumented rainfall record. If this can be validated, then a longer record may be extrapolated. The first instrumentally calibrated speleothem record used to provide a climate reconstruction over the last thousand years was from Madagascar (Brook et al. 1999), where lamina thickness was shown to correlate with the strength of the Southern Oscillation Index (SOI), with thicker layers being deposited in wetter, low SOI years. The reconstruction was shown to have a good correlation with other proxy archives which also reflect ENSO state, such as the Galapagos coral $\delta^{18}O$ archive (Dunbar et al. 1994). From the caves of Assynt, northwest Scotland, UK, Proctor et al. (2000) showed a correlation between annual lamina thickness in stalagmites and total annual precipitation and mean annual temperature, with the correlation with precipitation being stronger. This enabled a thousand-year reconstruction of total annual precipitation based on the detrended annual lamina thickness series.

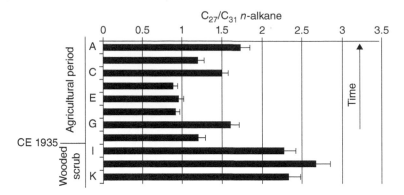

Figure 9.14 Comparison of changes in speleothem organic biomarkers, in this example, the ratio of C_{27}-C_{31} alkanes, and surface vegetation change. Source: From Woodhead et al. (2007) © 2019 Springer Nature.

The extraction of pollen from stalagmites, long held to be near-impossible, has recently been developed as a very useful tool for regions where conventional pollen traps, such as lakes, are scarce. Sniderman et al. (2019) extracted fossil pollen from radiometrically dated stalagmites collected in caves of southwest Western Australia. The pollen record, supported by 30 U-Th dates, reveals the vegetation response to Late Pleistocene climates between ∼34 and 14 000 years, through the Last Glacial Maximum. Today this region is the only part of southwest Western Australia that supports forests, yet for approximately 10 000 years between 28 000 and 18 000 years ago, eucalypt forests collapsed, and the dominant vegetation type appears to have been shrublands. The timing and duration of full glacial conditions in southwest Australia are very similar to patterns of glacial environmental change observed in other mid-high latitude regions of the Southern Hemisphere.

A promising proxy record may be gained from lipid biomarkers. In a modern stalagmite from Ethiopia, $\delta^{18}O$ and $\delta^{13}C$ were sampled at an annual resolution, and lipid biomarkers at approximately decadal resolution (Blyth et al. 2007) at a site where vegetation history was known (Figure 9.14). The lipid biomarkers showed a rapid response to surface vegetation change, explicable if the lipid signal bound in the stalagmite was dominated by an event-driven percolation of water through the epikarst, transporting the biomarkers rapidly from the soil to the stalagmite. In contrast, lipids transported more slowly in the karst aquifer may be biologically processed by bacteria and, therefore, retain a signal of the microbial activity in the karst water.

9.7 The Long Environmental History of the Nullarbor Plain, Australia

There has been much speculation over the age of the extensive caves on the Nullarbor Plain of Australia. Today the area is very arid, with annual rainfall at 250 mm or less. It is hard to explain the extensive tunnels and evidence of phreatic development without recourse to older, wetter climates. Early attempts at speleothem dating (Goede et al. 1990) indicated that the gypsum speleothems were likely to be at or beyond the limits of uranium-series dating.

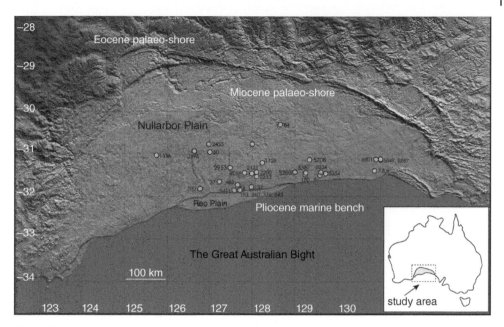

Figure 9.15 Caves sampled for speleothem dates on the Nullarbor Plain. This arid plateau is the world's largest area of exposed karst and hosts hundreds of shallow caves. Source: From Woodhead et al. (2019) © 2019 Springer Nature.

Recent U-Pb series dating by Woodhead et al. (2019) has for the first time provided data from 31 shallow caves scattered across 100 000 km^2 of the karst terrain (Figure 9.15). These clearly indicate caves and their contents have been preserved at the landscape scale since the Pliocene. In addition, the speleothems provide an invaluable archive of the long-term hydroclimatic history of the Pliocene Southern Hemisphere, including vegetation reconstructions that go a long way to explaining the rich marsupial fossil remains in the caves.

The speleothem ages tend to concentrate from 5 to 3 million years, with very few ages younger than 2.5 million years or older than 5.5 million years (Figure 9.16 and Table 9.2). Thus, the main phase of speleothem growth occurred during the early Pliocene and persisted until 2.5 Ma, near the beginning of the Pleistocene. The draining of the shallow phreatic caves and initiation of speleothem growth (Figure 9.17) must have occurred by the late Miocene, around 7–8 million years ago.

Pollen incorporated into the speleothem matrix was extracted from each stalagmite and analysed (Sniderman et al. 2016). As seen in Table 9.2, the yield of pollen from some speleothems was low but overall is statistically acceptable. The pollen grains were compared to contemporary vegetation types to identify individual taxa to Family or Genus level. The modern occurrence data for these taxa or clades, drawing on publicly accessible online databases of plant occurrence data, for Australia, New Zealand, and globally, was used to infer mean annual rainfall. These known environmental tolerances of these plant taxa were used in combination to develop a generalised additive model (GAM) in which rainfall is the dependent variable and the presence/absence of a taxon provides the dependent variable, assuming a binomial distribution and a cubic regression spline.

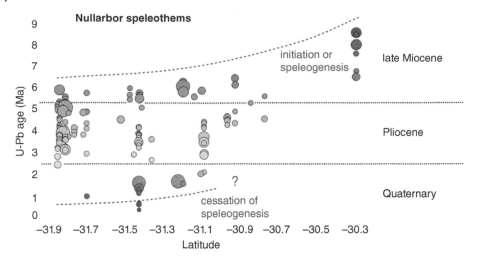

Figure 9.16 The distribution of Nullarbor (Australia) speleothem ages, plotted by latitude. The size of individual circles indicates the analytical uncertainty on each age determination while the colours code with age – Quaternary samples are orange while Pliocene and late Miocene samples are coloured blue. Source: From Woodhead et al. (2019) © 2019 Springer Nature.

Table 9.2 Age and pollen yield of Nullarbor speleothems.

Sample	Age, Ma ($\pm 2\sigma$)	Mass dissolved, (g)	Pollen count	Pollen grains (g^{-1})
2121-1	0.41 ± 0.07	215.48	401	1.9
645-15	3.47 ± 0.13	139.06	547	3.9
370-3	3.62 ± 0.14	311.68	105	0.3
370-1	3.63 ± 0.17	221.69	35	0.2
370-5	3.76 ± 0.12	198.75	221	1.1
645-13	4.14 ± 0.11	202.77	166	0.8
370-11	4.15 ± 0.12	302.09	256	0.8
2200-12.4	4.16 ± 0.12	57.8	389	6.7
2200-2	4.20 ± 0.14	68.47	279	4.1
483-9	4.89 ± 0.12	181.89	91	0.5
370-16	4.97 ± 0.12	760.1	60	0.1
370-17	5.34 ± 0.12	247.49	113	0.5
370-19	5.59 ± 0.15	240.83	152	0.6
		= 3148	= 2815	$\bar{x} = 1.7$

Source: From Sniderman et al. (2016).

Figure 9.17 Ancient speleothems in Witches Cave, Nullarbor Plain, Western Australia. These calcite formations are modified by spalling due to gypsum and salt deposition.

As seen in Figure 9.18, the Middle Pleistocene assemblage (B) is very similar to the composition of Late Pleistocene and Holocene pollen records (C) from Nullarbor cave/doline infills. These indicate a chenopod dominated shrubland similar to the vegetation today, confirming that the speleothem pollen assemblages register the surrounding vegetation in ways comparable to conventional fossil pollen records. The earlier Miocene to Pliocene assemblages show closer affinities to more closed vegetation types, indicative of shrubland and woodland in areas of higher rainfall. The genus Geniostoma is today found in shrubland to low forest in north-eastern Australia, Melanesia, Polynesia, and New Zealand. Doryanthes is a rosette plant found in near-coastal sclerophyll forests in northern New South Wales and southern Queensland. Genera such as Banksia and the Families Casuarinaceae and Myrtaceae, are widely distributed but generally associated with shrubland and woodland. The data suggest an abrupt onset of much warmer and wetter climates early within the Pliocene, driving complete biome turnover (Sniderman et al. 2016). The Pliocene warmth represents a discrete interval which reversed a long-term trend of cooling and aridification of the Australian continent. The mechanisms that drove the abrupt increase in Nullarbor precipitation and Southern Hemisphere sea surface temperatures are at present unknown.

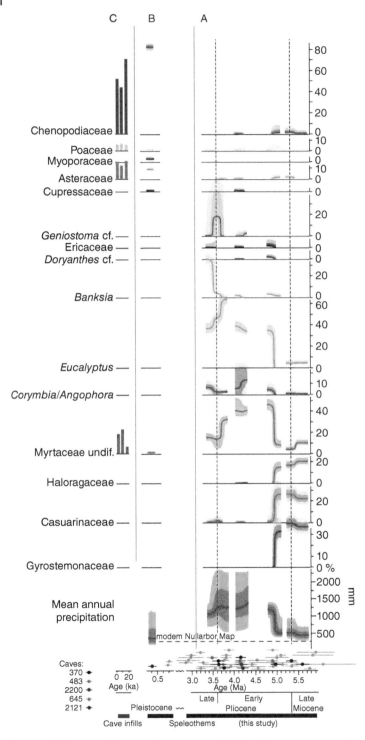

Figure 9.18 Late Miocene, Pliocene, and Middle Pleistocene vegetation change on the Nullarbor Plain, Australia. From Sniderman et al. (2016). Monte Carlo simulations of the late Miocene-Pliocene (a) and Middle Pleistocene (b) U-Pb−dated Nullarbor speleothem pollen record, accounting for Gaussian uncertainties in speleothem ages and in pollen percentage counts, and of the Nullarbor mean annual precipitation reconstruction derived from the pollen assemblages. Source: From Sniderman et al. (2016).

9.8 Some Speculations on the Future

It seems apparent that the age limits of U-Th dating have been reached, with the uncertainty of the ^{230}Th half-life the constraining factor preventing the determination of ages greater than 500 000 years. There is potential for other speleothem materials, such as gypsum, to be dated using this method, and this will expand the range of subjects that are possible.

New proxy records will be obtained with greatly increased temporal resolution. Annually resolved speleothem archives have been obtained for the Holocene, and these high-precision records may be extended to the Pleistocene and earlier. The long Quaternary records from Devils Hole, Soreq, and Hulu have all shown that exciting local to regional climate records can be obtained from speleothems. There may need to be a strategic approach to identify sites in critical geographic areas for understanding changes in regional or global circulation patterns.

Developments in cave monitoring equipment and techniques now make it possible to better define the relationships between surface proxies and cave speleothems. Understanding of the processes that transform speleothem feed water into calcite will also advance through this monitoring. Future research will increasingly focus on the use of multi-proxy records to obtain the best possible palaeoenvironmental information. Speleothem research has the potential to make great contributions to our understanding of the last thousand years of Earth's climate history.

There will be significant advances in the use of speleothem biomarkers. The ability to quantify several biomarkers is now established: future research will improve our understanding of the biomarker signal and its transfer from surface to cave, develop specific isotope analyses, and increase the biomarker toolbox to include proxies, such as tetraether lipids and rDNA.

References

Bakalowicz, M.J., Ford, D.C., Miller, T.E. et al. (1987). Thermal genesis of dissolution caves in the Black Hills, South Dakota. *Geological Society of America Bulletin* 99: 729–738.

Barker, S., Knorr, G., Edwards, R.L. et al. (2011). 800,000 years of abrupt climate variability. *Science* 334: 347–351.

Bar-Matthews, M., Ayalon, A., Gilmour, M. et al. (2003). Sea-land oxygen isotopic relationships from planktonic foraminifera and speleothems in the Eastern Mediterranean region and their implication for paleorainfall during interglacial intervals. *Geochimica et Cosmochimica Acta* 67: 3181–3199.

Blyth, A.J., Asrat, A., Baker, A. et al. (2007). A new approach to detecting vegetation and land-use change using high-resolution lipid biomarker records in stalagmites. *Quaternary Research* 68: 314–324.

Bowen, D.Q. (1978). *Quaternary Geology: A Stratigraphic Framework for Multidisciplinary Work*, 221. Oxford, Pergamon Press.

Brook, G.A., Rafter, M.A., Railsback, L.B. et al. (1999). A high-resolution proxy record of rainfall and ENSO since AD 1550 from layering in stalagmites from Anjohibe Cave, Madagascar. *The Holocene* 9: 695–705.

Caldwell, J.R., Davey, A.G., Jennings, J.N., and Spate, A.P. (1982). Colour in some Nullarbor Plain speleothems. *Helictite* 20: 3–10.

Cheng, H., Edwards, R.L., Broecker, W.S. et al. (2009). Ice age terminations. *Science* 326: 248.

Cheng, H., Edwards, R.L., Sinha, A. et al. (2016). The Asian monsoon over the past 640,000 years and ice age terminations. *Nature* 534: 640–646.

Coplen, T.B., Winograd, I.J., Landwehr, J.M., and Riggs, A.C. (1994). 500,000 year stable carbon isotopic record from Devils Hole, Nevada. *Science* 263: 361–365.

Cruz, F.W. Jr.,, Burns, S.J., Karmann, I. et al. (2006). A stalagmite record of changes in atmospheric circulation and soil processes in the Brazilian subtropics during the late Pleistocene. *Quaternary Science Reviews* 25: 2749–2761.

Denniston, R.F., Gonzalez, L.A., Asmerom, Y. et al. (1999). Evidence for increased cool season moisture during the middle Holocene. *Geology* 27: 815–818.

Dorale, J.A., Gonzalez, L.A., Reagan, M.K. et al. (1992). A high-resolution record of Holocene climate change in speleothem calcite from Cold Water Cave, Northeast Iowa. *Science* 258: 1626–1630.

Dulinski, M. and Rozanski, K. (2016). Formation of 13C/12C isotope ratios in speleothems: a semi-dynamic model. *Radiocarbon* 32: 7–16.

Dunbar, R.B., Wellington, G.M., Colgan, M.W., and Glynn, P.W. (1994). Eastern Pacific Sea surface temperature since 1600 AD: the δ18O record of climate variability in Galápagos corals. *Paleoceanography* 9: 291–315.

Fairchild, I.J. and Baker, A. (2012). *Speleothem Science: From Process to Past Environments*. Chichester: Wiley.

Fairchild, I.J., Smith, C.L., Baker, A. et al. (2006). Modification and preservation of environmental signals in speleothems. *Earth-Science Reviews* 75: 105–153.

Frappier, A.B., Sahagian, D., Carpenter, S.J. et al. (2007). Stalagmite stable isotope record of recent tropical cyclone events. *Geology* 35: 111–114.

Gascoyne, M. (1992). Palaeoclimate determination from cave calcite deposits. *Quaternary Science Reviews* 11: 609–632.

Genty, D., Blamart, D., Ouahdi, R. et al. (2003). Precise dating of Dansgaard-Oeschger climate oscillations in western Europe from stalagmite data. *Nature* 421: 833–837.

Goede, A., Harmon, R.S., Atkinson, T.C., and Rowe, P.J. (1990). Pleistocene climatic change in southern Australia and its effect on speleothem deposition in some Nullarbor caves. *Journal of Quaternary Science* 5: 29–38.

Haig, J., Nott, J., and Reichart, G.J. (2014). Australian tropical cyclone activity lower than at any time over the past 550-1500 years. *Nature* 505: 667–671.

Harmon, R.S., Schwarcz, H.P., and Ford, D.C. (1978). Stable isotope geochemistry of speleothems and cave waters from the Flint Ridge-Mammoth Cave system, Kentucky: implications for terrestrial climate change during the period 230,000 to 100,000 years B.P. *Journal of Geology* 86: 373–384.

Harmon, R.S., Schwarcz, H.P., and O'Neil, J.R. (1979). D/H ratios in speleothem fluid inclusions: a guide to variations in the isotopic composition of meteoric precipitation? *Earth and Planetary Science Letters* 42: 254–266.

Ingraham, N.L., Chapman, J.B., and Hess, J.W. (1990). Stable isotopes in cave pool systems: Carlsbad Cavern, New Mexico, U.S.A. *Chemical Geology: Isotope Geoscience* 86: 65–74.

Lachniet, M.S. (2009). Climatic and environmental controls on speleothem oxygen-isotope values. *Quaternary Science Reviews* 28: 412–432.

Lauritzen, S.-E. and Lundberg, J. (1999). Calibration of the speleothem delta function: an absolute temperature record for the Holocene in northern Norway. *The Holocene* 9: 659–669.

Lauritzen, S.E., Ford, D.C., and Schwarcz, H.P. (1986). *Humic Substances in Speleothem Matrix:* Paleoclimatic Significance. In: *Proceedings 9th International Congress of Speleology*, 77–79. Barcelona.

Liu, K.B., Shen, C., and Louie, K.S. (2001). A 1,000-year history of typhoon landfalls in Guangdong, southern China, reconstructed from Chinese historical documentary records. *Annals of the Association of American Geographers* 91: 453–464.

Maher, B.A. and Thompson, R. (2012). Oxygen isotopes from Chinese caves: records not of monsoon rainfall but of circulation regime. *Journal of Quaternary Science* 27: 615–624.

Mangini, A., Spötl, C., and Verdes, P. (2005). Reconstruction of temperature in the Central Alps during the past (2000) yr from a δ18O stalagmite record. *Earth and Planetary Science Letters*: 741–751.

Marčenko, E., Srdoč, D., Golubić, S. et al. (2016). Carbon uptake in aquatic plants deduced from their natural ^{13}C and ^{14}C content. *Radiocarbon* 31: 785–794.

Nott, J., Haig, J., Neil, H., and Gillieson, D. (2007). Greater frequency variability of landfalling tropical cyclones at centennial compared to seasonal and decadal scales. *Earth and Planetary Science Letters* 255: 367–372.

Proctor, C.J., Baker, A., Barnes, W.L., and Gilmour, M.A. (2000). A thousand year speleothem proxy record of North Atlantic climate from Scotland. *Climate Dynamics* 16: 815–820.

Sniderman, J.M.K., Woodhead, J.D., Hellstrom, J. et al. (2016). Pliocene reversal of late Neogene aridification. *Proceedings of the National Academy of Sciences*: 1999–2004.

Sniderman, J.M.K., Hellstrom, J., Woodhead, J.D. et al. (2019). Vegetation and climate change in southwestern Australia during the last glacial maximum. *Geophysical Research Letters* 46: 1709–1720.

Spate, A.P. and Ward, J.J. (1980). Preliminary note on the black speleothems at Jersey Cave, Yarrangobilly, NSW. In: *Proceedings of the 12th Biennial Conference*, 15–16. Australian Speleological Federation.

White, W.B. (1981). Reflectance spectra and color in speleothems. *National Speleological Society Bulletin* 43: 20–26.

White, W. B. and Brennan, E. S. (1990). Luminescence of speleothems due to fulvic acid and other activators. Budapest. pp. 212–214.

Winograd, I.J., Coplen, T.B., Landwehr, J.M. et al. (1992). Continuous 500,000-year climate record from vein calcite in Devils Hole, Nevada. *Science* 258: 255–260.

Winograd, I.J., Landwehr, J.M., Coplen, T.B. et al. (2006). Devils Hole, Nevada, δ18O record extended to the mid-Holocene. *Quaternary Research* 66: 202–212.

Woodhead, J., Hellstrom, J., Maas, R. et al. (2006). U–Pb geochronology of speleothems by MC-ICPMS. *Quaternary Geochronology* 1: 208–221.

Woodhead, J.D., Sniderman, J.M.K., Hellstrom, J. et al. (2019). The antiquity of Nullarbor speleothems and implications for karst palaeoclimate archives. *Scientific Reports* 9: 603.

Yonge, C.J. (1981). Fluid inclusions in speleothems as paleoclimate indicators. In: *Proceedings of the 8th International Congress of Speleology*, 301–304. Kentucky.

10

Cave Ecology

10.1 Introduction

Caves have attracted the interest of biologists and ecologists for over a hundred years (Schiner 1854; Vandel 1965). Physiological and evolutionary adaptations to cave living occur in many phyla of the animal kingdom, and also in some flowering plants and fungi. The principal changes are a loss of pigmentation, partial or total loss of eyes, an extension of sensory hairs or antennae, and changes in body part proportions (Culver 1982). This often produces somewhat bizarre-looking animals, which have thus attracted attention despite their cryptic habits. In this chapter, the basic characteristics of the cave ecosystem and its constituent trophic levels are described. This relates back to the basic concepts of mass and energy flow in the karst system. The role of external energy sources is critical for cave life. Cave biota are dependent on periodic inputs of nutrients, usually swept in by floods. They are also severely disadvantaged by quite minor disturbances. Thus, they have low resilience in the face of a change to the cave ecosystem. The spectrum of impacts on caves have serious consequences for their biology and ecology, and adequately conserving cave biota is a major challenge for protected area management.

10.2 Classification of Cave Life and its Function

The most obvious features of the cave environment are the reduced light levels and a near-constant temperature regime. Life in total darkness requires that other senses – principally those of touch and smell – become dominant. Thus, cave-adapted fauna have greatly enlarged antennae or bristles as well as specialised organs to detect vibration. These adaptations produce some bizarre morphologies in cave animals.

The most widely used classification of cave organisms was developed by Schiner (1854) who proposed an ecological classification based on habitat, which included a category for species that were limited to caves. Although this classification system, usually called the Schiner-Racoviţă system (Sket 2008), is based on the distribution of animals, it is the morphology that is the obvious attribute of many. Under this system, true cave dwellers or troglobites were obliged to live in the deep zone and show significant eye and pigment reduction. These creatures are relatively rare and unable to survive outside the cave environment. The often bizarre cave beetles with absent or residual non-functional eyes and long antennae provide one example of true cave dwellers. Those species which use the deep cave environment but show little eye and pigment reduction are termed troglophiles – facultative cave dwellers. They live and breed inside the cave, but on the basis of their morphology, it is assumed that they can live on the surface as well, usually in similar dark, humid microhabitats, such as the undersides of fallen logs. Many cave crickets, spiders, and millipedes fall into this group. Finally, those species often found in caves for refuge, but which leave to feed, are called trogloxenes. Bats are a good example of this group (Jefferson 1976). Some other species wander into caves accidentally but cannot survive there.

For aquatic organisms, there is a parallel classification. Those highly specialised animals living entirely in the groundwater environment, and absent in surface waters, are called stygobites. In contrast, stygophiles are found in both surface and underground waters without adaptation to subterranean life. Stygoxenes are organisms that appear rarely, and almost randomly, in underground waters but are essentially surface dwellers (Marmonier et al. 1993; Ginet and Juberthie 1988).

The discovery of morphologically similar animals outside caves – in scree slopes, in crevices, and in springs – has led to a radical revision of the Schiner-Racoviţă system. There has been a move away from an ecologically-based classification to a morphologically based system. Christiansen (1962) proposed a term to reflect this convergent morphology of subterranean-dwelling animals – troglomorphy. Terrestrial animals of this kind are termed troglobionts while their aquatic counterparts are termed stygobionts. The convergence is truly remarkable and makes troglomorphy easy to identify. It is also becoming clear that not all stygobionts and troglobionts are necessarily troglomorphic. There may be species which have had insufficient time in subterranean habitats to evolve troglomorphic features. For example, species living on guano piles in caves are often not troglomorphic (Culver 1982; Culver and Pipan 2009).

10.3 Adaptations and Modifications to Life in Darkness

The specific adaptations of organisms to life in the cold and dark are:

- Loss of pigmentation
- Reduction in eyes
- Extension of sensory structures
- Changes in feeding appendages
- Elongation of locomotory spines and claws
- Reduced size
- Changes to circadian rhythms
- Changes to reproductive biology and life cycle (Culver 1982)

These changes in morphology are shown in Figure 10.1. As well as dramatic changes in morphology, cave dwellers tend to have altered reproductive strategies involving reduced frequency of breeding and number of larval stages for invertebrates (Ueno 1987). Cave dwelling fish tend to have larger eggs and larvae coupled with a longer reproductive cycle. Adaptations to a life in darkness are not necessarily confined to cave dwellers. Troglobionts share adaptations with nocturnal animals, in that their non-visual senses tend to be well

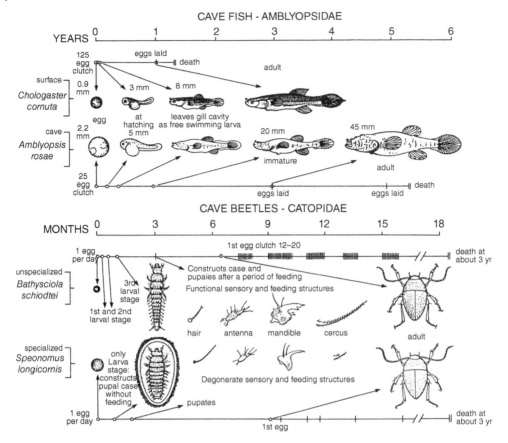

Figure 10.1 A comparison between surface and cave-adapted forms in fish of the family Amblyopsidae and beetles of the family Catopidae. Note changes in gross morphology and size, sensory organs and appendages, and reproductive cycles. Source: Adapted from Poulson and White (1969).

developed so that they can move, eat, and breed in darkness (Nevo 1979). However, their eyes tend to be reduced in contrast to the enlarged eyes or highly responsive retinas of nocturnal animals. Very long appendages, such as the antennae of cave crickets and scutigerid centipedes, are quite normal in troglobionts. These antennae can also function as chemoreceptors and may be sensitive to relative humidity (Jones et al. 1992). Echolocation is a well-developed trait in bats and some cave-dwelling birds such as swiftlets. Swiftlets click their tongues at low frequencies (1.5–5.5 kHz) and the sound echoes are interpreted to produce a "picture" of the surroundings. This enables them to navigate around cave passages but is inadequate for hunting their prey of small wasp-like insects, which must be sought in daylight. In contrast, insectivorous bat echolocation is highly complex, emitted at high frequencies (20–130 kHz) and very accurate, and allows hunting in total darkness for the bats' diet of moths and beetles.

As well as the loss of or reduction in eyes, many troglobiont invertebrates have lost cuticle pigments and wings. Some have also developed a thinner cuticle and a more slender body, as they are free of the desiccation of the surface environment. Foot modifications (hooks and bristles) to allow walking on wet or slippery surfaces are common, as are lower metabolic rates. These equip the individual organism to move more securely in the cave and to utilise the available energy sources most effectively.

Most animals have a clearly defined daily cycle of activity. Thus, nocturnal species are active at night, diurnal species during the day, and crepuscular species around dawn and dusk. Such cycles are clearly associated with light intensity changes and may be irrelevant to a cave community. There may well be other events that impose a daily rhythm on cave inhabitants. The departure and later return of surface-feeding bats or swiftlets cause a change to the rain of urine and faeces from the roof of the cave, a potent energy source for floor-dwelling organisms. Ectoparasites are temporarily deprived of their food source. Thus, there may be a daily cycle despite the lack of light, especially in tropical caves.

Providing a summary of the taxonomic breadth of subterranean fauna is daunting. Few generalisations are valid at a global scale, as species assemblages vary considerably within one karst region, within a country, and within a continent. The hexapod order Collembola has many troglobiontic species in French caves, very few in Slovenian caves, and is common in Australian caves. The order Syncarida, small aquatic crustaceans, is important in

Table 10.1 Invertebrate orders with more than 50 stygobionts and troglobionts.

Phylum	Class	Order
Platyhelminthes	Turbellaria	Tricladida
Annelida	Clitellata	Lumbriculida
		Tubificida
Mollusca	Gastropoda	Mesogastropoda
Arthropoda	Maxillopoda	Cyclopoida
		Harpacticoida
	Ostracoda	Podocopida
		Bathynellacea
	Malacostraca	Amphipoda
		Isopoda
		Decapoda
	Entognatha	Collembola
		Diplura
	Insecta	Coleoptera
		Homoptera
	Chilopoda	Lithobiomorpha
	Diplopoda	Chordeumatida (= Craspedosomida)
		Julida (= Iulida)
		Araneae
	Chelicerata	Opiliones
		Pseudoscorpionida

From Culver and Pipan (2009; 49).

Europe, southern Australia, and New Zealand, but not in North America. Many species have very restricted ranges and are known from only one or two sites. New species are being collected all the time, though their formal taxonomic description is a slower process. Finally, the subterranean fauna is relatively species-rich. Culver and Holsinger (1992) estimate that there would be approximately 50 000 species of stygobionts and troglobionts worldwide, if all species were known and described. The number of described species is much less but still large. Deharveng et al. (2009) reported 930 stygobiontic species from six European countries (Belgium, France, Italy, Portugal, Slovenia, and Spain) and Zagmajster et al. (2008) reported 282 troglobiontic beetle species in two families in the Dinaric karst of Italy, Slovenia, Croatia, Bosnia and Herzegovina, Serbia, Montenegro, and Albania. As a broad generalisation, there are 21 invertebrate orders with at least 50 stygobionts or troglobionts (Table 10.1) and two vertebrate groups with stygobionts – salamanders and fishes.

Culver and Pipan (2009; see Chapter 3) provide a very useful overview of the systematics of troglobionts and stygobionts, while more detail on systematics and biogeography can be found in the three-volume *Encyclopaedia Biospeologica*, edited by Juberthie and Decu (2001).

10.4   Life Zones within Caves

The entrance zone of caves is a highly variable habitat organised along a gradient of decreasing light and increasing humidity away from the entrance (Howarth 1983). Around the entrance overhang, the environment differs little from the surface, though more shelter is available and flows of mass and energy may be more focused. Light and warmth may be available for longer periods of time. The presence of collapse blocks and relatively coarse sediments may provide a wide range of interstitial spaces for colonisation. Greater amounts of organic matter, higher plant production, and a greater diversity of potential food for predacious animals may be present. Many of the organisms in this zone are not obligatory cave dwellers, and there may be many accidental visitors here, which have either been washed in or have strayed (Figure 10.2).

In the twilight zone, there is still some light, but a more stable temperature regime accompanied by higher humidity. There is less organic matter, but fungi may be important in the slow nutrient release by decomposition. Organisms start to show specific adaptations to cave-dwelling, but there are many accidental visitors in this zone as well as species which feed outside the cave.

The deep, truly dark zone of an individual cave is a rigorous to harsh environment for most surface- or soil-dwelling organisms. Within, the network of passages of varying sizes conditions are perpetually dark (except in tourist caves or frequently visited wild caves), with relatively constant temperatures close to the surface mean annual temperature. The substrate in many middle- to high-latitude caves is perpetually wet and the atmosphere close to saturated with water vapour, often at levels beyond the tolerances of terrestrial arthropods (Poulson and White 1969). Carbon dioxide concentrations may be high, and the seasonal cues – fluctuations in light and temperature – available to surface-dwelling organisms are lacking.

Figure 10.2 The upstream entrance of Deer Cave, Gunung Mulu NP, Sarawak, Malaysia provides an energy and light gradient from the surface rainforest to the perpetual dark of the cave. Organic matter is washed into the cave on a daily basis.

Temperatures are usually constant in caves on account of the buffering or insulating effects of cave walls and roofs and reduced air circulation. A cold trap effect may, however, be present in caves which descend steeply from the entrance, resulting in local drying on rock surfaces. Conversely, blind, upwards trending passages may trap warmer air and provide suitable habitats for bats, with static, saturated air. But most cave atmospheres conform to the mean annual temperature unless they lie close to the surface and are subject to local air circulation. Air movement may be seasonally variable, but there will be pockets of still air in which spiders can weave delicate webs to entrap their prey. Under such stable conditions, the approach of prey or potential predators can be detected by sudden air movements (Howarth 1983).

The humidity regime of many caves is more complex. Locally there may be films of water on rock surfaces, pools and active drips. There may be condensation at constrictions where air movement is pronounced. The constant high humidity in the deep parts of caves has led to the morphologies of troglobiontic arthropods being similar to those of aquatic arthropods (Howarth and Stone 1990). Higher carbon dioxide levels in the still air of deep caves may result in slowed metabolism as a physiologic response. Figure 10.3 illustrates the changes in key atmospheric parameters and distribution of animals along a section in Bayliss Cave, north Queensland, Australia. Both humidity and temperature increase away from the entrance, which is the normal trend in most caves. Carbon dioxide concentrations increase dramatically at the back of the cave, and troglomorphic species (TM) increase in this deep zone with foul air. The saturated air and high CO_2 levels (some 200 times

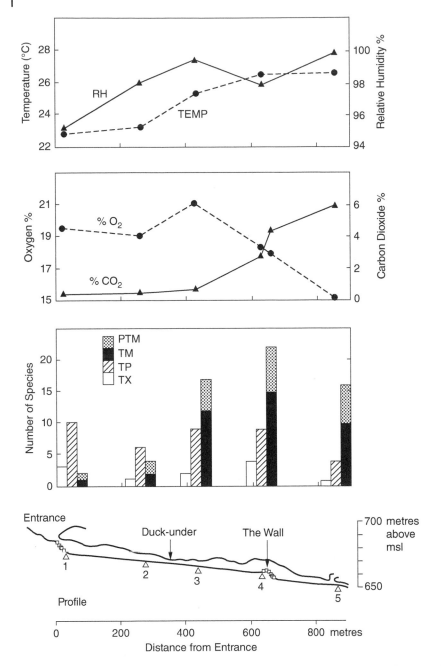

Figure 10.3 Atmospheric environments and animal distributions in Bayliss Cave, north Queensland, Australia. Bottom: profile view of Bayliss Cave with sampling points. On the bar graph, PTM = partly troglomorphic species; TM = troglomorphic species; TP = troglophilic species; TX = trogloxene species. Source: Adapted from Howarth and Stone (1990).

ambient concentration) support a highly diverse community of cave-adapted species tolerant of the conditions there, unlike surface and facultative cave species. Thus carbon dioxide concentrations may be significantly higher in caves than on the surface, owing to degassing from speleothems and the decomposition of organic detritus. This may play a role in the adaptation of cave invertebrates. Their adaptation to radiation sources, such as radon gas, remains unstudied.

Many troglobionts have fat stores to cope with intermittent food supplies. Although food sources may be very restricted, this is not always so in modified cave systems. But for many sites, the filtering action of cave passages may at the very least mean that inputs of food are very widely spaced in time. In such caves, periodic floods may be crucial for the introduction of energy reserves to maintain cave life.

10.5   The Cave as a Habitat

Biologists have traditionally regarded caves as finite environments whose boundaries are well defined (Barr 1968) and within which the resident flora and fauna can be censused, their trophic relationships defined and the flows of mass and energy measured. Thus, caves have been seen as ideal natural laboratories in which to test ideas about processes of adaptation, the structure, and function of simple ecosystems, and the reaction of ecosystems and their components to induced changes. Research (Holsinger 1988, Howarth 1991, Humphreys 1993) suggests, however, that the boundaries are not as finite as previously imagined. Speleologists can enter drained and some flooded cave conduits and examine their fauna and flora. It is rarely possible to examine the extensive network of fine cracks and the interstitial spaces in sediments which are normally associated with the conduits. These spaces represent a further habitat for cave organisms which have their own special characteristics and may connect with other conduits, groundwater bodies, and the surface environment. There is thus the possibility that caves are not closed environments, and that unexpected interchanges of biota, mass, and energy may occur in areas which are at present difficult to sample. This has especial relevance for the survival of small populations of organisms which have become locally extinct in a conduit as a result of some imposed changes in the habitat, such as pollution or increased sedimentation.

The most obvious change in the cave habitat as we move further inside is the gradual reduction in the intensity of light. Depending on the architecture of the passage, the total extinction of daylight may occur from a few metres to tens of metres into the cave. The absence of light deprives cave biota of one major source of energy and means that the production of biomass by the photosynthesis of green plants is limited to the entrance zone. Thus the base of the ecosystem structure – the primary producer – in most terrestrial and aquatic environments is largely lacking in caves. This is essentially true for nearly all caves, though a type of lichen comprised of gram-positive bacteria and blue-green algal cells coats cave walls in some tropical caves (Whitten et al. 1988). This colony has a high density of photosynthetic membranes (thylakoids) and may be able to use very low light levels. An exception also occurs when seedlings fall or are washed into caves, and produce sprouts from their stored starch reserves. Thus, pale or white seedlings may be found several hundred metres inside caves and may provide an extra energy source for cave biota.

This reduction in primary production by photosynthesis has important consequences for the types of organisms which live inside caves and their lifestyles. Most are, by necessity, detritivores or carnivores. The only common primary producers in such an ecosystem are bacteria and fungi. Bacteria may be divided into heterotrophs, breaking down organic debris and gaining nutrients and energy, or chemautotrophs, reducing mineral compounds such as iron sulphates and thereby gaining energy and nutrients (Dyson and James 1981). Irrespective of their mode of energy production, their total productivity under natural conditions is likely to be low. Their productivity is enhanced slightly by the commonly alkaline nature of cave sediments and waters, but low temperatures limit their metabolic processes. Fungi slowly break down organic debris and provide a further source of food for detritivores, but again their productivity is limited. This limited productivity means that the total biomass supportable in any cave is circumscribed, and thus cave biota are never numerous under unimpacted conditions.

All cave dwellers are largely dependent on food sources brought into the cave from the surface environment. The accession of food and energy from external sources becomes critical for the survival of viable populations of organisms comprising the ecology (Figure 10.4).

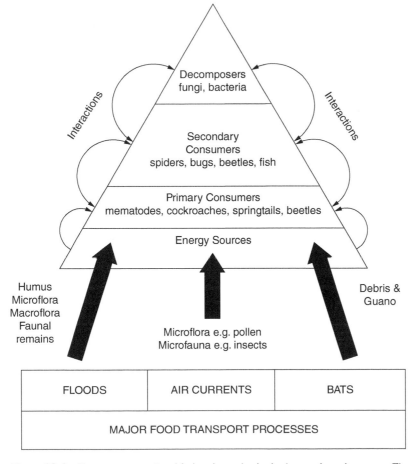

Figure 10.4 Energy sources, trophic levels, and principal organisms in a cave. There are many complex interactions between trophic levels which can only be hinted at here. Source: Adapted from Biswas (1992).

Figure 10.5 Tree roots provide an extra energy source and habitat in subtropical caves for animals such as this isopod in Yessabah Cave, New South Wales, Australia. Source: Photo by Stefan Eberhard.

The principal source is organic debris washed into the cave by running water, either as percolation water or as discrete cave streams. This material may be fine humus, which is readily utilised by cave biota, or coarser debris (twigs, leaves, and branches), which must first be broken down by bacteria and fungi to be useable. Sites within the cave that are rarely flooded may, therefore, be expected to be depauperate in fauna, while sites along main streams with a direct connection to the surface environment may be quite rich in species and in total numbers of organisms (Poulson 1977). Another significant source of external material is from air-fall processes below shafts or crack systems open to the surface. This is especially important for higher-level dry passages remote from the stream or for caves in dry climates. The penetration of tree roots into cave passages provides a very important energy source in both tropical and some temperate caves. Some animals feed directly on the plant roots (Figure 10.5) that penetrate the cave roof, while others feed on detritus that is washed in during floods or falls in from adjacent daylight holes. As well as the decomposition of bark and stem tissue to produce humus, the exudates from roots provide a source of energy (Howarth 1983).

A small amount of organic matter also percolates down through the epikarst and enters the cave via the small roof fissures. Bats and cave-nesting birds also provide food in a number of ways:

- Faeces or guano, which has high nutritional value and is fed on by various animals (coprophages) as well as providing a rich nutrient medium for fungi and bacteria.
- Moulted hair and skin fragments, which provide an additional proteinaceous food source.
- Their progeny may be susceptible to predators and parasites – for example, cave crickets may attack the eggs of swiftlets, and false vampire bats *(Macroderma* spp.) successfully predate on juvenile and adult bats and birds.
- They are host to a wide range of ecto- and endo-parasites.
- Their corpses provide a source of food for animals (necrophages).

There is thus a fairly clear distinction in caves between roof and floor communities (Figure 10.6), each with its own food web, but linked by detritus fall. As the understanding

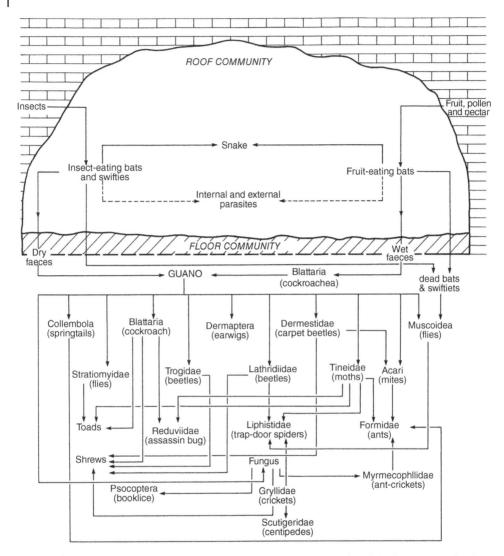

Figure 10.6 Simplified food web from a tropical cave system, showing links between roof and floor communities. Source: Adapted from Whitten et al. (1988).

of the cave community grows, the food web in Figure 10.6 tends to become more complex. The floor community is highly diverse, with each group of organisms strongly interlinked with others. The organic layer on the floor of most caves is composed of the waste products and corpses of animals. It is usually rich in nitrogen and phosphorus but often low in carbon because of the efficient breakdown of organic matter. On the cave floor, coprophages and necrophages are dominant, but this distinction may be blurred, with coprophages eating dead animals as well. Coprophages such as woodlice, flies, beetles, and Tineid moth larvae will readily exploit the guano of insectivorous bats (Hill 1981). The softer guano of fruit bats will be consumed by cockroaches and some cave crickets. Fungi and some bacteria also digest this food source and, in turn, are eaten by other organisms, such as

Figure 10.7 An Amblypigid predator from Gua Tempurung, Malaysia. This "whip scorpion" is actually related to spiders, and predates on insects, baby birds, and frogs. Total length, including antennae is 150 mm.

flies. In addition, intricate parasite or prey-predator relationships become more apparent with detailed research on population ecology. There are a great number of higher-order consumers (Figure 10.7) which prey on the animals that consume guano. These include spiders, assassin bugs, ants, crickets (Figure 10.8), and centipedes.

Elucidation of the food web can also lead to the construction of population pyramids (Figure 10.9).

The roof community is somewhat simpler and includes bats and swiftlets as well as all those animals that feed on or parasitize them. Bat hawks and pythons are often seen near cave entrances and have been observed feeding on emerging bats at dusk. Bats are host to a wide range of parasites that suck blood or are internal. These include nycteribiid bat flies, ticks, mites, bed bugs and internal roundworms, and flukes. Bat parasites show strong adaptation to their host and to the cave environment, with body flattening, grasping claws and spines, loss of wings, and reduced eyes.

In surface ecosystems, the numbers of organisms at the base of a food chain – the producers – are most numerous, while those at the top – the secondary or tertiary consumers – are least numerous. Where parasitism is present, a single host may have a large number of parasites in turn host to a larger number of hyper-parasites (both external and internal). These theoretical structures may also be expressed most usefully in terms of biomass. In surface environments, there will be a large biomass of producers and a steadily or exponentially reducing biomass of consumers at each energy or trophic level

Figure 10.8 The cave cricket Cavernotettix sp. from Australia shows the elongated appendages and reduced eyes typical of cave-adapted animals. Source: Photo by Stefan Eberhard.

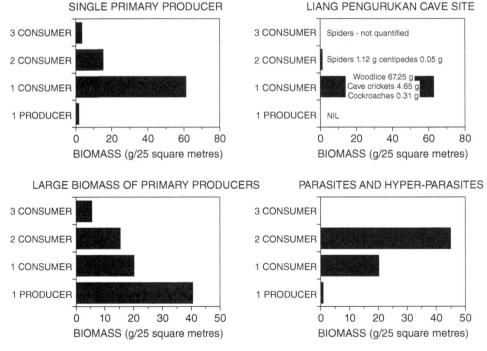

Figure 10.9 Simplified population pyramids for cave ecosystems with a single primary producer, a surface site with a large biomass of primary production, the tropical cave ecosystem of Liang Pengurukan, and a food web based on parasitism. Source: Data from Whitten et al. (1984).

(Figure 10.9). In cave ecosystems, there may be a large biomass of primary consumers but a drastically reduced biomass of secondary consumers. In the tropical cave Liang Pengurukan, Malaysia (Whitten et al. 1984), the total biomass in a 25 m^2 area of cave floor was 63.38 g (2.5 g·m^{-2}), of which 90% was contributed by a single species of woodlouse, which accounted for 99% of the animals found. The heaviest animal, a cave cricket, was 900 times heavier than the lightest, the woodlouse. A population pyramid based on numbers alone would overemphasise the woodlouse, but its importance becomes more apparent when the pyramid is constructed from biomass data. In cool climate caves, generally less productive, the differences between trophic levels become even more pronounced.

The frequency and magnitude of energy inputs into the cave ecosystem become very important for the maintenance of populations of organisms. In areas of cold climate with water movement restricted to the spring thaw, biological activity is phased to follow the major influx of water and organic matter, and at other times may be largely dormant. In areas with strongly seasonal precipitation, organisms may have to be adapted to survive desiccation for up to six months; perhaps longer if the climatic variability is high. Cave fauna in tropical areas is less restricted and may be active throughout the year, but reproduction may be phased to reduce competition for resources. Major changes to the frequency of water inputs may have serious consequences for the cave biology, and are common in rural areas where karst water is diverted or overutilised, or if surface changes such as vegetation clearance alter the quantity and quality of percolation water.

10.6 Energy Flows in Cave Ecosystems

We have already seen that cave ecosystems are likely to be limited by energy flows and that sources diminish away from any entrance. Periodic flood flows may be the only source deep inside any cave. In quantifying energy flows the scale of analysis is quite important (Figure 10.10). In stream ecology the stream reach is the fundamental unit of analysis; its underground equivalent would be a cave passage. These reaches could be integrated as the whole cave and beyond that the whole karst drainage basin.

Gibert (1986) studied a whole karst basin in France to quantify inputs and outputs, and made pioneering measurements of nutrient fluxes, showing that dissolved organic carbon (DOC) was the principal source of nutrients. Simon et al. (2007) compared the flux of nutrients in Organ Cave, Virginia, USA and Postojna Jama, Slovenia. They concluded that carbon was likely to be the limiting nutrient, not phosphorus or nitrogen. They considered two main inputs in karst basins: First, a very localised flow of particulate organic carbon (POC) and DOC through streamsinks and shafts. This may include leaves, wood, and fine detritus, as well as both invertebrate and vertebrate remains. Secondly, a diffuse flow of POC and DOC derived from overlying soils and transmitted through the epikarst (Table 10.2). Although POC is effectively filtered through the epikarst, Pipan (2005) has shown that there is a rain of POC in the form of copepods in epikarst drips and Simon et al. (2003) found that microbial films fuelled by DOC from soils were a primary food for animals in cave streams. These two principal types of organic matter are transformed and transported through the karst basin before eventually being lost, either through respiration or final export from the basin at springs (Figure 10.11). There may also be some production

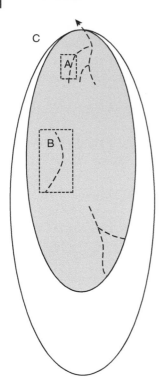

C

A

B

Figure 10.10 Schematic representation of the scale and extent of subterranean ecosystem models. The large ellipse represents the karst drainage basin, the shaded ellipse the extent of karst within the basin, the heavy dashed lines subterranean stream passages, and the arrow the exit of the water from the spring. The rectangles (A, B, and C) are possible scales of analysis. Source: From Culver and Pipan (2009).

Table 10.2 Estimates of dissolved organic carbon from Organ Cave and Postojna Planina Cave System (PPCS).

Cave Inflows	Organ Cave			Postojna-Planina Cave System		
	n	Mean, mg C L^{-1}	S.E.	n	Mean, mg C L^{-1}	S.E.
Sinking streams	3	7.67	1.03	2	4.36	0.46
Epikarst drips	20	1.10	0.15	99	0.70	0.04
Cave streams	6	1.08	0.32	3	4.75	1.57
Resurgence	3	0.90	0.17	2	2.67	0.80

Source: From Simon et al. (2007).

of carbon by chemautotrophs in cave streams and on rock surfaces; the latter biofilms may be an important source of nutrients for cave biota.

The authors made a number of measurements of organic carbon, especially dissolved organic carbon DOC, in epikarst drip water, cave streams, surface streams sinking into the cave, and at springs. These were also combined with other published data. In both caves, most of the epikarst derived organic carbon was DOC, at concentrations around 1 mg C L^{-1}. In both basins, sinking streams accounted for the majority of DOC input with concentrations around 4.4–7.7 mg C L^{-1}.

From a whole-ecosystem perspective, cave streams are very similar to rather unproductive surface streams. The concentration of DOC in streams of Organ Cave and Postojna

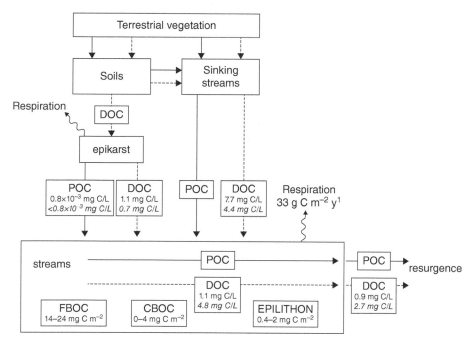

Figure 10.11 Schematic diagram of organic carbon flux in Organ Cave and the Postojna Planina Cave System (PPCS). Values in italics are PPCS data. Data are standing stocks of C except for respiration flux. POC, DOC, FBOC, and CBOC are particulate, dissolved, fine benthic, and coarse benthic organic carbon, respectively. Source: From Simon et al. (2007).

Planina Cave System (PPCS) are at the low end of the range ($0.1–36.6$ mg L^{-1}) reported for surface streams (Muholland 1997). In the Organ Cave basin, sinking streams drain 30% of the basin at a DOC concentration of 7.67 mg L^{-1} (Table 10.2), while the epikarst drains 70% of the basin with a DOC concentration of 1.10 mg L^{-1} (Table 10.2). The contribution of DOC from sinking streams is roughly three times higher than the supply from the epikarst. For PPCS, sinking streams drain about 93% of the basin and when combined with DOC concentrations, sinking streams account for 99% of the DOC entering the cave system. It is likely that considerable processing of organic carbon occurs within both caves, but measuring this in all areas of a cave system would be challenging. Simon et al. (2007) conclude that more detailed measures of organic carbon flux at both the basin and stream scale are needed.

The character and composition of dissolved organic matter (DOM) in sinking streams and epikarst drainage waters are fundamentally different (Simon et al. 2010). Epikarst waters contain humic-like and protein-like DOM, which is indicative of microbial uptake and release. In contrast, DOM in sinking streams contains humic and fulvic substances, which are more aromatic and are present in higher concentrations. The epikarst DOM is an important contributor to banding in speleothems and increases with flushing events. Simon (2013) has re-analysed Gibert's (1986) data to look at the role of water flux and flushing in the movement of organic detritus and cave biota. Gibert's data, collected over two years, remain the most comprehensive temporal datasets on the variations in nutrient flux and biological response anywhere. They provide a unique opportunity to examine

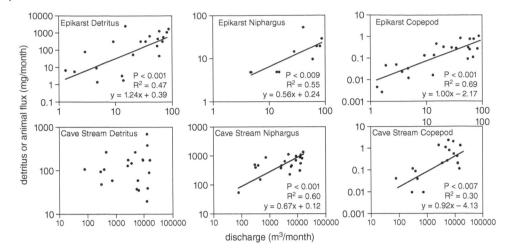

Figure 10.12 Relationship between detritus or animal flux and discharge over the two year study period in the epikarst seepage water and the Cormoran Cave stream (France). Lines and statistical values are results of linear regressions of log 10 transformed data. Source: From (Simon, 2013).

the relative magnitude of organic matter flux in dead and living forms (the amphipod *Niphargus* and Copepods) and to address factors that control it.

The monthly flux of detritus, *Niphargus* and copepods varied by at least two orders of magnitude over a year but there was no clear seasonal pattern, as might be expected. Variation in discharge explained 30–69% of the monthly variation of animals in both the epikarst and the cave stream and 47% of the variation in detritus flux from the epikarst. There were significant relationships between discharge and the flux of detritus, *Niphargus*, and copepods for the epikarst, but no significant relationship for detritus and discharge in the cave stream (Figure 10.12). Both *Niphargus* and copepods were significant for cave stream discharge. For detritus in the cave stream, peak monthly discharge provided a better predictor of organic matter transport. This may be related to the action of large flows that mobilise material along the stream bed from detritus traps in hollows.

The availability of instrumentation for in-stream monitoring of DOC and other water quality parameters will facilitate more intensive studies of carbon flux in karst basins and caves. The sampling of cave biota using automatic water samplers or rising stage samplers may also provide opportunities to quantify their contribution to living and dead detritus flux (Hartmann et al. 2018).

10.7 Cave Microbiology

Cave microbiology has been a very rapidly growing field, and the central role of microbes in nutrient cycling has now been realised. Most cave visitors have seen the small reflective white dots on moist rock. These are colonies of actinomycete bacteria (Northup and Lavoie 2004). These are but a very small component of the diverse microorganisms to be found in caves. Microbes from the groups Bacteria, Archaea, Fungi, and Protista are universally present in subterranean habitats. They may be common surface-dwelling bacteria or very exotic organisms with affinities to those found in deep-sea volcanic vents and lava tubes.

From the preceding section, caves are nutrient-limited environments. Given the lack of photosynthesis in caves, microorganisms are dependent on the heterotrophic decomposition of limited and episodic inflows of organic matter or are dependent on the release of nutrients through redox reactions facilitated by autotrophic microbes. Caves are also relatively static environments, buffered from change for thousands of years. In surface ecosystems, microbial community structure can be affected by changes in climate, vegetation, and human impact (Hershey and Barton 2018). The absence of these pressures in caves means microbial diversity is more likely to be dependent on the geologic context and especially the geochemical environment (Ortiz et al. 2013). Despite these limitations, tens of thousands of microbes have now been identified from caves, and they play a critical role in the breakdown of organic matter, in the production of organic matter (chemoautotrophy), and in the dissolution of rock and the deposition of minerals (Culver 2005; Hershey and Barton 2018).

Traditional microbiology derived from medical practice in which colonies of microbes are cultivated on Petri dishes using suitable nutrient media and controlled temperatures. Initial research on cave microbiology found relatively few cultivable microbes. It has now been estimated that over 99% of cave microbes are simply not able to be cultivated (Engel 2019). In part, this is due to the difficulty of matching the precise substrate and geochemical conditions under which they thrive. In the 1980s, the microbiologist and caver Norm Pace used the 16S small ribosomal subunit rRNA gene sequence as a genetic marker (phylotype) to distinguish previously uncultured species (Pace et al. 1986). This provided a revolutionary way of identifying microorganisms in the natural environment. These investigators used molecular phylogenetic approaches to examine the filamentous biofilms of a sulfidic stream within Sulfur River Cave, Kentucky, USA. One of the surprises from this study was the dominance of the *Epsilonproteobacteria*, which had previously only been seen in deep-sea volcanic vents. The study also demonstrated that microbial cave communities could be quite distinct from their surface counterparts (Angert et al. 1998). This development also led to the view that rather than supporting the idea that caves were dominated by a few specialised species adapted to nutrient limitation, caves appeared to be home to a diverse assemblage of species from multiple phyla, including the *Alpha-, Beta-, Gamma-, and Delta-proteobacteria, Chloroflexi, Planctomycetes, Bacteroidetes*, Acidobacteria, and Actinobacteria, with small but significant contributions from members of the *Nitrospirae, Gemmatimonadetes*, and *Verrucomicrobia* (Hershey and Barton 2018). However, the molecular approach was still limited to those microbes with sufficient biomass for gene sequencing. The next development was in optically based gene sequencing methods – collectively referred to as "next generation sequencing" (NGS) technologies. The dramatic increase in the number of bases that these technologies could sequence (>15 billion bases in as little as four hours) combined with their significant cost reductions has revolutionised the ability to sequence DNA from the natural environment.

The 13 phyla of microbes dominant in soils (Figure 10.13) may also be found in caves, and it is likely that soil microorganisms are introduced into the cave environment by both inflowing streams and from the epikarst. This view is supported by recent studies of speleothem dripwater (Hershey and Barton 2018).

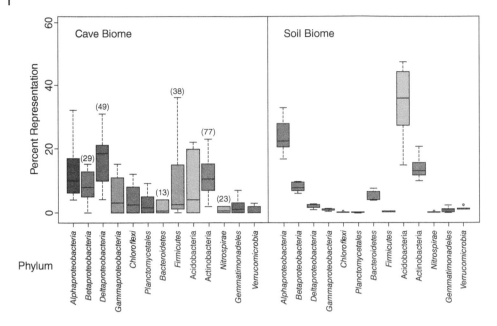

Figure 10.13 Boxplot comparison of soil microbial communities under alkaline conditions to those found in caves. The boundaries for the first and third quartile are shown, with the centreline representing the mean and whiskers representing the max/min values (outlier values for the cave biome data are shown in parentheses). Source: From Hershey and Barton (2018).

There may be selective pressures operating within caves that sort the introduced microbial species into cave populations. Ortiz et al. (2013) showed that less than 16% of the microorganisms found in caves share taxonomic identity with soil species.

Fungi are often present in caves, especially in the entrance zone. Fungi have been described from pristine cave environments, so they are natural components and not just introduced contaminants, while they may play an important role in mineral and speleothem chemistry. The growth of filamentous fungi across mineral surfaces creates microfabrics that support the growth of other microbial species. Together these diagenetic processes change the mineral matrix to form other deposits, such as calcite, goethite, halloysite, and montmorillonite, all of which have been detected in caves. Fungi may also play a role in breaking down organic matter, especially coarse woody debris. The effective ability of fungi to breakdown these recalcitrant carbon sources could subsequently provide a pool of nutrients for the growth of other microorganisms (Barton 2015).

There are a range of chemoautotrophic processes identified in predominantly reducing environments within caves. These include iron oxidation, sulphur oxidation, iron reduction, ammonia oxidation, methanogenesis, and methanotrophy. Engel (2005) and Northup and Lavoie (2001) provide details on the metabolic functions and phylogenetic position of the various subterranean chemoautotrophic microbes.

Early in Earth's history conditions were probably too hot for autotrophy and chemoautotrophy was probably dominant (Engel 2005). On Mars today it is probably too cold (−80 °C) for liquid water to be stable on the surface of the planet and so subterranean environments may offer potentially favourable conditions for life as we know it. Caves such as Peştera Movile in Romania may be a suitable model of the kind of environment where extra-terrestrial life might occur. The cave is only 240 m long with no natural entrances (entered originally by drilling for a power station); the smell of sulphur dioxide and elevated CO_2 levels make it a most unpleasant environment for humans. The cave air is enriched in CO_2 up to 3.5% and depleted in O_2 down to 7.0%. Despite this, it is one of the most diverse caves with 48 stygobionts and troglobionts (Culver and Sket 2000; Lascu 2004). Water surfaces in the cave have microbial mats with sulphur oxidising microbes.

10.8   Origin and Dispersal of Cave-Dwelling Animals

The isolation of species in caves may be the result of a variety of causes, including climatic change, sea-level change, and geomorphic processes. Whatever the cause, isolation promotes evolutionary changes that make cave faunas of special interest to ecologists and evolutionary biologists, and pose special problems for conservation biology. It is clear from recent taxonomic work that many families and genera of cave-adapted animals have a long evolutionary history, and have dispersed from centres of origin in the supercontinents Laurasia and Gondwana (Humphreys 1993). Thus, cave fauna in the Cape Range, Western Australia, show strong affinities with those in western India, while cave amphipods in North America and northern Europe are closely related. It is clear that many caves are of great age and transcend the Quaternary, the Tertiary, and possibly the Mesozoic. Thus, there has been a considerable time – 200 million years or so – for evolutionary adaptation of cave-dwelling animals to occur.

There are many examples of markedly disjunct distributions in cave animals. The aquatic amphipod *Stygobromus* – a small crustacean – is widely distributed in caves of North America, including some several hundred kilometres north of the limit of continental ice during the Last Glacial (Barr 1973; Holsinger 1981). Many of its habitats are widely separated by non-karstic terrain, the species attaining something like a continuous distribution only in the extensive karst of the Appalachians. Local extinction of populations of this species are now occurring because of water pollution, chiefly from increased nitrate levels owing to herbicide and fertiliser use. Those populations north of the ice limit survived in montane refugia, inhabiting caves in nunataks or sheltered south-facing valleys. The survival of an aquatic species implies that cave streams and groundwater remained unfrozen despite the proximity of ice several kilometres thick (Holsinger and Culver 1988).

There are two main theories of cave colonisation and troglobiont evolution: 1) the Pleistocene effect theory (Barr 1968; Peck 1981; Vandel 1965); p. 2) the adaptive shift theory (Howarth 1993). The Pleistocene effect theory has been the most widely accepted model for the evolution of terrestrial troglobionts until quite recently. During cold glacial climates, the cooler, wetter conditions south of the continental ice masses of Europe, Asia, and America favoured the spread of invertebrates inhabiting both temperate forest ecosystems and caves. With the amelioration of climate, those taxa that survived were those living in caves as the forest ecosystems changed radically. Ultimately geographic and genetic isolation in these cave refugia produced adaptive radiation and the evolution of distinct troglobionts. In favour of this theory are the close affinities between closed forest and cave taxa (many forest species becoming accidental cave visitors or trogloxenes), the present

Figure 10.14 The cave-adapted beetle Idacarabus troglodytes is endemic to the Ida Bay karst of southern Tasmania's World Heritage area, Australia. Source: Photo by Stefan Eberhard.

distributions of taxa in mountain areas separated by deep valleys, and former wider distributions evidenced from the fossil record. Thus, the evolution of the troglobitic beetles of the genus *Pseudanophthalmus,* today represented by some 240 species in the eastern United States, may be on account of this mechanism. The surface ancestors of this genus may have spread out from forest habitats in the Allegheny Plateau during cooler glacial periods, and have become a series of isolated populations in the Appalachians and western plateaux, such as the Cumberland Plateau of Tennes (Barr and Holsinger 1985). Similarly, the cave beetle fauna of southern Australia (Figure 10.14) evolved from a widespread ancestral stock, but the eastern and western populations became isolated by the formation of the arid Nullarbor Plain, an effective barrier to movement (Moore 1964). Cave beetles with affinities to both groups survive on the arid karst today. Overlapping distributions of two troglobitic species may suggest invasion during different glacial cycles, such as that provided by the cave harvestmen *Hickmanoxyomma cristatum* and *H. clarkei* in southern Tasmania, Australia (Kiernan and Eberhard 1990), which inhabit the deep and entrance zones of caves.

The second theory, that of adaptive shift, was advanced by Howarth (1983) to explain the origin of troglobionts in the Hawaiian lava tubes, USA, but may have much wider applicability. This theory does not rely on climate change; rather, it proposes that partially adapted ancestral species moved into cave niches almost continuously. These may have been species out-competed in surface environments. Thus, the availability of food is the keystone of this theory, and troglobiont evolution has been continual rather than episodic. Caves are buffered against climatic change, and individual taxa may survive for a very long time.

In the seasonally humid tropics, climate change may also have been important, the contrast being wet–dry cycles rather than warm–areool cycles. But the emerging picture of major Pleistocene changes in low latitudes (to which cave studies have made a major contribution) may have more implications for troglobiont biogeography than has been realised. There are a very large number of tropical troglobionts, and finding a theory which explains their presence is a major challenge. It may well be that no single theory of troglobiontic evolution will be valid in all circumstances.

For some trogloxene animals, such as forest cockroaches and centipedes, the cave is an acceptable habitat but only one of their potential living places. But many cave animals are more or less confined to caves, and often to particular cave microhabitats. Caves to them represent islands of biological opportunity in a sea of inhospitable habitats or barely hospitable habitats inhabited by inhospitable competitors or predators (Howarth 1983). Dispersal can, therefore, be a serious problem to some cave dwellers. In West Virginia's Greenbrier County, USA, seven blocks of limestone are set among other non-karstic rocks or are divided from each other by rivers – effective barriers to dispersal (Culver et al. 1973). Cave-dwelling animals can move within a block but not between blocks. The number of terrestrial arthropods is correlated with block size, but there is no such correlation for the aquatic arthropods. Presumably, they can move in groundwater and some surface waters. Thus, the terrestrial animals are vulnerable in the face of habitat change – either climatic or humanly induced – while the aquatic species have more of a chance of survival by dispersal or staying in small refugia.

The network of small fissures and enlarged joints is of great importance for cave-dwelling animals. They are relatively stable, protected habitats where small populations may survive local extinction caused by adverse conditions in a main cave conduit. Many of these fissures are unenterable by people, and thus the recolonisation of the major trunk passage by small populations comes as a pleasant surprise to biologists and cave managers. This has happened in Horse Cave, Kentucky, where the removal of pollution sources and some rehabilitation has permitted the resettlement of the main cave passage by fauna which survived in small tributary fissures (Lewis 1993).

Similarly, the zone of saturated coarse sediments in a cave stream may be an important refuge for stygobionts (Figures 10.15 and 10.16). In surface environments, the interstitial spaces in saturated coarse alluvium – the hyporheic zone of stream ecologists – is an important habitat, allowing taxa to survive the drying of stream beds and to escape surface impacts (Allan and Castillo 2007). The spelean equivalent has the advantage of near-constant temperature and usually buffered water-level change. It is, therefore,

Figure 10.15 A new species of aquatic Syncarid (Psammaspides sp.) from Wellington Caves, New South Wales, Australia. Such organisms are highly adapted and are vulnerable to changed water quality. Source: Photo by Peter Serov.

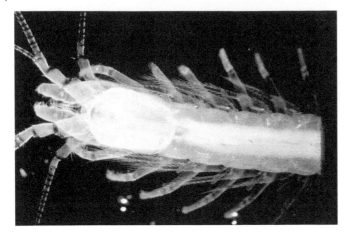

Figure 10.16 A closer view of the anterior of the aquatic Syncarid from Wellington Caves, New South Wales, Australia. The animal has strong mandibles, numerous sensory hairs and vestigial eyes. Source: Photo by Peter Serov.

important that the coarse open texture of many cave stream sediments be maintained, and excessive fine sedimentation be controlled. Where there are connections between cave conduits and gravel-bed surface streams, the hyporheic zone may allow stygobionts to move between caves. Similarly, they may move between caves in deep groundwater as long as oxygenation of the water is adequate.

In common with occurrences in surface environments, cave-adapted fauna may be moved between sites accidentally. It is very easy to move microbiota (bacteria and fungi especially) between caves on boots, rope, and clothing. Washing these items between caves is, therefore, a wise precaution. Stygobionts may be flushed from cave passages following drainage diversion by cavers or pulse wave release for dye tracing. Bats and birds may act as agents of dispersal for ecto- and endo-parasites; this is especially so in tropical karst regions. Domestic pets, such as dogs, may also pick up cave entrance invertebrates and move them around.

10.9 Threats to Cave Fauna

There are many obvious threats to cave-dwelling animals:

- Caves are used for legal or illegal dumping of refuse. Although relatively few animals can survive in refuse or garbage, cave faunas are especially sensitive. Increased resource levels may allow non-cave dwelling species to thrive and outcompete the obligate cave species.
- Inadequate sewage treatment may pollute groundwater and sediments, leading to local extinction and/or replacement of cave species by surface species.
- Cave entrances are vulnerable to closure on account of landfills, housing developments, or road construction.
- Deforestation or land-use change may affect water flow and sedimentation rates.
- Pesticides and herbicides in an agricultural area may impact both cave vertebrate and invertebrate populations.

- Mining and quarrying remove cave habitats.
- Human visitors to caves impact their fauna in many ways.

Many cave species are described from one or two specimens, and extensive searching often reveals only a few individuals. The largest recorded population of large troglobionts is 9090 *Orconectes* sp. crayfish in Pless Cave, Indiana, USA (Culver 1986). For most caves, low numbers of endemic species are very vulnerable to pollution by sewage (Iliffe et al. 1984). Serious problems may also arise from pollution by heavy metals and hydrocarbons (Holsinger 1966). The increased bacterial activity in sewage or contaminated water results in loss or great reduction in levels of dissolved oxygen. The animals literally smother in their polluted habitat. Toxins in sewage-contaminated water (endotoxins, heavy metals, phosphates, and nitrates) may directly poison cave invertebrates, especially those higher in the food chain. Hamilton-Smith (1970) discussed the extinction of cave invertebrate faunas by trampling, water pollution, and changes in cave microclimates. Pesticides and herbicides may have serious effects on cave invertebrates. There are very few studies of nitrate (or other chemical) pollution on caves and karst in Australia. In the porous karst aquifer of south-eastern South Australia, Waterhouse (1984) reported background groundwater nitrate-nitrogen levels of 2 ppm (parts per million) and up to 300 ppm at sites grossly contaminated by dairy and piggery wastes. There are also anecdotal accounts of sewage pollution in Tasmania (Kiernan 1988; pp. 8–9); no measured nitrate or phosphate values are given for these cases. Similar levels to the South Australian study were recorded by Quinlan and Ewers (1985) for Horse Cave, Kentucky, where sewage, creamery waste, and heavy metal effluent had been detected at 56 springs.

Wheeler et al. (1989) investigated the impacts of fertiliser and herbicide use on the degraded water quality of the 16 km long Coldwater Cave, Iowa, USA. Average nitrogenous fertiliser application rates on farms nearby were 113–170 kg ha^{-1}; atrazine herbicide was being applied at 2.2–3.3 kg ha^{-1}. Nitrate-nitrogen concentrations in groundwater averaged about 5.8 ppm in the cave stream from 1985 to 1987, with a range of 7.4–3.5 ppm This can be contrasted with a recorded level of 4.1 ppm from 1973 to 1974. Background levels in farm water supply wells were less than 0.1 ppm Thus, the higher recorded values were approaching the Environmental Protection Agency (EPA) maximum standard for drinking water, which is 10 ppm as nitrate-nitrogen (Field 1988). Atrazine was detected in 78% of cave sites, and concentrations ranged from 99 to 1980 ppt (parts per trillion). Thus, most of the observed nitrate in cave streams was derived from fertiliser. At these levels, observable water-quality decline had occurred (algal blooms, odours).

Vandike (1982) reported on the effects of a fertiliser (ammonium nitrate–urea) spill on karst groundwater in Missouri, USA. Water-quality degradation began at the spring within eight days of the spill: dissolved oxygen levels decreased from the normal 7–8 mg L^{-1} to 0.2 mg.L^{-1}. Nitrate plus nitrite nitrogen peaked at 4.2 ppm from a background level of 0.05 ppm Total spill volume was estimated at 24100 gal (109 400 L) and the effects on water quality were observable for five months after the accident.

The effects of urban runoff on cave invertebrates were investigated by Pride et al. (1989) for three cave spring sites in Tennessee, USA. The authors measured a range of water-quality parameters and surveyed benthic macroinvertebrate populations. All sites had some history of pollution. The direct storm fed, hydrologically open caves had lower biodiversity indices and significantly higher levels of pollutants. The increased presence

Table 10.3 Water quality parameters and biodiversity indices for three polluted caves in Tennessee, USA.

Site	Dissolved oxygen (mg L^{-1})	Nitrate-N (ppm)	Orthophosphate (ppm)	Shannon biodiversity index (H′)
Canal Cave	7.9	0.7	39.4	0.32
Ament Cave	7.5	0.8	55.5	0.25
City Cave	8.9	3.1	56.6	0.49

Source: From Pride et al. (1989).

of Dipterans and Oligochaetes in both cave streams suggests organic pollution (in this case, from sewage). The cave fed by diffuse recharge had higher biodiversity and better water quality on account of the filtering effect of the karst system; it still contained isopod populations which are regarded as sensitive to pollution. Table 10.3 gives some relevant data from this study. Lowering of dissolved oxygen levels by bacterial uptake will disadvantage aquatic macroinvertebrates, and it is clear that the nutrient concentrations cited are maximal values beyond which the major decline of the aquatic fauna would occur.

Increased cave sedimentation may also adversely impact fauna. This may occur in two ways. First, loss of habitat for organisms living or seeking refuge in coarse-textured sediments (sands and gravels) may occur if fine-textured sediments, such as silts and clays, are deposited. Secondly, fine sediments with significant organic matter content may contain bacterial colonies, which will also deplete dissolved oxygen, leading to the smothering of cave life.

During their long evolutionary history, most troglobionts have reduced their capacity to withstand environmental perturbations. Many troglobionts cannot withstand desiccation or temperature fluctuations. Cave faunas are thus at especial risk from environmental change.

The general sequence of use of caves by people may be summarised as local use, discovery by speleologists, increased recreational caving, scientific study, physical and biological resource depletion and, finally, protection of what remains. Cave conservation plans must, therefore, take account of the impact of visitors on subterranean faunas. The biological impact of repeated cave visits may be far more severe than the physical effects of sediment compaction, flowstone muddying, and erosion. Many cave animals have perished under the boots of cavers without the latter being even aware of their existence. The compaction of sediments by repeated visits to a cave removes potential habitat and may result in increased water flows and rill erosion in areas downslope of flowstones and drips. Hamilton-Smith (1970) recorded an important invertebrate site in the unconsolidated floor sediments of a lava tube in western Victoria, Australia; this potentially important study site was wrecked because the landowner encouraged (for a fee) unmanaged wild caving, which reduced the floor to a slick, compacted surface with no fauna at all. Around entrances, the increased foot traffic may accelerate the inwashing of fine sediments, as well as reducing the carrying capacity of an energy-rich zone of importance to many obligatory or accidental cave dwellers. These impacts are probably more severe for low-energy caves – dry systems with infrequent energy inputs – than for high-energy systems subject to repeated flooding and detritus transfer (Parsons 1990).

The introduction of extra energy sources is one way in which cavers may drastically or subtly alter the cave ecosystem. It is instructive to spread out a plastic sheet and consume a caving snack while seated on it. Around your body will be spread a ring of small portions of the food – chocolate or Mars bar fragments, biscuit crumbs, soup drops – which represent a major local increase in available energy. These items become potential sites for the colonisation of fungi and bacteria or food for cave animals. Even in cold, wet caves, such food fragments will be colonised by a fine fur of fungi in a week or so. The best solution is either to avoid eating in caves or to sit on a plastic sheet which can be carefully brushed afterwards into a bag and carried out. Similarly, human body wastes are potential energy and pollution sources which should either not be deposited inside the cave or in extremis deposited in secure containers for removal and disposal off the karst (Poulson 1977; Tercafs 1992). The sudden appearance of *Mondmilch* (moonmilk) in some caves may be related to the introduction of bacteria foreign to the cave by speleologists (Derek Ford, pers. comm.). In both wild and tourist caves, there is the potential for these bacteria to be introduced on the fine rain of lint and skin cells normally shed by people every day. Such bacteria may also serve as extra energy sources for cave biota unaccustomed to this rich bounty.

Spent carbide is a potent agent for the demise of cave fauna, especially in low-energy systems where the reaction products may remain for long periods. Calcium carbide reacts to produce acetylene gas and lime and may contain considerable impurities such as sulphides and some metals. These substances may leach into cave streams or groundwater to pollute them. The unsightly carbide dumps on ledges or in alcoves may remain for periods of several years, and particles may adhere to the appendages of cave animals who walk over them. Although some spent carbide may be flushed by flooding, some always remain in the cave passage, forming unsightly crusts. Fortunately, the widespread use of effective and cheap light-emitting diode (LED) cavers' lights has rendered carbide lighting mostly obsolete.

The widespread practice of opening entrances by digging or using explosives causes meteorological changes, which may result in the progressive desiccation of rock and sediment surfaces, as well as a reduction in relative humidity. The fumes from explosives also constitute a potential toxin for cave animals. Artificially opened or enlarged entrances should be sealed effectively between visits to minimise changes to air flows and potential desiccation.

The social effects of visitors on bat colonies, especially maternity sites, are a major concern for cave biology (Figure 10.17). There is no question that disturbances as trivial as briefly entering a maternity area with lights can result in decreased survival of young bats and possible abandonment of the site (McCracken 1989). Visits to maternity sites early in the breeding season can result in pregnant females moving to other, less favourable sites. The general level of activity by bats is also raised, and this results in a greater expenditure of energy and use of fat reserves. This may cause adult mortality in extreme cases and certainly reduces the growth of the young bats. Young bats may also become dislodged from the walls and ceiling, falling to the cave floor, where they may perish or be consumed. Clustering of bats in maternity sites has thermoregulatory advantages for the colony. If the size of the colony decreases, by death or by abandonment, then a critical threshold of energy conservation may be reached below which the young cannot be raised successfully.

Figure 10.17 The evening flight of hundreds of thousands of insectivorous wrinkle-lipped bats (*Tadarida plicata*) from Deer Cave, Gunung Mulu NP, Sarawak, Malaysia.

Hibernating bats are very vulnerable as well. The energy reserves that bats accumulate prior to hibernation are often close to the level necessary for over-winter survival. Disturbance during hibernation can cause bats to arouse prematurely, expending their energy reserves in flight and raising their body temperatures. The bats may go back into torpor after disturbance, but may not have sufficient energy to survive the winter. This may not be apparent to the casual visitor, but a return in spring will often reveal a carpet of dead bats or corpses still hanging on the cave walls.

Human activities outside the cave may also have a negative effect on bat colonies. Vegetation modification by clearing, from controlled burns or from stock grazing, may reduce food sources for bats. There may be other implications for biological conservation in karst regions. In deserts and in tropical areas many plants are pollinated by bats. Their decline or local extinction may affect plant regeneration, especially for cacti. Insectivorous bats are very vulnerable to pesticide poisoning (Clark 1981, 1986), although the reduction of their resource base (insects) and visitor disturbance may be of greater significance. There are several cases of bat colonies in caves of Alabama and Tennessee, USA, being badly polluted with toxic chemicals (Dichlorodiphenyltrichloroethane DDT and Polychlorinated Biphenyls PCBs), and making population recoveries once the caves were protected (Tuttle 1986).

The ecological effects of tracks and lighting in tourist caves have been well documented, are reasonably well understood and are discussed further in Chapter 13. Developed caves add a new dimension to impacts on cave biology because of the greater physical modification of the cave environment and the addition of extra energy in the forms of light, heat, and food.

10.10   Conservation of Biological Diversity in Caves

How many species of troglobionts are there? We can provide only a crude estimate based on extrapolation from finite, well-studied karst areas to continents and then the globe. With the possible exception of birds and mammals, there is no group of organisms on earth for which the total number of species is precisely known. So we are forced to make estimates based on known species diversity from air-filled caves, which may be but a small part of the subterranean habitat. For Virginia, USA, Culver and Holsinger (1992) have listed 160 known species of terrestrial troglobionts collected from over 500 caves, a good sample. This covers 15 taxonomic groups, not all of which are described; for amphipods alone, only 29 of the estimated 43 taxa are formally described. Including aquatic troglobionts, this total number comes to 549 for Virginia, and perhaps 1000 for the Appalachian karsts. Excluding the diverse fauna of the Hawaiian lava tubes, the US estimate is 6000 species, based on area-species curves. The global estimate, based on proportional area, is about 60 000 species, but the estimates may vary from 50 000 to 100 000. Many of these are known from one cave site only, and many occur in caves subject to pollution. Cave faunas are a fragile resource and the special conservation problems of caves become critical for their biota.

There is, thus, a great need for surveys of cave faunas at local and regional levels. In several regions of North America and Europe, the tireless efforts of individuals have resulted in a good estimate of cave biodiversity and, for certain taxonomic groups, some knowledge of their ecology. But, for large regions of the world, cave faunal survey has been restricted to short visits by scientists. These visits have served only to demonstrate the great potential diversity of cave life, especially in the humid tropics. The studies by Howarth and his co-workers (Howarth 1987, 1988) have demonstrated the great diversity of lava tube faunas in Hawaii and Australia, and similar species richness is being recorded for tropical limestone caves (Humphreys and Adams 1991).

Given the key role of periodic inputs of food for cave dwellers during floods, there is also a need for seasonal surveys of individual caves and their region. Weinstein and Slaney (1995) have demonstrated that trapping techniques may cause significant variation in species recovery. In particular, the use of wet litter traps, simulating the detritus carried in by floods, may greatly increase the yield of individual sites. Thus, our current knowledge of cave organisms is probably only a fraction of the species present. Unfortunately, habitat modification may cause their demise before they are sampled, let alone formally described and conserved. Despite the energetic efforts of a growing number of cave biologists, definitive studies of cave ecology at the community level are relatively few. Studies of human impact on cave ecology are limited and should be a high research priority.

The most widespread classification used for assessing the vulnerability of species is the IUCN classification employed in the Red Data Book (IUCN Conservation Monitoring Centre 1986). There is a hierarchy of categories, as follows:

- Extinct
- Extinct in the wild
- Critically endangered
- Endangered
- Vulnerable
- Conservation dependent
- Low risk
- Data deficient
- Not evaluated

Extinct species are those for which there is no reasonable doubt of their extinction, while *extinct in the wild* species are those which survive only in cultivation, in captivity, or as a naturalised population well outside their natural range. A taxon is *critically endangered* when it is facing an extremely high risk of extinction in the wild in the immediate future, as evidenced by severe population decline over the last decade, or when its extent of occurrence is less than $100\,km^2$. An *endangered* species is not critically endangered, but is facing a very high risk of extinction in the wild in the near future, evidenced by severe population decline of 50% over the last decade, or when its extent of occurrence is less than $5000\,km^2$. A *vulnerable* taxon faces a high risk of extinction in the wild in the medium future, evidenced by a severe population decline of 50% over the last 20 years, or when its extent of occurrence is less than $20\,000\,km^2$ or its population is fewer than 1000 individuals. *Conservation dependent* taxa must be the focus of a specific conservation programme, the cessation of which would result in its being reclassified into one of the three higher categories. *Low-risk* species are those that are close to qualifying for the above, are abundant or are of less concern. *Data deficient* taxa are those for which there are inadequate data to make a meaningful evaluation but may be listed as threatened when more data become available.

The IUCN Red Data Books provide a useful perspective on the threats to cave and karst faunas, but the listings therein are only a small proportion of cave fauna at risk. They do, however, provide a good idea of the range of problems facing cave dwellers. The IUCN global review of living resources (World Conservation Monitoring Centre [WCMC] 1992), does not list caves as a habitat type but has a section devoted to subterranean fishes (WCMC 1992; pp. 121–2, 131–5) which indicates at least 47 species of fish that are either cave-adapted or have cave-adapted populations and provides useful data on species, location, and adaptations. Population sizes are generally unknown and distributions are localised. It notes that the waters in which these species live or have evolved are the final sumps for water-soluble chemicals used on land, and thus their habitats may be under threat. There is not a comparable review of cave invertebrates, but a planned volume on worldwide cave biology (Hamilton-Smith, pers. comm.) may go some distance to achieving this. Accordingly, the Cave Invertebrate Species Specialist Group of IUCN is now compiling information on threatened cave communities, with the eventual aim of a Red Data Book on caves and their fauna.

The IUCN classification is a useful starting point for consideration of the vulnerability of cave fauna, but by its nature, it is species-oriented rather than designed to preserve the whole community and its habitat. In addition, it was designed to cope with species that are really extensive (such as large herbivores) and whose population dynamics are either well known or amenable to study. The present Red Data Book demonstrates only the minute tip of the iceberg that is the endangered cave fauna of the globe. Despite numerous legal efforts and legislation aimed to give sanctuary to particular cave-dwelling species by designating them as protected fauna, in many cases, there is little or no parallel effort to adequately safeguard their habitat or to ensure that ecological studies of species requirements are made to provide a sound basis for conservation. This situation is slowly changing, but in the time taken to enact new legislation, many species will become at least locally extinct.

Some major problems with existing legislation are:

- That token gestures such as listing particular species as rare or endangered are futile unless their habitat is also protected and studies of their ecology are conducted.
- That scientific collecting is likely to have a minimal effect on their populations compared with ongoing habitat modification or destruction.
- That species may not have been selected for legislative protection by any obvious rational scientific procedure related to their scarcity or their biological peculiarity.
- That the legislation may at times be hard to enforce because of the difficulty of recognising certain species by non-specialist personnel.

10.11 Caves and Ecosystem Services

Karst terrains contain many natural resources and provide valuable ecosystem services, such as fresh water for human consumption; aquatic ecosystems and agricultural irrigation; great biodiversity both at the land surface and in the underground environment; landscapes and caves with high recreational and cultural value; and soils that provide the basis for agricultural production. Karst terrains also act as natural sinks for carbon dioxide (CO_2), thus helping to mitigate climate change. All these resources and ecosystem services cannot be considered in an isolated way but are intensely interconnected. Because of these complex feedback mechanisms, impacts on isolated elements of the karst ecosystem can have unexpected impacts on other elements or even on the entire ecosystem.

Freshwater from karst is probably the most significant as well as the safest source of drinking water. Ford and Williams (2007) have estimated that karst aquifers supply drinking water for about 25% of the world's population while the karst itself occupies about 15% of the ice-free land area. In Austria and Slovenia, about 50% of the water supply comes from the karst (Ravbar and Goldscheider 2009). In fact, the city of Vienna, with its two million inhabitants, is entirely supplied by karst water. In Italy, many areas and cities are also supplied by karst waters, including Rome with its 2.8 million inhabitants, which has been predominantly been supplied by water from several large karst springs since Roman times (Kresic and Stevanovic 2009). In the USA, the Edwards Aquifer of Texas is a karst groundwater resource supplying millions of people, including big cities, such as San Antonio (Wong et al. 2012). Finally, China is a country where more than 100 million people rely on karst water resources (Lu 2007).

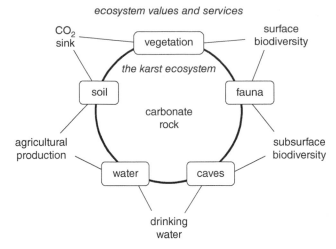

ecosystem values and services

Figure 10.18 Generalised presentation of an undisturbed karst ecosystem and its natural resources that represent a variety of values and provide ecosystem services. Source: From Goldscheider (2012) © 2012 Springer Nature.

Where karst aquifers deliver drinking water of excellent quality, a number of factors combine to maintain this. Favourable hydrogeological settings involve thick overlying layers, absence of sinking streams and dolines as entry points for contaminants, a thick unsaturated or epiphreatic zone, or deep regional flow paths. Healthy karst ecosystems with undisturbed soils and vegetation also provide natural water filtration and purification (Figure 10.18). In turn, clean groundwater emerging from karst springs provides the basis for healthy aquatic ecosystems (Bonacci et al. 2009).

Soils in karst areas provide a valuable resource for agriculture and forestry, as well as being the main source of carbon dioxide for the karst hydrological cycle. Soil thickness is generally more variable in karst terrains due to the density of fissures and the propensity of karst soils to enter unstable, positive feedback loops in which soil is lost down those fissures (Gillieson et al. 1986). Over-use of karst soil resources depleting soil nutrients can also lead to loss of soil aggregate stability and subsequent soil erosion (Figure 10.19). Mechanical action by cattle, agricultural machines or other activities can easily damage the topsoil and leave nothing but naked limestone. Similarly, removal or degradation of the vegetation can cause rapid soil erosion by intense rainfall.

The mineral content of soils on karst generally derives from the insoluble residue, which for purer limestone ranges from 1% to 10%. Thus, soils tend to be thin and accumulate slowly. Exceptions are found where there has been an accession of aeolian dust (loess) in China or volcanic ash, as in southern Italy and Papua New Guinea. Rough estimates suggest that limestone dissolution generates 0.1–10 mm of residual minerals in 1000 years. Thus, soil erosion on karst is irreversible on a human timescale, although organic humus soil matter can form much more rapidly in cooler climates. The cost of remediation of karst soils is very high as measures to prevent the abstraction of soil down fissures involve engineering works and restoring the hydrology (Gillieson and Houshold 2000).

General subterranean ecosystems possess a high number of rare and endemic species (Achurra and Rodriguez 2008) because of their high degree of isolation. TM may be

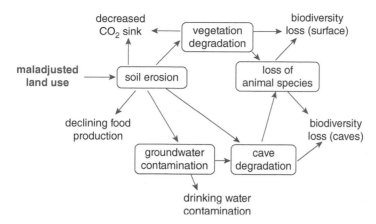

Figure 10.19 Effects of poor land-use practices on ecosystem services in karst terrains. Source: From Goldscheider (2012) © 2012 Springer Nature.

confined to one cave or one karst region. New species are regularly discovered, but their taxonomy may take many years to be finalised, during which time they may become locally extinct. The destruction or contamination of karst habitats is very likely to lead to the extinction of unknown species. The epikarst also provides habitats for specifically adapted species assemblages (Pipan et al. 2008).

Karst landscapes are often more diverse, with greater internal topographic variation. They, thus, offer a greater variety of potential habitats than non-karst landscapes and are often relatively isolated from their surroundings, for example in tower karst landscapes. They often host great biodiversity of animal and plant species, including rare and endemic species (Clements et al. 2006). In Vietnam Delacour's langur (*Trachypithecus delacouri*), an endangered primate species, is endemic to some Vietnamese karst areas (Tuyet 2001; Workman 2010). In the extensive karst on the border between Vietnam and Laos, large blocks of limestone terrain are separated by rivers, which provide effective barriers to species dispersal. There are at least six species of Leaf-eating langurs (*Trachypithecus* spp), each endemic to a specific block of limestone (Nadler et al. 2007). Similarly, in Guanxi Province, China habitat fragmentation separates populations of the White-headed langur (*Trachypithecus leucocephalus*) (Huang et al. 2008). Whitten (2009) has provided a useful summary of the issues facing the conservation of biological resources in Chinese caves, while Clements et al. (2006) has provided a review of the biodiversity values of caves and karst in Southeast Asia.

Karst areas are potentially effective sinks for carbon dioxide, due to the gas being central to the geochemical cycle. In bare karst, such as in alpine or arctic areas, CO_2 only comes from the atmosphere. CO_2 partial pressures in the atmosphere have been steadily increasing from 316 ppm in 1959 to 409 ppm in August 2019 (https://www.esrl.noaa.gov/gmd/ccgg/trends/monthly.html). For a given CO_2 partial pressure in the air, the equilibrium concentration in water depends inversely on temperature.

With well-vegetated karst with healthy soil, most of the CO_2 comes from the microbial and fungal decomposition of organic matter in the soil. The concentration of CO_2 depends on many factors, such as soil structure, type and content of organic matter, pH, moisture content, and temperature. Soil CO_2 partial pressures range between atmospheric levels (0.041%) and 10% with 0.5–5% being a normal range. Some of this CO_2 dissolves in soil water and moves underground through the epikarst, where it reacts with carbonate bedrock to form dissolved calcium (Ca^{++}) cations and bicarbonate (HCO_3^-) anions in the groundwater. As a result, karst systems covered with soil and vegetation are more efficient as CO_2 sinks than bare carbonate rock outcrops, for three main reasons:

1) photosynthetic CO_2 uptake by the vegetation,
2) carbon storage in organic-rich soils, and
3) increased microbial CO_2 production in the soil and subsequent neutralisation by carbonate rock dissolution.

Liu et al. (2010) estimated that karst processes account for 10% of the total anthropogenic CO_2 emission or 29% of the "missing CO_2 sink." Recent studies suggest that the role of carbonate rock weathering as a CO_2 sink had previously been underestimated by a factor of three (Liu et al. 2011).

Karst and caves have very high scenic and recreational values. There at least 58 cave and karst sites inscribed on the World Heritage List for their outstanding universal values, and these attract millions of visitors each year. Aspects of cave tourism and management are explored further in Chapters 12 and 13.

Thus, karst systems contain significant natural resources, host very high biodiversity, and deliver valuable ecosystem services, largely unrecognised. These resources and services are very vulnerable to human impacts. Impacts on relatively isolated elements of the whole karst ecosystem can have unexpected impacts on other elements of the karst ecosystem (Goldscheider 2019). For example, groundwater contamination can lead to the extinction of endemic and yet undiscovered species in the karst aquifer and, thus, to a loss of biodiversity. Soil erosion can also cause groundwater contamination and decrease the effectiveness of the karst system to act as a natural sink for carbon dioxide. Therefore, a holistic approach is needed to understand and protect karst groundwater, biodiversity, natural resources, and ecosystem services in karst terrains.

10.12 White Nose Syndrome

White-nose syndrome is a very infectious fungal disease that has killed millions of cave-dwelling bats in North America since its first appearance in 2006 (Figure 10.20). It is caused by the fungus *Pseudogymnoascus destructans*, which has been identified on bats in both Europe and China without causing population declines. The fungus thrives at temperatures below 15 °C while growth ceases above 20 °C. Preferring high humidity (Warnecke et al. 2012), it grows on and adversely impacts hibernating cave-dwelling bats while they are in torpor. Visible symptoms include fuzzy white patches on the bat's nose and white patches on the body and wings.

White-nose syndrome may lead to:

- Increased consumption of fat/energy reserves
- Increased frequency and duration of arousals

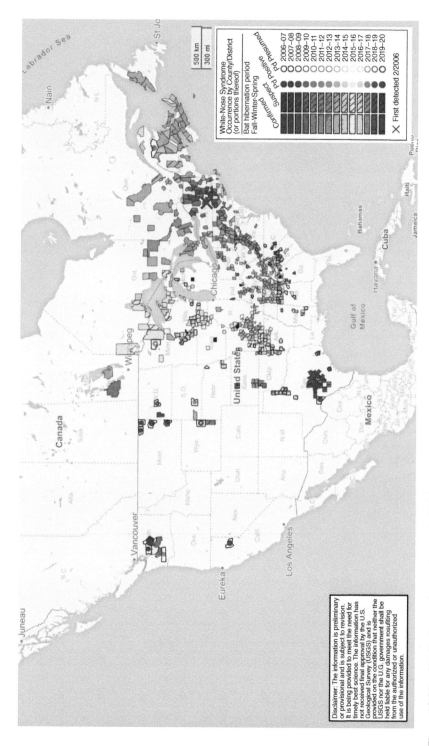

Figure 10.20 Distribution and annual spread of White-nose syndrome in North America and Canada. Source: White-nose Syndrome Response Team 2019.

- Damaged wing membrane increasing evaporation and water loss
- Altered behaviour, including spending more time grooming to remove the fungus, and flying around looking for water and food when it is not available
- Increased CO_2 levels in blood, leading to respiratory acidosis and hyperkalaemia (high potassium in blood), which can also cause heart failure
- Exhaustion of energy reserves and subsequent death (Verant et al. 2014).

The first evidence of White-nose syndrome in northern America (Frick et al. 2016) was provided in 2006 by a caver photographing a pile of dead bats in Howes Cave, New York State, USA, a popular tourist attraction with hundreds of thousands of visitors a year (Frick et al. 2016). The fungus is believed to have been introduced by someone with infected caving gear from Europe into North America. Humans can spread the fungus from one hibernation cave to another by accidentally carrying the fungus on shoes, clothing, or caving gear (National Parks Service 2019). Tourists visiting show caves may also spread the disease widely within and outside a country. Procedures such as boot/shoe decontamination stations have now been established at show caves, such as Mammoth Cave in Kentucky.

The fate of the Little Brown Bat (*Myotis lucifugus*) is instructive. It was once a very common bat species in North America, numbering in the millions. Prior to White-nose syndrome, it would have been considered a very resilient species, using a wide variety of habitat types and overwintering in very cold humid caves rejected by most other bat species. The population has now been greatly reduced by White-nose syndrome with millions of bats dead, and is now regarded as threatened in many areas.

Little Brown Bats have a number of characteristics and behaviours that make them particularly susceptible to White-nose syndrome, with fatal consequences:

- Very small bat (9 g prior to torpor)
- Very long winter torpor (Sept to April or May, depending on gender)
- Long torpor periods (average 16 days)
- Tight clustering roosting behaviour
- Summer breeding swarming
- Hibernates in very cold, humid caves

These characteristics can be used as a benchmark for susceptibility and are a useful tool when looking at the potential vulnerability and susceptibility of other bat species (Langwig et al. 2015).

The WNS Decontamination Team (2019) recommends the following procedures to decontaminate caving gear and equipment:

For items submersible in water:

- Thoroughly clean caving gear by removing all dirt and grime
- Immerse in hot water maintaining a temperature of greater than 55 °C for a minimum of 20 minutes

For items not submersible in water:

- Disinfect using 6% hydrogen peroxide spray or isopropanol disinfectant wipes.
- Boots should be scrubbed to remove all mud and dirt, then sterilised as above.

- Any gear that has been taken into potentially infected caves, and cannot be treated using the appropriate decontamination procedures, should NOT be taken to other cave areas or caves in other countries.

White-nose syndrome is currently believed to be absent from Australia. Seven cave-dwelling bat species in southern Australia are at risk from contracting the disease with potentially devastating consequences. The Southern Bent-wing Bat (*Miniopterus orianae bassanii*) was assessed as having a high risk and the Eastern Bent-wing Bat (*Miniopterus orianae oceanensis*) was assessed as having a medium risk, while the following species are regarded to be at low risk: Eastern Horseshoe Bat (*Rhinolophus megaphyllus*), Chocolate Wattled Bat (*Chalinolobus morio*), Large-eared Pied Bat (*Chalinolobus dwyeri*), Large-footed Myotis (*Myotis macropus*), and Finlayson's Cave Bat (*Vespadelus finlaysoni*) (Holz et al. 2016).

10.13   Unravelling the Secrets of the Carrai Bat Cave

Caves provide homes for a variety of organisms and afford challenges to the ecologists who study them. As we have seen, cave biota are rarely numerous, especially at higher trophic levels, and the cave environment is not an easy one in which to study biology.

Bats roosting in large numbers provide a rich bounty for cave invertebrates. Large supplies of guano build up only in caves which satisfy the microclimatic or social requirements of the bat colony and where there is a low chance of flooding or of active cave streams. Caves or specifically cave chambers which satisfy these requirements are seasonally occupied by many thousands of bats and may be used for over-wintering, mating, or maternity sites. Such sites may be used for many years, and the guano heaps in these caves support resident communities of invertebrates of ecological complexity.

One such cave is located in the dense rainforests of north-eastern New South Wales, Australia. High on a ridge overlooking the deep valley of Stockyard Creek, the Carrai Bat Cave has been the object of study by vertebrate and invertebrate zoologists (Harris 1970) interested in the dynamics of cave communities. The main outer chamber of the cave is colonised by large numbers of cave crickets, which feed on the leaf litter and humus washed into the cave by slopewash. An inner, enclosed chamber with a domed ceiling is home to between one and three thousand bent-winged bats *(Miniopterus schreibersii),* which roost there from late January to June, and again from October to early December, each year (Figure 10.21).

These regular visits have been repeated with little variation for at least 20 years. The guano heap under their roost is about 150 cm high and 200–300 cm across (Figure 10.22). It is inhabited by bacteria, fungi, protozoans, nematodes, mites, flies, moths, and spiders. The bacteria and fungi occur in largest numbers in the surface layers, as do the protozoans and nematodes. Guano mites occur in very large numbers: in particular, the small mite *Uroobovella coprophila* (Figure 10.23) has an astounding density of 33 700 000 per square metre! Guano flies (*Cypelosoma australis)* and spiders are found close to the top of the heap.

When bats are absent from the cave, there is no food source for the guano community. Either the organisms on the heap become inactive, neither growing nor reproducing (as in

Figure 10.21 Bent-winged bats (*Miniopterus schreibersii*) in flight in a cave in northern New South Wales, Australia.

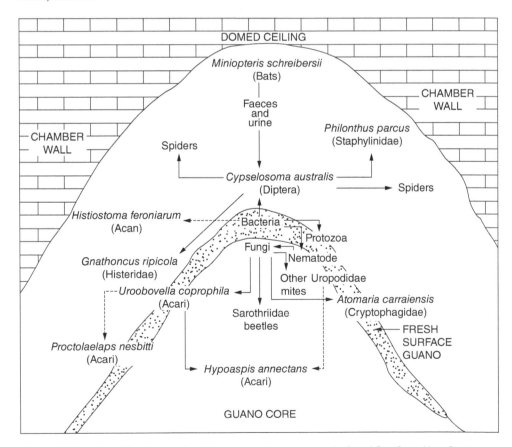

Figure 10.22 Simplified food web of the guano pile ecosystem in Carrai Bat Cave, New South Wales, Australia. Source: From Harris (1970).

Figure 10.23 In Carrai Bat Cave, Kemspey, New South Wales, Australia, there is a high density (33 700 000 animals.m^{-2}) of the guanophile mite Urobovella sp. Tineid moths also occur in this nutrient-rich environment. Source: Photo by Stefan Eberhard.

the case of the guano mite), or their population declines markedly (as in the case of the guano flies). Resting histerid beetle larvae occur 30 cm into the heap. Only the staphylinid beetles disappear; presumably, they migrate to somewhere else in the cave. The organisms appear to be unable to feed on the core of red, decomposing guano, which is mostly the chitinous body parts of moths and beetles.

When the bats arrive in October, fresh dung and urine reinvigorate the community. Microorganisms colonise the guano and their numbers grow rapidly. The guano mites feed, grow, and reproduce speedily. Adult flies lay their eggs in the guano, and the mature larvae of the histerid beetles pupate and adults emerge at the surface. The activity of the guano pile is intense for up to two months, and the temperature of the heap increases markedly from 14 °C (the cave average annual temperature) to 24 °C. In early December, the majority of the female bats vacate the cave to give birth at a cave six kilometres away. However, there seems to be enough fresh guano to carry the system until the bats return in late January. From then until June, the numbers of organisms appear stable as a state of equilibrium is attained. This series of changes appear to be a regular annual cycle.

Why do the guano-dwelling organisms not migrate elsewhere in the cave when living gets hard? The central guano core provides not only food, but also a large surface area for colonisation and a depth of deposit into which vulnerable larvae can burrow and hide. So those soft-bodied species can be in position for the next input of food in a relatively constant and safe environment. The presence of the core seems necessary for the maintenance of the very large numbers present.

Thus, the bat guano community is made up of a small number of species that have evolved to exploit a very specific niche, where the environment is characterise by a regular input of food – the guano – and complex physical conditions where living space is available. Caves are a further demonstration of the ever-changing nature of ecosystems, forming, developing for a while, then collapsing in response to changing energy inputs. Threatening processes that alter these flows of energy include forest clearance in the cave catchment, altered slope hydrology washing too much sediment into the cave, and disturbance to the bat colonies. Fortunately, Carrai Bat Cave is now in the Werrikimbee National Park, and its catchment is largely free from the logging and quarrying that affects other cave sites in the region.

References

Achurra, A. and Rodriguez, P. (2008). Biodiversity of groundwater oligochaetes from a karst unit in northern Iberian Peninsula: ranking subterranean sites for conservation management. *Hydrobiologia* 605: 159–171.

Allan, J.D. and Castillo, M.M. (2007). *Stream Ecology: Structure and Function of Running Waters*. Springer Science & Business Media.

Angert, E.R., Northup, D.E., Reysenbach, A.L. et al. (1998). Molecular phylogenetic analysis of a bacterial community in Sulphur River, Parker cave, Kentucky. *American Mineralogist* 83: 1583–1592.

Barr, T.C. (1968). Cave ecology and the evolution of Troglobites. In: *Evolutionary Biology*, vol. 2 (eds. T. Dobzhansky, M.K. Hecht and W.C. Steere). Boston, MA: Springer US.

Barr, T.C. (1973). Refugees of the ice age. *Natural History* 72: 26–35.

Barr, T.C. and Holsinger, J.R. (1985). Speciation in cave faunas. *Annual Review of Ecology and Systematics* 16: 313–337.

Barton, H.A. (2015). Starving artists: bacterial oligotrophic heterotrophy in caves. In: *Microbial Life of Cave Systems* (ed. A. Summers Engel), 79–104. Berlin: Walter de Gruyter.

Biswas, J. (1992). Kotumsar cave ecosystem: an interaction between geophysical, chemical and biological characteristics. *National Speleological Society Bulletin* 54: 7–10.

Bonacci, O., Pipan, T., and Culver, D.C. (2009). A framework for karst ecohydrology. *Environmental Geology* 56: 891–900.

Christiansen, K.A. (1962). Proposition pour la classification des animaux cavernicoles. *Spelunca* 2: 76–78.

Clark, D.R. (1981). *Bats and environmental contaminants: a review*. Special Scientific Report Wildlife No. 235, Washington, D.C.: US Fish & Wildlife Service, 27pp.

Clark, D.R. (1986). Toxicity of methyl parathion to bats: mortality and coordination loss. *Environmental Toxicology and Chemistry* 5: 191–195.

Clements, R., Sodhi, N.S., Schilthuizen, M., and Peter K.L., Ng (2006). Limestone karsts of Southeast Asia: imperiled arks of biodiversity. *BioScience* 56: 733–742.

Culver, D.C. (1982). *Cave Life: Ecology and Evolution*. Cambridge, MA: Harvard University Press.

Culver, D.C. (1986). Cave faunas. In: *Conservation Biology: The Science of Scarcity and Diversity* (ed. M.E. Soule). Sunderland, Massachusetts: Sinauer Associates.

Culver, D.C. (2005). Microbes. In: *Encyclopedia of Caves* (eds. D.C. Culver and W.B. White). Amsterdam, The Netherlands: Elsevier/Academic Press.

Culver, D.C. and Holsinger, J.R. (1992). How many species of troglobites are there? *National Speleological Society Bulletin* 54: 79–80.

Culver, D.C., Holsinger, J.R., and Baroody, R. (1973). Towards a predicitive cave biogeography: the Greenbrier Valley as a case study. *Evolution* 27: 689–695.

Culver, D.C. and Pipan, T. (2009). *The Biology of Caves and Other Subterranean Habitats.* Oxford: Oxford University Press.

Culver, D.C. and Sket, B. (2000). Hotspots of subterranean biodiversity in caves and wells. *Journal of Cave and Karst Studies* 62: 11–17.

Deharveng, L., Stoch, F., Gibert, J. et al. (2009). Groundwater biodiversity in Europe. *Freshwater Biology* 54: 709–726.

Dyson, H.J. and James, J.M. (1981). The incidence of iron bacteria in an Australian cave. In: *Proceedings of the 8th International Congress of Speleology*, 79–81. Kentucky: Department of Geography and Geology, Western Kentucky University.

Engel, A.S. (2005). Chemoautotrophy. In: *Encyclopedia of Caves* (eds. D.C. Culver and W.B. White). Amsterdam, The Netherlands: Elsevier/Academic Press.

Engel, A.S. (2019). *Microbes. Encyclopedia of Caves*, 691–698. Academic Press.

Field, M.S. (1988). *Vulnerability of Karst Aquifers to Chemical Contamination.* Washington, DC: Office of Health and Environmental Assessment.

Ford, D.C. and Williams, P.W. (2007). *Karst Hydrogeology and Geomorphology.* Chichester, England: Wiley.

Frick, W.F., Puechmaille, S.J., and Willis, C.K.R. (2016). White-nose syndrome in bats. In: *Bats in the Anthropocene: Conservation of Bats in a Changing World* (eds. C.C. Voigt and T. Kingston), 245–262. Cham: Springer.

Gibert, J. (1986). Ecologie d'un systeme karstique jurassien. Hydrogéologie, dérive animale, transits de matières, dynamique de la population de Niphargus (Crustacé Amphipode). *Mémoires de Biospéologie* 13: 1–379.

Gillieson, D.S. and Houshold, I. (2000). Rehabilitation of the Lune River quarry, Tasmanian wilderness world heritage area, Australia. In: *Karst Hydrogeology and Human Activities: Impacts, Consequences and Implications* (eds. H. Holtzl and D. Drew). Vienna: International Association for Hydrogeology.

Gillieson, D.S., Oldfield, F., and Krawiecki, A. (1986). Records of prehistoric soil Erosion from Rockshelter sites in Papua New Guinea. *Mountain Research and Development* 6: 315–324.

Ginet, R. and Juberthie, C. (1988). Le peuplement animal des karsts de France. *Karstologia* 11: 61–71.

Goldscheider, N. (2012). A holistic approach to groundwater protection and ecosystem services in karst terrains. AQUA mundi Am06046: 117–124.

Goldscheider, N. (2019). A holistic approach to groundwater protection and ecosystem services in karst terrains. *Carbonates and Evaporites* 34: 1241–1249.

Hamilton-Smith, E. (1970). Biological aspects of cave conservation. *Journal of the Sydney Speleological Society* 14: 157–164.

Harris, J.A. (1970). Bat-guano cave environment. *Science* 169: 1342–1343.

Hartmann, A., Luetscher, M., Wachter, R. et al. (2018). Technical note: GUARD – an automated fluid sampler preventing sample alteration by contamination, evaporation and gas exchange, suitable for remote areas and harsh conditions. *Hydrology and Earth System Sciences* 22: 4281–4293.

Hershey, O.S. and Barton, H.A. (2018). The microbial diversity of caves. In: *Cave Ecology* (eds. O.T. Moldovan, L'. Kováč and S. Halse), 69–90. Cham, Switzerland: Springer.

Hill, S.B. (1981). Ecology of bat guano in Tamana Cave, Trinidad, WI. In: *Proceedings of the 8th International Congress of Speleology*, 243–246. Kentucky: Department of Geography and Geology, Western Kentucky University.

Holsinger, J.R. (1966). A preliminary study on the effects of organic pollution of Banners Corner cave, Virginia. *International Journal of Speleology* 2: 75–89.

Holsinger, J.R. (1981). *Stygobromus canadensis*, a troglobitic amphipod crustacean from Castleguard Cave, with remarks on the concept of cave glacial refugia. In: *Proceedings of the 8th International Congress of Speleology*, 93–95. Kentucky: Department of Geography and Geology, Western Kentucky University.

Holsinger, J.R. (1988). Troglobites: the evolution of cave-dwelling organisms. *American Scientist* 76: 146–153.

Holsinger, J.R. and Culver, D.C. (1988). The invertebrate cave fauna of Virginia and a part of eastern Tennessee: zoogeography and ecology. *Brimleyana* 14: 1–164.

Holz, P., Hufschmid, J., Boardman, W. et al. (2016). Qualitative risk assessment: White-nose syndrome in bats in Australia. www.wildlifehealthaustralia.com.au/Portals/0/Documents/ ProgramProjects/WNS_Disease_Risk_Analysis_Australia.pdf (accesssed 23 December 2020).

Howarth, F.G. (1983). Ecology of cave arthropods. *Annual Review of Entomology* 28: 365–389.

Howarth, F.G. (1987). The evolution of non-relictual tropical troglobites. *International Journal of Speleology* 16: 1–16.

Howarth, F.G. (1988). Environmental ecology of North Queensland caves: or why are there so many troglobites in Australia? In: *Proceedings of the 17th Biennial Conference Australian Speleological Federation*, 1988. pp. 76–84.

Howarth, F.G. (1991). Hawaiian cave faunas: macroevolution on young islands. In: *The Unity of Evolutionary Biology* (ed. E.C. Dudley). Portland, Oregon: Dioscorides Publishers.

Howarth, F.G. (1993). High-stress subterranean habitats and evolutionary change in cave-inhabiting arthropods. *The American Naturalist* 142: S65–S77.

Howarth, F.G. and Stone, F.D. (1990). Elevated carbon dioxide levels in Bayliss cave, Australia: implications for the evolution of obligate cave species. *Pacific Science* 44: 207–218.

Huang, C., Li, Y., Zhou, Q. et al. (2008). Karst habitat fragmentation and the conservation of the white-headed langur (*Trachypithecus leucocephalus*) in China. *Primate Conservation* 23: 133–140.

Humphreys, W.F. (1993). The significance of the subterranean fauna in biogeographical reconstruction: examples from Cape Range peninsula, Western Australia. *Records of the Western Australian Museum Supplement* 45: 165–192.

Humphreys, W.F. and Adams, M. (1991). The subterranean aquatic fauna of the North West Cape peninsula, Western Australia. *Records of the Western Australian Museum* 15: 383–411.

Iliffe, T.M., Jickells, T.D., and Brewer, M.S. (1984). Organic pollution of an inland marine cave from Bermuda. *Marine Environmental Research* 12: 173–189.

IUCN Conservation Monitoring Centre (1986). *Red List of Threatened Animals*. Gland, Switzerland: International Union for Conservation of Nature and Natural Resources.

Jefferson, G.T. (1976). Cave fauna. In: *The Science of Speleology* (eds. T.D. Ford and C.H. Cullingford). London: Academic Press.

Jones, R., Culver, D.C., and Kane, T.C. (1992). Are parallel morphologies of cave organisms the result of similar selection pressures? *Evolution* 46: 353–365.

Juberthie, C. and Decu, V. (2001). *Encyclopaedia biospeologica, 3 vols.* Moulis, France: Société Internationale de Biospéologie.

Kiernan, K. (1988). *The Management of Soluble Rock Landscapes: an Australian Perspective.* Sydney, Sysdney: Speleological Society. 61 pp.

Kiernan, K. and Eberhard, S. (1990). Karst resources and cave biology. In: *Tasmanian Wilderness World Heritage Values* (eds. S.J. Smith and M.R. Banks), 28–37. Hobart: Royal Society of Tasmania.

Kresic, N. and Stevanovic, Z. (2009). *Groundwater Hydrology of Springs: Engineering, Theory, Management, and Sustainability.* Butterworth-Heinemann.

Langwig, K.E., Frick, W.F., Reynolds, R. et al. (2015). Host and pathogen ecology drive the seasonal dynamics of a fungal disease, white-nose syndrome. *Proceedings of the Royal Society B: Biological Sciences*: 20142335.

Lascu, C. (2004). *Movile Cave, Romania. Encyclopedia of Caves and Karst Science*, 528–530. New York/London: Fitzroy Dearborn/Taylor and Francis.

Lewis, J.J. (1993). The effects of cave restoration on some aquatic cave communities in the central Kentucky karst. In: *Proceedings of the 1991 Cave Management Symposium*, 346–350. Kentucky: Bowling Green.

Liu, Z.H., Dreybrodt, W., and Liu, H. (2011). Atmospheric CO_2 sink: silicate weathering or carbonate weathering? *Applied Geochemistry* 26: S292–S294.

Liu, Z.H., Dreybrodt, W., and Wang, H.J. (2010). A new direction in effective accounting for the atmospheric CO_2 budget: considering the combined action of carbonate dissolution, the global water cycle and photosynthetic uptake of DIC by aquatic organisms. *Earth-Science Reviews* 99: 162–172.

Lu, Y. (2007). Karst water resources and geo-ecology in typical regions of China. *Environmental Geology* 51: 695.

Marmonier, P., Vervier, P., Giber, J., and Dole-Olivier, M.-J. (1993). Biodiversity in ground waters. *Trends in Ecology & Evolution* 8: 392–395.

McCracken, G.F. (1989). Cave conservation: special problems of bats. *National Speleological Society Bulletin* 51: 49–51.

Moore, B.P. (1964). Present day cave beetle fauna in Australia: a pointer to past climate change. *Helictite* 1: 3–9.

Mulholland, P.J. (1997). Dissolved organic matter concentration and flux in streams. *Journal of the North American Benthological Society* 16(1): 131–141.

Nadler, T., Thanh, V.N., and Streicher, U. (2007). Conservation status of Vietnamese primates. *Vietnamese Journal of Primatology* 1: 7–26.

National Parks Service (2019). *White nose syndrome and Mammoth Cave.* https://www.nps.gov/maca/whitenose.htm (Accessed 14 April 2019).

Nevo, E. (1979). Adaptive convergence and divergence of subterranean mammals. *Annual Review of Ecology and Systematics* 10: 269–308.

Northup, D.E. and Lavoie, K.H. (2001). Geomicrobiology of caves: a review. *Geomicrobiology Journal* 18: 199–222.

Northup, D.E. and Lavoie, K.H. (2004). Microbial processes in caves. In: *The Encyclopedia of Caves and Karst Science* (ed. J. Gunn). New York: Fitzroy Dearborn.

Ortiz, M., Neilson, J.W., Nelson, W.M. et al. (2013). Profiling bacterial diversity and taxonomic composition on speleothem surfaces in Kartchner Caverns, AZ. *Microbial Ecology* 65: 371–383.

Pace, N.R., Stahl, D.A., Lane, D.J., and Olsen, G.J. (1986). *The Analysis of Natural Microbial Populations by Ribosomal RNA Sequences. Advances in microbial ecology*. Boston, MA: Springer.

Parsons, P.A. (1990). The metabolic cost of multiple environmental stresses: implications for climatic change and conservation. *Trends in Ecology & Evolution* 5: 315–317.

Peck, S.B. (1981). The geological, geographical and environmental setting of cave faunal evolution. *Proceedings of the 8th International Congress of Speleology*: 501–502.

Pipan, T. (2005). *Epikarst: A Promising Habitat*. Ljubljana, Slovenia: Založba ZRC.

Pipan, T., Navodnik, V., Janzekovic, F., and Novak, T. (2008). Studies of the fauna of percolation water of Huda luknja, a cave in isolated karst in northeast Slovenia. *Acta Carsologica* 37: 141–151.

Poulson, T.L. (1977). Ecological diversity and stability: principles and management. In: *Proceedings of the National Cave Management Symposium*. Albuquerque, NM: Speleobooks.

Poulson, T.L. and White, W.B. (1969). The cave environment. *Science* 165: 971–981.

Pride, T.E., Ogden, A.E., and Harvey, M.J. (1989). Biology and water quality of caves receiving urban runoff in Cookville, Tennessee, USA. In: *Proceedings of the 9th International Congress of Speleology*, 27–29. Budapest: Hungarian Speleological Society.

Quinlan, J.F. and Ewers, R.O. (1985). Ground water flow in limestone terranes: strategy, rationale and procedure for reliable, efficient monitoring of ground water quality in karst areas. In: *Proceedings of the National Symposium and Exposition on Aquifer Restoration and Ground Water Monitoring*, 197–234. Worthington, Ohio: National Water Well Association.

Ravbar, N. and Goldscheider, N. (2009). Comparative application of four methods of groundwater vulnerability mapping in a Slovene karst catchment. *Hydrogeology Journal* 17: 725–733.

Schiner, J.R. (1854). Fauna der Adelsberger-, Lueger- und MagdalenGrotte. *Verhandlungen der Zoologisch-Botanischen Gesellschaft in Wien* 3: 1–40.

Simon, K.S. (2013). Organic matter flux in the epikarst of the Dorvan karst, France. *Acta Carsologica* 42: 237–244.

Simon, K.S., Benfield, F.E., and Macko, S.A. (2003). Food web structure and the role of epilithic films in cave streams. *Ecology* 84: 2395–2406.

Simon, K.S., Pipan, T., and Culver, D.C. (2007). A conceptual model of the flow and distribution of organic carbon in caves. *Journal of Cave and Karst Studies* 69: 279–284.

Simon, K.S., Pipan, T., Ohno, T., and Culver, D.C. (2010). Spatial and temporal patterns in abundance and character of dissolved organic matter in two karst aquifers. *Fundamental and Applied Limnology/Archiv für Hydrobiologie* 177: 81–92.

Sket, B. (2008). Can we agree on an ecological classification of subterranean animals? *Journal of Natural History* 42: 1549–1563.

Tercafs, R. (1992). The protection of the subterranean environment: conservation principles and management tools. In: *The Natural History of Biospeleology* (ed. A.I. Carmacho). Madrid, Spain: Monografias Museo Nacional de Ciencias Naturales.

Tuttle, M.D. (1986). Endangered gray bat benefits from protection. *Bats* 4: 1–4.

Tuyet, D. (2001). Characteristics of karst ecosystems of Vietnam and their vulnerability to human impact. *Acta Geologica Sinica-English Edition* 75: 325–329.

Ueno, S. (1987). The derivation of terrestrial cave animals. *Zoological Science* 4: 593–606.

Vandel, A. (1965). *Biospeleology: The Biology of Cavernicolous Animals*. Oxford: Pergamon Press.

Vandike, J.E. (1982). *Hydrogeologic Aspects of the November 1981 Liquid Fertilizer Pipeline Break on Groundwater in the Maramec Spring Recharge Area*, vol. 25, 93–101. Phelps County, Missouri: *Missouri Speleology*.

Verant, M.L., Meteyer, C.U., Speakman, J.R. et al. (2014). White-nose syndrome initiates a cascade of physiologic disturbances in the hibernating bat host. *BMC Physiology*: 10.

Warnecke, L., Turner, J.M., Bollinger, T.K. et al. (2012). Inoculation of bats with European Geomyces destructans supports the novel pathogen hypothesis for the origin of white-nose syndrome. *Proceedings of the National Academy of Sciences* 109: 6999–7003.

Waterhouse, J.D. (1984). Investigation of pollution of the karstic aquifer of the Mount Gambier area in South Australia. In: *Hydrogeology of Karstic Terrains: International Contributions to Hydrogeology* (eds. A. Burger and L. Dubertret). Hannover: Heise.

Weinstein, P. and Slaney, D. (1995). Invertebrate faunal survey of Rope Ladder Cave, northern Queensland: a comparative study of sampling methods. *Australian Journal of Entomology* 34: 233–236.

Wheeler, B.J., Alexander, E.C., Adams, R.S., and Huppert, G.N. (1989). Agricultural land use and groundwater quality in the Coldwater cave Groundwater Basin, upper Iowa River karst region, USA: part II. In: *Resource Management in Limestone Landscapes: International Perspectives Special Publication. Department Geography & Oceanography* (eds. D. Gillieson and D. Ingle Smith). University College, Australian Defence Force Academy.

White-Nose Syndrome Response Team (2019) White-nose syndrome response team. U.S. Fish and Wildlife Service. Available: https://www.whitenosesyndrome.org (Accessed 14 April 2019).

Whitten, A.J., Damanik, S.J., Anwar, J., and Hisyam, N. (1984). *The Ecology of Sumatra, Yogyakarta, Gadjah Mada*. University Press.

Whitten, A.J., Mustafa, M., and Henderson, G.S. (1988). *The Ecology of Sulawesi*. Bogor: Java, Gadjah Mada University Press.

Whitten, T. (2009). Applying ecology for cave management in China and neighbouring countries. *Journal of Applied Ecology* 46: 520–523.

Wong, C.I., Mahler, B.J., Musgrove, M., and Banner, J.L. (2012). Changes in sources and storage in a karst aquifer during a transition from drought to wet conditions. *Journal of Hydrology* 468: 159–172.

Workman, C. (2010). Diet of the Delacour's Langur (*Trachypithecus delacouri*) in Van Long nature reserve. Vietnam. *American Journal of Primatology* 72: 317–324.

World Conservation Monitoring Centre (WCMC) (1992). *Global Biodiversity: Status of the Earth's Living Resources*. London: Chapman & Hall.

Zagmajster, M., Culver, D.C., and Sket, B. (2008). Species richness patterns of obligate subterranean beetles (Insecta: Coleoptera) in a global biodiversity hotspot–effect of scale and sampling intensity. *Diversity and Distributions* 14: 95–105.

11

Cave Archaeology

11.1 Introduction

Archaeological deposits in caves form a large part of the surviving record for many prehistoric periods (Straus 1997). Early in human history, caves provided shelter and also facilitated the accumulation of sediments and cultural material. The long geologic and archaeological record within caves provides the original and basic data of prehistoric archaeology in many parts of the world. Caves provide access to water sources and also provide ready-made natural structures without the need of any significant modification; thus, they were convenient and useful for a variety of purposes (Skeates 1997; Straus 1997). Caves provide not only natural shelter and protection, but also a memorable and confined living space with a sense of durability and familiarity, probably precursors of the modern idea of 'home'. Caves may have been the first 'home base' sites during the late Middle Pleistocene (Rolland 2004).

Caves were used in various ways by hominids, certainly for disposal of the dead and probably for shelter. The Rising Star cave within the Cradle of Humankind in South Africa is but one site where important early hominid fossils have been found (Berger et al. 2015; Pickering and Kramers 2010). Widespread, regular cave occupation was certainly a feature of Neanderthal societies from at least 400 000 years ago, and caves were being widely exploited as occupation sites in the last 80 000 years. The regular use of fire made deep caves available for specialised ritual functions, as is indicated by the Upper Palaeolithic art caves of France and Spain. During the Neolithic there was an expansion of seasonal cave occupation by pastoral groups, a use which continues today in mountainous areas of Europe and Central Asia (Tolan-Smith and Bonsall 1997).

The birth of archaeology as a discipline in the nineteenth century was essentially founded on excavations in caves. Investigations by inquisitive individuals, geologists and antiquarians revealed the bones of animals and humans in a stratigraphic context, often associated with stone tools and other artefacts (Simek 2004). At that time, the biblical account of creation, asserted by Archbishop Ussher to be at 10 : 30 a.m. on the 23 October 4004 BCE, was not questioned by enlightenment scientists. The cave excavations started to reveal the bones of large extinct mammals associated with the Ice Age. The eminent French scientist Georges Cuvier argued, in 1795, that the extinctions of mammoth, woolly rhinoceros, and giant elk were due to catastrophic floods that produced stony deposits ('diluvium') in which their bones came to rest. He believed that such events could have occurred in the last 6000 years and that human agency was not involved. In 1822, the geologist William Buckland excavated Paviland Cave in Wales and found both extinct animals and human bones, but denied any association. This opened the way for others to question the biblical account. Kent's Cavern near Torquay, England was excavated by Rev. John McEnery in 1825. Underneath flowstone floors he found animal bones and flint tools in stratigraphic association. His scientific mind summarised that the antiquity of man must go back to a time earlier than biblical chronology, but his religious beliefs, and the views of those he turned to for advice, including Buckland, persuaded him otherwise. These flint hand axes could only have been placed in the cave by modern visitors and thoughts about them being contemporary with extinct animals were ill-founded.

We now know that McEnery had discovered proof for the antiquity of man, but his lack of scientific endeavour to support this proposition and a reluctance to publish papers left others to develop his theories. McEnery's unpublished work inspired William Pengelly, a teacher from Looe, Cornwall, who had developed a fascination for natural history in his later years and gained a reputation in Torquay as a tutor to prominent Victorian families. In 1865, he set about excavating and recording the sediments lying between the stalagmite floors and the Devonian limestone base of Kents Cavern. It would take 15 years to complete the task, a work he conducted with great exactitude. He pioneered a three-dimensional recording system that produced a level of recording accuracy that would not be repeated anywhere else for decades. Most museums founded prior to the 1870s are likely to have artefacts from Kents Cavern in their collections today thanks to Pengelly's correspondence. This includes the Smithsonian, in Washington, USA, the Natural History Museum at the Jardin des Plantes, in Paris, France, and most museums in Britain, including the British Museum and Natural History Museum, in London.

This evidence was initially rejected by the scientific establishment, and it took another 20 years for the antiquity of humanity and its association with extinct mammals to be accepted. At Engis (Belgium) in 1832, the physician Schmerling excavated archaic human skulls associated with the bones of rhinoceros and elephants; these were, in fact, Neanderthal remains but were not recognised as such at the time (Shaw 2004). The weight of evidence proved so overwhelming that both Buckland and the eminent geologist Charles Lyell had to admit the possibility of human prehistory prior to the deluge. The pace of archaeology in caves quickened, and the discipline spread to other continents.

11.2 Prehistoric Uses of Caves

Around the world, humans have long used the natural shelter afforded by caves to process food, cook, make tools and clothing, and many other social activities. While the daylight zone of most caves is used, there is compelling evidence that humans have gone far beyond the limits of daylight to mine minerals and resources such as flint, to draw or engrave art, to place or bury their dead, and to perform rituals.

In France, Neanderthal people created two ring-shaped structures from broken stalagmites deep inside Bruniquel Cave (Jaubert et al. 2016). The structures are over 300 m from the cave entrance, and some show evidence of the use of fire. There are over 400 stalagmites in the assemblage, some of which have grown on top of broken stalagmites. Dating

Figure 11.1 Broken stalagmite assemblage in Bruniquel Cave, France, created by Neanderthal people. Source: From Jaubert et al. (2016) © Springer Nature.

both by Uranium series methods provides an opportunity to obtain both a minimum and maximum age for the assemblage. On average it was created 176 000 years ago. The assemblage may have had either a ritual function or a domestic function, though the site is very wet (Figure 11.1).

11.3 Cave Faunas and Hominids

In southern Africa, the Cradle of Humankind World Heritage Site includes caves that preserve a range of hominid fossils (Berger et al. 2010, 2015; Bruxelles et al. 2014). The hominid remains are generally found in cemented sediments within caves developed in dolomite bedrock. Sedimentological and taphonomic analyses of several notable fossil sites (Brain 1981; Dirks et al. 2010; Pickering and Kramers 2010) suggest that the fossils were preserved in caves as a result of a range of processes including death traps, scavenging, mudflows, and predation. In most cases, these sites were close to the critical resource of water.

Over the last 3 million years, cave sediment traps have collected fossils and provide a unique window into Pliocene-Pleistocene ecosystems, and how they changed with Pleistocene glacial periods and associated landscape change (Bruxelles et al. 2014; Dirks and Berger 2013; Granger et al. 2015; Pickering and Kramers 2010). The hominid-bearing deposits formed in broadly similar settings involving debris cone accumulations near cave openings. The debris cones resulted from both breakdown of ceilings and walls and allogenic sand to clay-sized sediments either blown or washed into the cave.

In general, the hominid fossil record of the Middle Pleistocene between 780 000 and 130 000 years ago in Africa is quite limited. There are perhaps a half a dozen partial mandibles and a somewhat larger number of postcranial fragments or dental remains. However, many of these were found prior to 1960 and lack both good provenance and reliable dating. In some cases, a date has been estimated on the basis of their morphology or the associated faunal remains. This was not the case with *Homo naledi*. More than 1500 fossils representing at least 15 individuals of this species were unearthed from the Rising Star cave system in South Africa between 2013 and 2014. Found deep underground in the Dinaledi Chamber, the *H. naledi* fossils are the largest collection of a single species of an ancient human relative discovered in Africa (Berger et al. 2017).

The Rising Star cave system lies in the Bloubank River valley, 2.2 km west of Sterkfontein cave. It lies in the core of a gently west-dipping open fold in dolomite with five thin chert marker horizons that have been used to evaluate the relative position of chambers within the system. The network of cavities is developed along northerly and north-westerly trending fractures and joints. The Dinaledi Chamber is about 30 m below the surface and was found during speleological surveys. The only identified access point into the Dinaledi Chamber involves an exposed, 15 m climb up the side of a sharp-edged dolomite block that has dislodged from the roof (the Dragon's Back). From the top of the Dragon's Back, the Dinaledi Chamber is accessed via a narrow vertical fissure involving a 12 m climb down, with squeezes as tight as 20 cm, to reach the floor.

Dating of *H. naledi* fossils from the Dinaledi Chamber (Berger et al. 2015) indicates that they were deposited between about 236 000 and 335 000 years ago (Dirks et al. 2017), placing *H. naledi* firmly in the later Middle Pleistocene. Prior to this, only larger-brained modern humans or their close relatives had been demonstrated to exist at this time in Africa, but the fossil evidence in southern Africa was limited. It now seems more likely that there was a diversity of hominid types in the region, with some lineages contributing DNA to living humans. It may be that *H. naledi* represents a survivor from the earliest stages of diversification within the genus *Homo*. The late presence of this hominid also suggests that archaeological stone tools from the contemporaneous Middle Stone Age may not necessarily be due to modern humans.

The Dinaledi collections of hominid remains displays taphonomy indicative of several stages of burial with some breakage patterns consistent with repeated reworking of at least part of the bones. The distribution of the bone material and skeletal part representation is indicative of limited winnowing showing that that the fossils of *H. naledi* must have found their way into the chamber via a difficult route that precluded any other large vertebrates from disturbing the bones (Figure 11.3). The presence of articulated, functional anatomical units that conventionally disassociate early in the decomposition sequence also suggests that bodies were fresh or in the early stages of decomposition when they entered the chamber and were encapsulated in the matrix (Haglund 1993; Lyman and Lyman 1994). The lack of green fractures on any of the elements in the assemblage suggests that the bodies did not enter the chamber due to catastrophic accident, such as falling into the chamber or due to flooding.

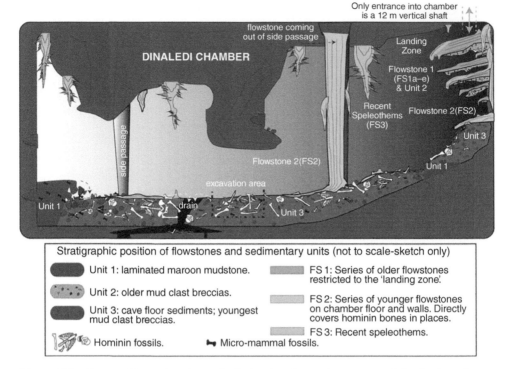

Figure 11.2 Cartoon illustrating the geological and taphonomic context and distribution of fossils, sediments, and flowstones within the Dinaledi Chamber, South Africa. The distribution of the different geological units and flowstones is shown together with the inferred distribution of fossil material. Source: From Dirks et al. (2017) © eLife Sciences Publications Ltd.

The traditional 'Out of Africa' single dispersal of modern humans (*Homo sapiens*) about 60 000 years ago is now being questioned in the light of new archaeological and geomorphological evidence. Much of this new evidence has been gained from careful analysis of cave deposits over the last decade. The discovery of archaeological sites from 46 000 to 65 000 years old in Southeast Asia and Australia suggest that earlier dispersal from Africa must have occurred to allow the spread of modern humans into regions remote from Africa and their interbreeding with Neanderthal and Denisovan peoples in Central Asia. Inconvenient though this might be, there is a growing body of evidence supporting multiple phases of dispersal and perhaps multiple sites of origin for modern humans.

The earliest morphologically modern humans – *Homo sapiens* – have been reported from two sites in the Horn of Africa at 160 000 and 195 000 years ago, while the cave deposits of Jebel Irhoud in Morocco are dated to 310 000 years ago. The sites of Skhul and Qazzeh in Israel date to between 90 000 and 120 000 years ago. This raises questions about the central, pivotal role of the East African Rift in the emergence of modern humans. A single dispersal model of modern humans out of Africa at around 60 000 years ago has been supported by modern genetic studies of human population variation around the world. However, multiple dispersals in the Late Pleistocene pre-60 000 years ago are now supported by the growing archaeological evidence and emerging genetic evidence from both skeletal material and environmental DNA in cave sediments.

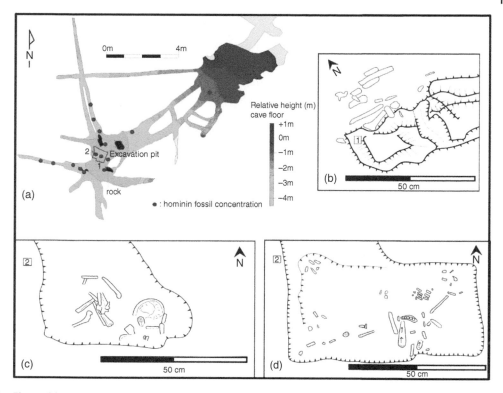

Figure 11.3 Map of the cave chamber showing the distribution of hominin fossils. (a) distribution of concentrations of bone fragments of *Homo naledi* along the floor of the Dinaledi Chamber, South Africa. The positions of maps (b), (c), and (d) are shown relative to survey pegs 1 and 2, respectively. (b) Concentration of long bone fragments encountered next to a rock embedded in Unit 3 sediment. (c) Distribution of fossils in the excavation pit at the start of the excavations in November 2013 (~5 cm below surface). (d) Distribution of fossils in the excavation pit during excavations in March 2014 (~15 cm below surface); the long- bone in the central part of the pit is in a near-vertical position. Source: From Dirks et al. (2017) © eLife Sciences Publications Ltd.

Migration pathways out of Africa have been hypothesised (Figure 11.4) as being via the Sinai Peninsula from northern Egypt and by a short water crossing of the Bab el Mandab Strait to Yemen. This latter crossing is about 20 km now and would have been from 5 to 15 km wide during glacial low sea levels. From there, coastal dispersal around the Indian Ocean into Asia may have been possible. The interactions between dispersing modern humans and Neanderthal people are now emerging from genetic evidence. Neanderthal people are well represented in cave deposits in Europe and Central Asia from about 400 000 years ago to 50 000 years ago, with quite sophisticated technology and art (Figure 11.5). Archaeological evidence of interbreeding between Neanderthals and modern humans is now widespread, with modern populations hosting between 1% and 4% Neanderthal DNA. A modern human skeleton from Romania with a minimum age of 40 000 years has 9% Neanderthal DNA, suggesting an ancestor of that species four to six generations back. Recent genetic identification of Denisovan ancestry for about 5% of modern Melanesian DNA also suggests interbreeding between

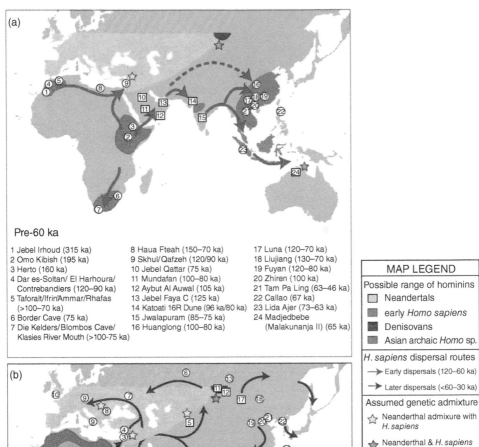

Pre-60 ka

1 Jebel Irhoud (315 ka)
2 Omo Kibish (195 ka)
3 Herto (160 ka)
4 Dar es-Soltan/ El Harhoura/
   Contrebandiers (120–90 ka)
5 Taforalt/Ifrin'Ammar/Rhafas
   (>100–70 ka)
6 Border Cave (75 ka)
7 Die Kelders/Blombos Cave/
   Klasies River Mouth (>100-75 ka)

8 Haua Fteah (150–70 ka)
9 Skhul/Qafzeh (120/90 ka)
10 Jebel Qattar (75 ka)
11 Mundafan (100–80 ka)
12 Aybut Al Auwal (105 ka)
13 Jebel Faya C (125 ka)
14 Katoati 16R Dune (96 ka/80 ka)
15 Jwalapuram (85–75 ka)
16 Huanglong (100–80 ka)

17 Luna (120–70 ka)
18 Liujiang (130–70 ka)
19 Fuyan (120–80 ka)
20 Zhiren (100 ka)
21 Tam Pa Ling (63–46 ka)
22 Callao (67 ka)
23 Lida Ajer (73–63 ka)
24 Madjedbebe
   (Malakunanja II) (65 ka)

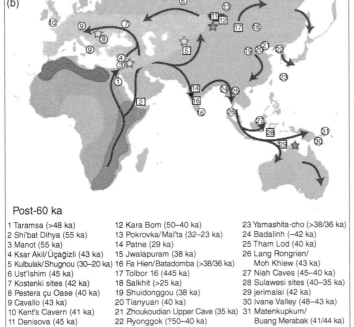

Post-60 ka

1 Taramsa (>48 ka)
2 Shi'bat Dihya (55 ka)
3 Manot (55 ka)
4 Ksar Akil/Üçağizli (43 ka)
5 Kulbulak/Shugnou (30–20 ka)
6 Ust'Ishim (45 ka)
7 Kostenki sites (42 ka)
8 Pestera çu Oase (40 ka)
9 Cavallo (43 ka)
10 Kent's Cavern (41 ka)
11 Denisova (45 ka)

12 Kara Bom (50–40 ka)
13 Pokrovka/Mal'ta (32–23 ka)
14 Patne (29 ka)
15 Jwalapuram (38 ka)
16 Fa Hien/Batadomba (>38/36 ka)
17 Tolbor 16 (445 ka)
18 Salkhit (>25 ka)
19 Shuidonggou (38 ka)
20 Tianyuan (40 ka)
21 Zhoukoudian Upper Cave (35 ka)
22 Ryonggok (?50–40 ka)

23 Yamashita-cho (>38/36 ka)
24 Badalinh (–42 ka)
25 Tham Lod (40 ka)
26 Lang Rongrien/
   Moh Khiew (43 ka)
27 Niah Caves (45–40 ka)
28 Sulawesi sites (40–35 ka)
29 jerimalai (42 ka)
30 Ivane Valley (48–43 ka)
31 Matenkupkum/
   Buang Merabak (41/44 ka)

MAP LEGEND

Possible range of hominins
☐ Neandertals
▨ early *Homo sapiens*
■ Denisovans
▨ Asian archaic *Homo* sp.

H. sapiens dispersal routes
⟶ Early dispersals (120–60 ka)
⟶ Later dispersals (<60–30 ka)

Assumed genetic admixture
☆ Neanderthal admixure with
   H. sapiens
★ Neanderthal & *H. sapiens*
   admixure with Denisovans

① Site with *H. sapiens* fossils
⬜28 Site with archaeology only

Figure 11.4 Distribution of sites and possible migratory pathways for modern humans dispersing across Asia in the late Pleistocene. Source: From Bae et al. (2017).

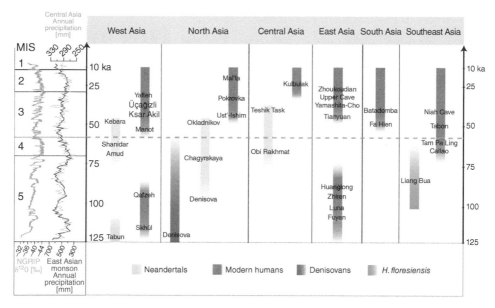

Figure 11.5 Age ranges for the presence of major hominin taxa (modern humans, Neanderthals, Denisovans, and *H. floresiensis*) in major regions of Asia. Source: From Bae et al. (2017).

species. Thus, the relationships between *Homo sapiens* and other hominid species are undergoing major revision and some species, such as *Homo floresiensis*, may be hybrids.

The traditional view of migration out of Africa envisaged modern humans carrying with them a set of behaviours involving stone tools, such as blades and microblades, art, and symbolism. However, these 'modern' artefacts appear in sites where there are no modern human remains, and these artefacts have also been excavated in association with Neanderthal and Denisovan people, and possibly other hominid taxa. For example, a set of perforated teeth, ostrich eggshell, and bone pendants were excavated from levels of Denisova Cave (Siberia) associated with Neanderthal and Denisovan people.

Douka et al. (2019) have applied a modelling approach that combines chronometric (radiocarbon, uranium series, and optical ages), stratigraphic and genetic data to probabilistically calculate the age of the human fossils at the site. Their modelled estimate for the age of the oldest Denisovan fossil suggests that these people were present at the site as early as 195 000 years ago. All Neanderthal fossils – as well as Denisova 11, the daughter of a Neanderthal and a Denisovan – date to between 80 000 and 140 000 years ago. The youngest Denisovan dates to between 52 000 and 76 000 years ago. Direct radiocarbon dating of Upper Palaeolithic tooth pendants (Figure 11.6) and bone points yielded the earliest evidence for the production of these artefacts in northern Eurasia, between 43 000 and 49 000 years before present. On the basis of current archaeological evidence, it may be assumed that these artefacts are associated with the Denisovan population and not modern humans.

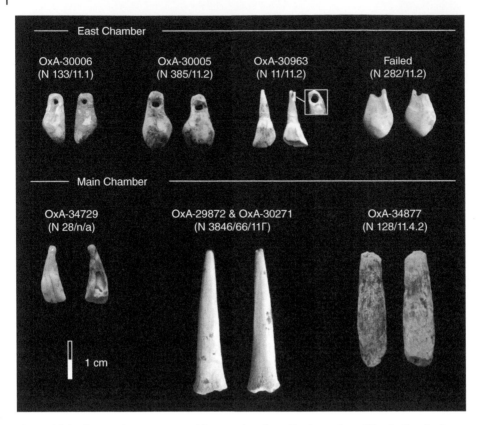

Figure 11.6 Personal ornaments and bone points from Denisova Cave, Siberia, Russia that were sampled for radiocarbon dating. Source: From Douka et al. (2019) © Springer Nature.

11.4 Cave Art in Context

The majority of prehistoric art is found in shallow caves and rock shelters, but there is nevertheless a lot of cave art within deep caves. The best-known examples come from Europe, but there are many other examples from caves in Southeast Asia, the Americas, and Australasia (David 2017). The discovery of many sites of prehistoric art in the nineteenth century paralleled the scientific discoveries in palaeontology, but it was not until stone tools were found in association with art that their antiquity was recognised (Bahn 2004). Just as the various phases of prehistoric archaeology were based on stone tools types and pottery series, the various styles of cave art were based on theoretical constructs of complexity. The development of direct dating of rock art by Bednarik (1995, 2004) led to a more objective basis for art typologies and refutation of earlier concepts. It is very clear that certain stylistic elements can be seen across an entire region; for example, in the Dordogne region of France, contemporaneous cave art about 15 000 years old features mammoth, woolly rhinoceros, horse, bison, and ibex (Desdemaines-Hugon 2010). Similarly, in northern Australia, the *wandjina* figures are seen over a geographic

Figure 11.7 Wandjina figures in cave entrance, Leopold Range, Kimberleys, Australia.

range of 500 km and date from the Last Glacial Maximum through to the recent past as they were re-painted by traditional owners over time (Figure 11.7).

In Europe, cave art ranges in age from c. 48 000 years ago (Bahn 2004) to medieval times, but most attention has been focussed on art ranging from 35 000 to 15 000 years ago. There are at least three hundred known cave art sites in Europe, and no doubt more will be found. For example, the Cosquer Cave site contains exceptional art dating to the Last Glacial Maximum and was discovered by cave divers in 1985. The tunnel leading into the cave is currently 37 m below sea level, so the site was clearly used during glacial low sea levels. The art contains many hand stencils dated to 19 000 years ago and animals such as bison, ibex, and horses, but also marine animals, such as auks (Clottes and Courtin 1996). Figure 11.8 provides a schematic time sequence for French cave art.

In the Western Hemisphere, cave art ranges from about 9000 years ago in Brazil (Caverna da Pedra Pintada) to historic times. Handprints in the Cueva de las Manos in Patagonia have been dated to 7300 years ago (Stone 2004). Probably the most abundant cave art is found in Mexico and the Caribbean islands of Cuba, Puerto Rico, and Jamaica (Greer and Greer 1999; Stone 2004).

Robert (2017) has analysed the distribution of Palaeolithic art in caves in relation to both substrate architecture and composition. He concludes that the nature and architecture of the cave wall plays an integral role in the design of the art. The careful use of natural hollows, edges, and overhangs on a cave wall add three-dimensionality to the engraved

Site	Example Image	Age Before Present (years)	Culture
Le Portel		11,600	Magdalenian
Trois-Freres		13,000	
Rouffignac		13,000	
Niaux		14,000	
Le Cap Blanc		15,000	
Altamira	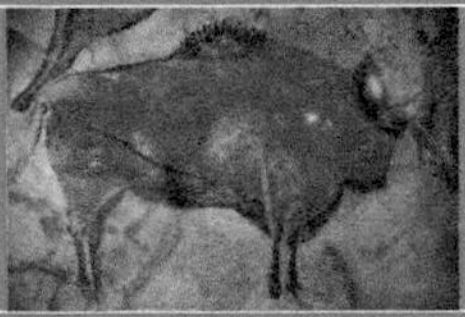	17,000	
Cosquer (Phase 2)		19,000	Solutrean
Lascaux		20,000	
Le Placard		21,000	
Cougnac	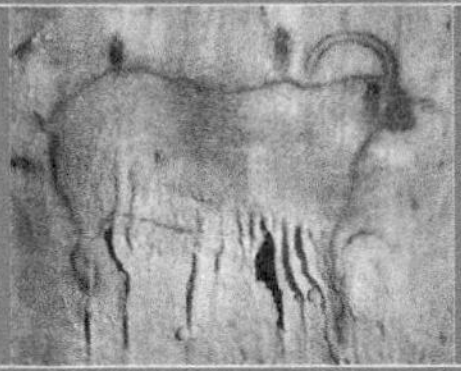	25,000	Gravettian
PechMerle		25,000	
Gargas		27,000	
Cosquer		27,000	
Chauvet		32,000	Aurignacian

Figure 11.8 Chronology of parietal art at several French cave art sites. Source: Images from Ministère de la Culture/Centre National de la Préhistoire

or painted art. Given that the prehistoric artists could not erase any mistakes, the design and placement of the art must have been conceived and visualised before being executed in a series of very precise brush strokes. Furthermore, the location of the wall within the cave and its accessibility plays a role in the design, especially in the choice of the figures or symbols employed (Figures 11.9 and 11.10).

Most of these figures are found in small chambers or at the end of tight passages; in the past access to these locations would have been difficult and involved crawling and climbing. Many of the Palaeolithic art sites in the Dordogne have had their passage floors lowered by excavation in the nineteenth and twentieth centuries, so access is now very easy. One might conclude that access to the sites in prehistory may have been part of some ritual or rite of passage.

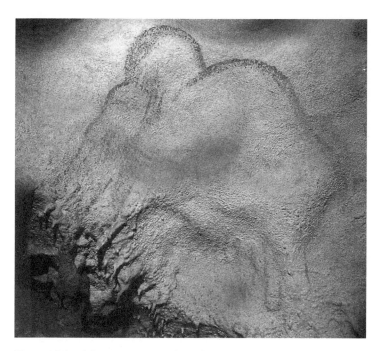

Figure 11.9 Painted mammoth from Grotte de Bernifal, Dordogne, France. The artist has carefully painted the animal using the rock wall topography to create a three-dimensional effect.

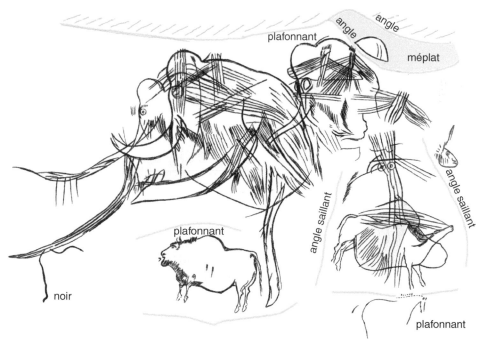

Figure 11.10 A frieze of animals from Bernifal Cave, Dordogne, France. The figures are carefully placed in the landscape of the cave wall, so as to add realism to their form. It should be noted that these figures are barely visible in direct light, but are clearly seen and have a sense of movement in oblique flickering light. Plafonnant = reaching upper limit, angle saillant = projecting rib of rock, meplat = flat. Source: From Robert (2017) © 2017b Elsevier.

11.5 Depositional Environments in Caves

The nature of rockshelter sediments was briefly outlined in Chapter 7, while a more detailed treatment of the internal facies of cave sediments was also provided there. The external facies of these sediment deposits are derived from surface wash, gravity fall, and in situ deposition, and may be subject to a range of geomorphic processes including slumping, erosion by water at the dripline, and collapse. These result in complex depositional patterns and sediment stratigraphy that is subject to greater fluctuations in moisture and temperature (O'Connor et al. 2016). In contrast, the internal facies of cave sediments is in an environment where geomorphic processes are less active and may be confined to episodic floods and some collapse. All of these disturbance processes operate at varying rates over time and at different intensities as a result of the climatic and tectonic regime.

In terms of the sediments normally associated with cave archaeology, we can usefully distinguish between active and passive karst settings (Table 11.1). Active or humid caves are generally linked to a major internal cavern or water source. Thus, at the earliest stages of its occupation, Nombe Cave (Papua New Guinea) acted as a cave spring, and the sediments were dominantly clays and silts, reworked by erosive peak flows from the emerging cave stream (Gillieson and Mountain 1983). In such humid environments, chemical alteration or diagenesis is rapid, and there is active deposition of calcite as flowstone. This can serve

Table 11.1 Some of the characteristics of active karst settings and passive karst settings for rockshelter and cave mouth environments in limestone terrains.

Active (Humid) karst setting	Passive (dry) karst setting
Linked to an internal cavern or conduit system	No significant links with an internal conduit system
Dripping vadose waters	Dry site without flowing or dripping water
Seasonal water flows and ponding	Limited or no inwashing of sediments via karstic cavities
Range of hydrological pathways	Highly localised or no chemical precipitation
Precipitation of calcite and other minerals	Limited vegetation growth in the site
Inwashing of fine sediments via conduits in the host bedrock	Desiccation of macroscopic plant remains
Development of vegetation within the site	Limited chemical diagenesis and mineral alteration
Mineralisation of macroscopic plant remains	Import of fine sediments through the shelter opening may be dominant
Strong chemical diagenesis and mineral alteration	Limited host rock weathering by solution
Humidity may encourage host rock breakdown by frost action	Subaerial processes are dominant
Evidence of erosion and sediment removal by invasive karst waters	

Source: Adapted from Woodward and Goldberg (2001).

to seal occupation deposits from further erosion and also provides a means of dating the deposits using uranium series methods (Chapter 8).

In contrast, a passive setting or dry cave environment has no flowing water and may only have seasonally active drips. There is very little inwashing of sediments from karst conduits, and the main processes of sedimentation are from roof fall, windblown sediments and more biogenic deposition from middens and hearths. Chemical diagenesis is much more limited, and there may be desiccation of plant and animal remains. The passive setting is the one most commonly encountered by archaeologists excavating in cave entrances. There may be substantial erosion and slumping of the entrance deposits at the entrance dripline where rain flows off the rock surface in a 'teapot' effect. This may affect occupation deposits as the sunny outer part of the cave entrance is the likely locus of most human activities. Analysis of archaeological cave sediments should take account of the changing depositional environments on the site over the full length of its occupation.

11.6 Cave Deposits and Biological Conservation

The role of palaeontology in conservation has been highlighted by several authors (Hadly and Barnosky 2009). The use of contemporary data only to predict faunal change fails to take into account the range of natural variation experienced by ecosystems over ecological and evolutionary time (Hadly and Barnosky 2009). Future conservation managers must broaden their strategies beyond preserving species within their current range. It will become increasingly necessary to develop tools for predicting where species can exist in light of rapid climate change, human expansion, and habitat fragmentation (Hadly and Barnosky 2009). The fossil record provides the only means for assessing long-term patterns of faunal change against climate and supplying meaningful data for such predictive models (Reed 2012).

Palaeontology has been used effectively in Yellowstone National Park, USA, where palaeoecological data from Holocene cave deposits were used to assess elk populations and restore wolves to the park (Hadly and Barnosky 2009). Other studies have revealed striking results that highlight limitations to our current understanding of ecological niches. Bilney et al. (2010) used sub-fossil deposits of the Greater Sooty Owl (*Tyto tenebricosa*) from cave and rockshelter deposits as a tool for understanding the diversity of the small mammal palaeocommunity. These results were compared to the contemporary Greater Sooty Owl diet from the same geographical region of south-eastern Australia to investigate the degree of small mammal decline. Of the 28 mammal species detected in sub-fossil deposits and considered prey items of the Greater Sooty Owl at the time of European settlement, only 10 species were detected in the contemporary Greater Sooty Owl diet. Numerous small mammal species have not only recently suffered severe declines in distribution and abundance, but have also recently undergone niche contraction, as they occupied a greater diversity of regions and habitats at the time of European settlement.

In order to investigate patterns of landscape change and faunal response over time, fossil deposits must meet a few important criteria (Reed 2012). Caves are a good place to start as they typically contain deep, stratified sediment beds which are well preserved within the relatively sheltered confines of the cave (Reed and Gillieson 2003). The site must have good stratigraphy with little evidence of reworking and post-depositional disturbance. It must also contain material that can be dated to provide a timeline for the deposit and a good assemblage of fossil bone material to allow statistical inferences. It is important to consider the taphonomic (site formation) history of the cave to ensure comparison between sites with a similar depositional history.

11.7 Taphonomy of Cave Deposits

Some of the most detailed work on the post-depositional alteration of organic remains in caves has been carried out in South Australia. Naracoorte Caves was declared a World Heritage site for its Pleistocene marsupial remains and has subsequently become a major site for cave palaeontology, environmental reconstruction, and for science education. There are three broad types of cave deposit at Naracoorte (Reed 2003). The first is a 'classic' pitfall trap with a vertical narrow solution pipe entrance in the cave roof, which accumulates sandy sediment cones below it. Small and large animals fall into the cave via these well-concealed entrances. Subsequent low energy water flows and some slumping lead to coarsely stratified deposits. Large mammals, such as kangaroos, are well represented and usually dominated by groups susceptible to entrapment (Reed 2003, 2008). This type of cave is typically quite warm and humid with limited light penetration. Carcasses falling into the cave decompose rapidly, leaving bone deposits which may be partially articulated (Reed 2009).

The second type of cave at Naracoorte has a large window entrance due to roof collapse. Blanche Cave is a good example of this type and has extensive light zones with vegetation growing on sediment cones, fluctuating temperature and humidity, and generally drier conditions within the cave. Sediment accumulates as a cone beneath the entrance and is reworked downslope by water flow during storms, resulting in very finely laminated sediment beds (Reed 2003). Decomposition is much slower, with animal remains and other organic material often becoming desiccated (Reed 2009). While some animals do become trapped in these caves via pitfall, the major accumulation type is owl pellet deposition, with some contribution from regular cave inhabitants.

The final type of cave forms from the roof window type where subsequent collapse removes any vertical component so that animals can walk or hop into the cave. These caves serve as dens and shelter for both predators and herbivores alike (Reed 2003). In the past, animals such as Tasmanian Devils would have used these caves as dens, bringing their prey back to the cave to consume and thus allowing bones to accumulate. Therefore, the type of cave is important in determining the nature of processes operating, sediment stratification, and the types of animal remains that accumulate there. Scientific research projects are usually based on a series of key questions and hypotheses. The range of data considered and methods of collection must be appropriate to the questions being asked.

Many owl species roost in caves for extended periods, leaving behind the remains of their prey in the form of regurgitated pellets. At Nettle Cave, Jenolan, Australia, Greater Sooty Owls have roosted for at least 16 000 years (see Table 11.2; Morris et al. 1997). Owls search the surrounding valleys for prey, mostly small mammals and reptiles, and thus provide a sample of the local fauna during each hunting trip. Over time, owls concentrate thousands of bones, which become incorporated into cave sediments and are thus available for

Table 11.2 Proposed correlation of fauna and environmental conditions at Nettle Cave, Australia.

Years before present	*20 000–15 000*	*15 000–10 000*	*10 000–present*
Levels of the deposit	68–44 cm	44–41 cm	41–0 cm
Mammals in deposit	Small mammals: *Burramys parvus*, *Cercartetus lepidus*, *Antechinus swainsonii*, *Mastacomys fuscus*	*Burramys parvus* and *Cercartetus lepidus* locally extinct, *Mastacomys fuscus* and *Antechinus swainsonii* numbers reduced, *Rattus fuscipes and arboreal species increased*	Small mammal fauna similar to that of present, increased arboreal species such as *Petauroides volans, Petaurus breviceps, Petaurus australis* reduced native murids
Events at Nettle Cave	Erosion of topsoil above cave	Water ponding, increased silt and clay	Increased humidity
Surface vegetation	Dominated by subalpine grasses, woodland at lower altitudes	Open forest and woodland by 11 500 years ago	Wet and dry eucalypt forest similar to present
Inferred climate	Very cold and dry, maximum aridity 16 000 years ago	Wetter, warmer conditions with humid period about 14 000 years ago	Period of humidity about 7500 years ago, drier conditions followed

Source: Adapted from Morris et al. (1997).

study. The large sample size makes this a useful resource for environmental reconstruction using statistical techniques. Assemblages based on owl pellets tend to reflect the prey range and size of the predator, so are biased towards small vertebrates. At Jenolan, the 1-m deep deposit covers the transition from the end of the Last Glacial, when open subalpine grassland extended through the region, to the arrival of tall forest communities as climate warmed. So, the deeper deposits contain bones of mountain pygmy possum (*Burramys parvus*), an alpine endemic species, while the more recent deposits contain bones of greater gliders (*Petauroides volans*), sugar gliders (*Petaurus breviceps*), yellow-bellied gliders (*Petaurus australis*), and other mammals inhabiting tall eucalypt forests today.

Larger mammals that fall into caves undergo a process of decomposition that is very dependent on the temperature and humidity at the site. In very dry caves the soft tissues of a carcass may desiccate and become mummified, sometimes providing exceptional preservation over long periods of time e.g. a 5000 year old mummified Thylacine carcass found in Thylacine Hole on the Nullarbor Plain in Australia (Archer et al. 1991). Terrell-Nield and MacDonald (1997) experimentally placed rat carcasses in two different caves in England to observe insect activity in relation to decomposition. They report that skeletonisation and disarticulation was complete by 1300 days in one cave and 730 days in a slightly warmer cave, suggesting within cave environmental conditions influence the rate of decomposition. Reed (2009) presented results of a study of decomposition and disarticulation of kangaroo carcasses in caves at Naracoorte, South Australia. Carcasses were placed in two caves and observed over a period of nearly three years. Decomposition progressed rapidly within the caves with almost immediate infestation by blowflies and

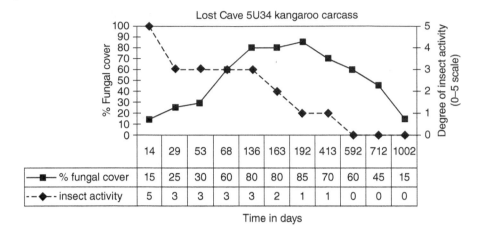

Figure 11.11 Degree of fungal cover (expressed as a % of the total carcass covered) and insect activity (expressed as a value from 0 to 5, with 0 indicating no activity and 5 indicating extreme activity), for the Lost Cave, Naracoorte, South Australia (5 U34) carcass during the study period. Source: From Reed (2009).

fungi (Figure 11.11). Invertebrate activity had ceased by 28 months; however, fungal colonisation continued for the course of the study period. Decomposition, skeletonisation, and disarticulation were complete by 600 days. The position of the carcass within the cave also influences the degree of dispersal of remains in pitfall caves.

Immediate invasion of the kangaroo carcasses by blowflies and house flies (*Calliphora* sp. and *Musca domestica*) during the fresh stage was recorded for both caves at Naracoorte, and this is typical for this decay stage. Fungi also appear to be important decomposers in caves at Naracoorte, and colonisation occurs rapidly in the fresh stage of the decomposition sequence, *Zygomycetes* fungi require warm and moist conditions for germination, relying on wind, rain, or animal dispersal of spores. They are saprophytic and are usually widespread in soil and dung; some forms are parasitic. In the kangaroo carcasses, fungal growth was largely concentrated on the hair and some nearby wood material.

Once decomposed, the long-term preservation of bone in caves is dependent upon rapid burial. This is most likely due to water flow or wind-blown deposition, but roof-fall may also contribute. In more humid sites, rapid deposition of calcite flowstone may also seal in bone deposits for posterity. Vertebrate remains, being essentially large clasts, are prone to movement in caves, particularly during flooding and mass movement events. However, steep energy gradients in cave passages, a result of variable flood events and the geometry of the passage (Gillieson 2004) localise erosional events and can result in translocation of clasts not far from their source. Within a cave, sediments and fossils can also be deposited in a single, fluidised, self-perpetuating sliding mass, resulting in deeply penetrating 'sliding bed' facies; a result of pipe-full conditions or mass movement events. Such facies have been recorded for the highland caves of New Guinea (Gillieson 1986) and some of the sedimentary units in Niah Caves, Borneo, Malaysia (Gilbertson et al. 2013).

The primary source of biogenic sediments in caves is excrement from animals, with bat guano being the dominant source but also from other cave-dwelling animals such as bears, hyenas, rodents, and some marsupials. The decay of this material, whatever its source, leads to the chemical alteration or diagenesis of other geogenic and biogenic sediments. This can result in the formation of a range of phosphate minerals (Karkanas et al. 2002), which can either form discrete deposits or infiltrate artefacts, such as bone, shell, or wood. In extreme cases, this may lead to the complete loss of bone from a deposit.

The activities of people in caves can produce a variety of sediments and artefacts, but burnt remains are usually dominant. Burning produces abundant sediment as ash, charcoal, and charred remains, and their structure can survive intact despite chemical alteration. These deposits are generally well preserved in limestone caves and their constituents, microstructure, and spatial pattern can provide useful evidence relating to the type and intensity of occupation over time. The use of fire in prehistoric times is of great interest because it is one the most important elements of human evolution (Rolland 2004). Farrand (2001) reported sedimentary accumulation at a rate of as high as 250 cm per 100 years in Mesolithic deposits in Franchthi Cave, Greece. Human activities significantly favour the long-term preservation of vertebrate bone by increasing sedimentation rates near entrances and altering cave entrance topographies, particularly slope, so that subsequent erosion is minimised.

Specific combustion features can be well preserved and may thus provide a clear picture of the burning activities which formed them. In Kebara Cave, Israel, field observations and micromorphology revealed a variety of features including massive accumulations and small patches of charcoal and ashes, intact hearth structures, and diffuse ashy lenses. The composition and microstructure of each of these types of deposits provided information on specific activities such as dumping, trampling, and cleaning that can provide a context for other artefacts, such as bone and stone tools. (Goldberg et al. 2007; Meignen et al. 2007). The Middle Pleistocene cave of Qesem, Israel, contains sediments that appear in the field as a light reddish-brown, lithified massive deposit that archaeologists recognise as a common fill of angular rock fragments and matrix called 'cave breccia'. However, micromorphological analysis revealed that a considerable proportion of these sediments consist of recrystallised wood ash derived from prehistoric fires (Karkanas et al. 2007).

11.8 Archaeology of Liang Bua Cave, Flores (the Hobbit Cave)

Liang Bua means 'cool cave' in the local language on Flores, an eastern island of Indonesia (Figure 11.12). The site is now famous for the discovery of deeply stratified deposits (more than 17 m) containing stone artefacts and faunal remains (Morwood et al. 2004), and the discovery of an almost complete skeleton of a new species of diminutive human named *H. floresiensis* (Brown et al. 2004). Flores has never been connected to Southeast Asia and is separated from Australia by deep oceanic trenches. This has limited major migrations of animals even during low Pleistocene sea levels.

The large cave of Liang Bua (Figure 11.13) was first excavated by the Dutch priest Father Theodor Verhoeven (1950–1965) revealing a wealth of Neolithic burials with grave goods (Verhoeven 1953). In 2001, a joint Australian-Indonesian team led by Mike Morwood extended the west wall excavations and were able to explore deeper levels using innovative shoring techniques to guard against collapse as lower levels of the cave were explored. During this excavation, the skeleton of a mature female was exposed, nicknamed 'Hobbit'

Figure 11.12 Plan and location of Liang Bua in Indonesia. Source: From Sutikna et al. (2016).

by the team members. This skeleton represented a new species of human with a tiny brain and standing a metre tall, with limb proportions similar to apes.

Liang Bua initially formed as a large phreatic chamber in Miocene limestone, later breached and exposed by downcutting of the Wae Racang river about 190 000 years ago (Westaway et al. 2009a). The cave is at the same level as the highest of three alluvial terraces (510 m asl), and coarse fluvial deposits are found at the rear of the cave.

The interplay of fluvial activity and karst processes has produced a complex stratigraphy reflecting the results of slopewash processes, pooling of water, cut and fill deposits, and extensive flowstone formation. At the earliest stages of the cave's evolution, the area available for human occupation would have been very limited until at least 100 000 years ago. From 74 000 to 61 000 years ago a dominant zone of occupation was established in the middle of the cave (Figure 11.14). Within these deposits, nine main sedimentary units have

Figure 11.13 Looking to the northwest at the front of Liang Bua, Indonesia. The cave has accumulated a wealth of archaeological material during its 100 000-year occupation. Source: From Westaway (2017) © Springer Nature.

been identified containing stone artefacts, plant and animal remains, pottery, metal items, and human skeletal remains. This evidence was protected from sheetwash and channel processes by an extensive flowstone that caps this occupation zone (Westaway et al. 2009b).

The skeletal material was associated with stone tools and hearths, and it was initially thought that the 'hobbits' had persisted on the site until about 12 000 years ago. More recent analysis and dating of the complex stratigraphy has led to a revision which indicates that *H. floresiensis* became extinct around 50 000 years ago. The new study dated layers of volcanic ash and calcite directly above and below the fossils.

The *H. floresiensis*-bearing deposits consist of multiple layers of fine-grained sediment interstratified by layers of weathered limestone, speleothem, and loose gravel. These deposits are conformably overlain by an ~2 m thick sequence of five tephras (referred to here as T1-T5), separated by clastic sediments and flowstones. The revision is based on the recognition of a pedestal of stratified deposits at the site which had been partially eroded and thus surrounded by other layered transported sediments. This pedestal extends ~12 m laterally from the eastern wall to the cave centre and is at least 6 m long from north to south. The identification and dating of five distinct tephra layers clarified the stratigraphic relationships (Figure 11.15). Sediment samples for infrared stimulated luminescence (IRSL) and thermoluminescence (TL) dating were collected from the *H. floresiensis*-bearing deposits directly underlying T1. These samples gave statistically

Figure 11.14 View of the rear of Liang Bua, Indonesia, taken from the northwest corner looking southeast. The height of the cave and shape of the domed front chamber are apparent, along with the fluvial gravel deposit on the right that represents the river's first exposure of a subterranean chamber. Excavations have been conducted mostly on the flat front section of the cave where the evidence for occupation was concentrated. From Westaway (2017) © Springer Nature.

indistinguishable IRSL and TL ages of 65 ± 5 and 71 ± 13 kyr respectively, for the time since sand-sized grains of feldspar and quartz were last exposed to sunlight. A TL age of 89 ± 7 kyr was obtained for the basal *H. floresiensis*-bearing deposits, while two further samples from the immediately underlying, gravel layer gave IRSL and TL ages of 128 ± 17 and 113 ± 9 kyr.

Sediments between T2 and T3 yielded a weighted mean IRSL age of 66 ± 9 kyr and a TL age of 59 ± 13 kyr. Interbedded flowstones and a small stalagmite in the same area gave $(^{234}U)/(^{230}Th)$ ages of between 66.1 ± 0.3 and 54.4 ± 0.3 kyr.

Three *H. floresiensis* ulnae and seven *Stegodon* bones were sampled for $(^{234}U)/(^{230Th})$ dating. The *H. floresiensis* ulnae have modelled $(^{234}U)/(^{230}Th)$ ages ($\pm 2\sigma$) for individual laser-ablation tracks that range from 86.9 ± 7.9 to 71.5 ± 4.3 kyr for LB1, 71.4 ± 1.1 to 66.7 ± 0.8 kyr for LB2, and 66.0 ± 4.3 to 54.6 ± 2.1 kyr for LB6. The *Stegodon* bone samples span a modelled age range of 80.6 ± 11.3 to 40.5 ± 2.0 kyr, with the youngest minimum age deriving from a bone recovered from the same sediments and depth as LB6.

The bones of *H. floresiensis* thus range in age from about 100 000 to 60 000 years old (Figure 11.16). Stone tools in the cave used by the 'hobbit' are from 190 000 to 50 000 years old (Sutikna et al. 2016). In 2010 and 2011, archaeologists discovered two hominin teeth in

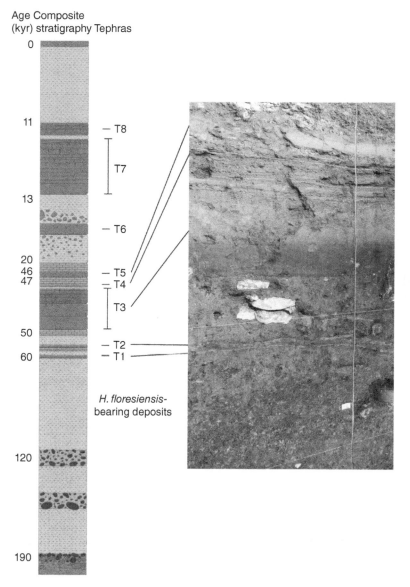

Age Composite
(kyr) stratigraphy Tephras

Figure 11.15 Composite stratigraphy of Liang Bua (Indonesia) showing tephra layers and location of *H. floresiensis* bearing deposits. Source: From Sutikna et al. (2016).

the cave that did not come from *H. floresiensis*. According to Sutikna et al. (2016), the teeth date to around 46 000 BP and are likely to have come from *Homo sapiens*. It is possible that modern humans played a role in the extinction of the 'hobbits', as potential prey animals such as giant storks (*Leptoptilos robustus*), vultures (*Trigonoceps*), and miniature elephants called stegodons (*Stegodon florensis insularis*) vanish from the cave's sediment layers after 46 000 BP. But the final fate of the 'hobbits' remains an open question to be answered by Indonesian and collaborating Australian archaeologists.

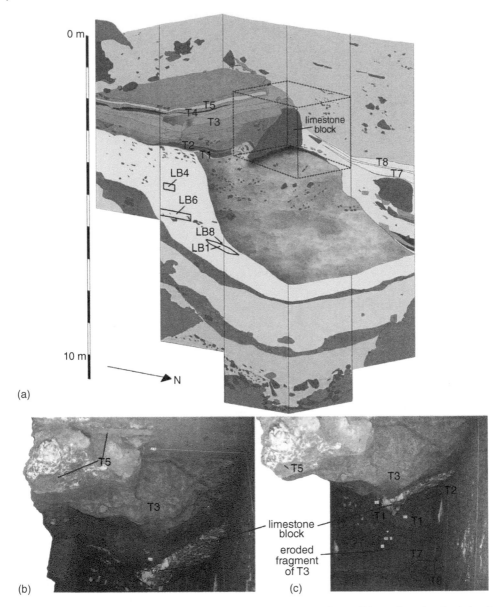

Figure 11.16 Illustration of the erosional surface and the locations of Homo floresiensis skeletal material LB1, LB4, LB6, and LB8 below the boundary. The deposits that unconformably overlie the pedestal are shown in the south and west baulks. The stippled cube outlines the photographed area (in Sector XV) shown in b and c. Both photographs taken from above, with north towards the bottom of the page. Source: From Sutikna et al. (2016) © Springer Nature.

References

Archer, M., Hand, S.J., and Godthelp, H. (1991). *Riversleigh. The Story of Animals in Ancient Rainforests of Inland Australia*. Sydney: Reed Books.

Bae, C.J., Douka, K., and Petraglia, M.D. (2017). On the origin of modern humans: Asian perspectives. *Science* 358: 1269.

Bahn, P. (2004). Art: cave art in Europe. In: *The Encyclopedia of Caves and Karst Science* (ed. J. Gunn). New York: Fitzroy Dearborn.

Bednarik, R.G. (1995). Refutation of stylistic constructs in Palaeolithic rock art. *Comptes Rendus de l'Académie des Sciences* 321: 817–821.

Bednarik, R.G. (2004). Art: cave art in Australasia. In: *The Encyclopedia of Caves and Karst Science* (ed. J. Gunn), 88–90. New York: Fitzroy Dearborn.

Berger, L.R., De Ruiter, D.J., Churchill, S.E. et al. (2010). Australopithecus sediba: a new species of homo-like Australopith from South Africa. *Science* 328: 195–204.

Berger, L.R., Hawks, J., De Ruiter, D.J. et al. (2015). Homo naledi, a new species of the genus Homo from the Dinaledi chamber, South Africa. *eLife* 4: e09560.

Berger, L.R., Hawks, J., Dirks, P. et al. (2017). Homo naledi and Pleistocene hominin evolution in subequatorial Africa. *eLife* 6: e24234.

Bilney, R.J., Cooke, R., and White, J.G. (2010). Underestimated and severe: small mammal decline from the forests of south-eastern Australia since European settlement, as revealed by a top-order predator. *Biological Conservation* 143: 52–59.

Brain, C.K. (1981). *The Hunters or the Hunted? An Introduction to African Cave*. Taphonomy: Chicago University of Chicago Press.

Brown, P., Sutikna, T., Morwood, M.J. et al. (2004). A new small-bodied hominin from the late Pleistocene of Flores, Indonesia. *Nature* 431: 1055–1061.

Bruxelles, L., Clarke, R.J., Maire, R. et al. (2014). Stratigraphic analysis of the Sterkfontein StW 573 Australopithecus skeleton and implications for its age. *Journal of Human Evolution* 70: 36–48.

Clottes, J. and Courtin, J. (1996). *The Cave beneath the Sea: Paleolithic Images at Cosquer*. Harry Abrams: New-York.

David, B. (2017). *Cave Art*. Thames & Hudson.

Desdemaines-Hugon, C. (2010). *Stepping-Stones: A Journey through the Ice Age Caves of the Dordogne*. New Haven, Conneticut: Yale University Press.

Dirks, P.H.G.M. and Berger, L.R. (2013). Hominin-bearing caves and landscape dynamics in the cradle of humankind, South Africa. *Journal of African Earth Sciences* 78: 109–131.

Dirks, P.H.G.M., Kibii, J.M., Kuhn, B.F. et al. (2010). Geological setting and age of Australopithecus sediba from southern Africa. *Science* 328: 205–208.

Dirks, P.H.G.M., Roberts, E., Hilbert-Wolf, H., and Berger, L.R. (2017). The age of Homo naledi and associated sediments in the Rising Star Cave, South Africa. *eLife Sciences* 6: e24231.

Douka, K., Slon, V., Jacobs, Z. et al. (2019). Age estimates for hominin fossils and the onset of the upper Palaeolithic at Denisova cave. *Nature* 565: 640–644.

Farrand, W.R. (2001). Sediments and stratigraphy in rockshelters and caves: a personal perspective on principles and pragmatics. *Geoarchaeology* 16: 537–557.

Gilbertson, D., Mclaren, S., Stephens, M. et al. (2013). The cave entrance sequences and environmental change. In: *Rainforest Foraging and Farming in Island Southeast Asia* (ed. B. Barker). Cambridge: McDonald Institute for Archaeological Research.

Gillieson, D.S. (1986). Cave sedimentation in the New Guinea Highlands. *Earth Surface Processes and Landforms* 11: 533–543.

Gillieson, D.S. (2004). Sediments: Allochthonous Clastic. In: *The Encyclopedia of Caves and Karst Science* (ed. J. Gunn). New York: Fitzroy Dearborn.

Gillieson, D.S. and Mountain, M.J. (1983). The environmental history of Nombe Rockshelter. Papua New Guinea Highlands. *Archaeology in Oceania* 18: 53–62.

Goldberg, P., Laville, H., and Meignen, L. (2007). Stratigraphy and geoarchaeological history of Kebara cave, Mount Carmel. In: *Kebara Cave, Part 1* (eds. O. Bar-Yosef and L. Meignen), 49–89. Cambridge, MA: Peabody Museum of Archaeology and Ethnology, Harvard University.

Granger, D.E., Gibbon, R.J., Kuman, K. et al. (2015). New cosmogenic burial ages for Sterkfontein member 2 Australopithicus and member 5 Oldowan. *Nature* 522: 85–88.

Greer, J. and Greer, M. (1999). Dark zone and twilight zone pictographs in U-Bar cave, southwestern New Mexico. In: *Rock Art Papers*, vol. 14 (ed. K. Hedges), 11–19. San Diego: San Diego Museum of Man.

Hadly, E.A. and Barnosky, A.D. (2009). Vertebrate fossils and the future of conservation biology. In: *Conservation Paleobiology: Using the Past to Manage for the Future*. The Paleontological Society Papers, vol. 15 (eds. G.P. Dietl and K.W. Flessa), 39–59.

Haglund, W.D. (1993). Disappearance of soft tissue and the disarticulation of human remains from aqueous environments. *Journal of Forensic Science* 38: 806–815.

Jaubert, J., Verheyden, S., Genty, D. et al. (2016). Early Neanderthal constructions deep in Bruniquel cave in southwestern France. *Nature* 534: 111–114.

Karkanas, P., Rigaud, J.-P., Simek, J.F. et al. (2002). Ash, bones and guano: a study of the minerals and phytoliths in the sediments of Grotte XVI, Dordogne, France. *Journal of Archaeological Science* 29: 721–732.

Karkanas, P., Shahack-Gross, R., Ayalon, A. et al. (2007). Evidence for habitual use of fire at the end of the lower Paleo-lithic: site-formation processes at Qesem cave, Israel. *Journal of Human Evolution* 53: 197–212.

Lyman, R.L. and Lyman, C. (1994). *Vertebrate Taphonomy*. Cambridge: Cambridge University Press.

Meignen, L., Goldberg, P., and Bar-Yosef, O. (2007). The hearths at Kebara cave and their role in site formation processes. In: *Kebara Cave, Part 1* (eds. O. Bar-Yosef and L. Meignen). Cambridge, MA: Peabody Museum of Archaeology and Ethnology, Harvard University.

Morris, D.A., Augee, M.L., Gillieson, D., and Head, J. (1997). Analysis of a late Quaternary deposit and small mammal fauna from Nettle Cave, Jenolan, New South Wales. In: *Proceedings of the Linnean Society of New South Wales*, 135–162.

Morwood, M.J., Soejono, R.P., Roberts, R.G. et al. (2004). Archaeology and age of a new hominin from Flores in eastern Indonesia. *Nature* 431: 1087–1091.

O'Connor, S., Barham, A., Aplin, K., and Maloney, T. (2016). Cave stratigraphies and cave breccias: implications for sediment accumulation and removal models and interpreting the record of human occupation. *Journal of Archaeological Science* 77: 143–159.

Pickering, R. and Kramers, J.D. (2010). Re-appraisal of the stratigraphy and determination of new U-Pb dates for the Sterkfontein hominin site, South Africa. *Journal of Human Evolution* 59: 70–86.

Reed, E.H. (2003). Vertebrate Taphonomy of large mammal bone deposits, Naracoorte Caves World Heritage Area. Unpublished Ph.D thesis. Flinders University of South Australia. 557 pp.

Reed, E.H. (2008). Pinning down the pitfall: entry points for Pleistocene vertebrate remains and sediments in the fossil chamber, Victoria fossil cave, Naracoorte, South Australia. *Quaternary Australasia* 25: 2–8.

Reed, E.H. (2009). Decomposition and disarticulation of kangaroo carcasses in caves at Naracoorte, South Australia. *Journal of Taphonomy* 7: 265–283.

Reed, E.H. (2012). Of Mice and Megafauna, new insights into Naracoorte's fossil deposits. *Journal of the Australasian Cave and Karst Management Association* 86: 7–14.

Reed, L. and Gillieson, D. (2003). Mud and bones: cave deposits and environmental history in Australia. In: *Beneath the Surface: A Natural History of Australian Caves* (eds. B. Finlayson and E. Hamilton-Smith), 89–110. Sydney: University of NSW Press.

Robert, E. (2017). The role of the cave in the expression of prehistoric societies. *Quaternary International* 432: 59–65.

Rolland, N. (2004). Was the emergence of home bases and domestic fire a punctuated event? A review of the middle Pleistocene record in Eurasia. *Asian Perspectives* 43: 248–280.

Shaw, T. (2004). Archaeologists. In: *The Encyclopedia of Caves and Karst Science* (ed. J. Gunn), 158–166. New York: Fitzroy Dearborn.

Simek, J.F. (2004). Archaeology of caves: history. In: *The Encyclopedia of Caves and Karst Science* (ed. J. Gunn). New York: Fitzroy Dearborn.

Skeates, R. (1997). The human uses of caves in east-Central Italy during the Mesolithic, Neolithic and copper age. In: *The Human Use of Caves* (eds. C. Bonsall and C. Tolan-Smith). Oxford: Archaeopress.

Stone, A. (2004). Art: cave art in the Americas. In: *The Encyclopedia of Caves and Karst Science* (ed. J. Gunn), 190–195. New York: Fitzroy Dearborn.

Straus, L.G. (1997). Convenient cavities: some human uses of caves and rockshelters. In: *The Human Use of Caves* (eds. C. Bonsall and C. Tolan-Smith). Oxford: Archaeopress.

Sutikna, T., Tocheri, M.W., Morwood, M.J. et al. (2016). Revised stratigraphy and chronology for Homo floresiensis at Liang Bua in Indonesia. *Nature* 532: 366–369.

Terrell-Nield, C. and Macdonald, J. (1997). The effects of decomposing remains on cave invertebrate communities. *Cave and Karst Science* 24: 53–63.

Tolan-Smith, C. and Bonsall, C. (1997). The human use of caves. In: *The Human Use of Caves* (eds. C. Bonsall and C. Tolan-Smith). Oxford: Archaeopress.

Verhoeven, T. (1953). Eine Mikrolithenkultur in Mittel- und West-Flores. *Anthropos* 48: 597–612.

Westaway, K.E. (2017). Liang Bua. In: *Encyclopedia of Geoarchaeology*, 473–476. Springer, Springer Nature.

Westaway, K.E., Roberts, R.G., Sutikna, T. et al. (2009a). The evolving landscape and climate of western Flores: an environmental context for the archaeological site of Liang Bua. *Journal of Human Evolution* 57: 450–464.

Westaway, K.E., Sutikna, T., saptomo, W.E. et al. (2009b). Reconstructing the geomorphic history of Liang Bua, Flores, Indonesia: a stratigraphic interpretation of the occupational environment. *Journal of Human Evolution* 57: 465–483.

Woodward, J.C. and Goldberg, P. (2001). The sedimentary records in Mediterranean rockshelters and caves: archives of environmental change. *Geoarchaeology* 16: 327–354.

12

Historic Uses of Caves

12.1 Introduction

Caves and rockshelters have been used by humans throughout history and indeed prehistory, often on an opportunistic basis as convenient cavities. Some caves have been used for the same purposes over a very long time, while others have seen major changes in their use. Many caves have been used by people, then abandoned and used by carnivores and raptors as dens. We can discern several major themes in the human use of caves. Caves have always been used as shelters, as living spaces, and as refuges in times of conflict. They are used as shrines or temples – as sacred spaces that engender feelings of awe and veneration, and facilitate religious observances by being places set apart from daily living. Caves are important sources of raw materials used in industry, such as saltpetre for gunpowder, guano for fertiliser, rare minerals, and some foodstuffs. They have been used as factories or storage spaces for high-valued commodities, such as cheese, wine, and preserved foods. Finally, there has been a tremendous growth in cave tourism over the last three centuries to feed the demand for novel experiences amongst travellers.

12.2 Caves as Shelter

Caves are ecologically-friendly houses. They are the most ecologically-sensitive form of construction and could be combined with an alternative energy system to be almost completely sustainable. They maintain a constant indoor temperature with natural earth and rock insulation (Figure 12.1), which also keeps them very quiet (Earth Homes Now 2018). They do not require large inputs in terms of bricks, concrete, mortar, metal, or wood, and do not demand costly synthetic or ecologically-questionable construction materials. In the Loire Valley of France, there are 45 000 cave homes (according to Association CATP – Carrefour Anjou Touraine Poitou – an organisation set up to promote and preserve the troglodytic heritage of the area). In Granada, Spain, there are cave dwellings dating back to Moorish times before the fifteenth century CE. The Albayzin and Sacromonte neighbourhoods are adjacent hillside cave-dwellings, both of which form part of a UNESCO World Heritage Area (WHA) along with the Alhambra Palace.

Caves are used as temporary dwelling places by many people. During transhumance in alpine areas, caves, and rockshelters are used as seasonal bases and may be modified by

Figure 12.1 Modern cave house in Cotignac, France. Source: https://www.airbnb.com.au.

rock walls, fireplaces, internal floor modifications, and rough benches. They may also be used to stable goats and other livestock. Many caves and rockshelters are used for camping and for the secret trysts of young people, away from disapproving parental gaze.

Caves have long been used for defensive purposes. Many fossil river caves in the Guilin tower karst, China have partially walled entrances. There are many medieval fortified caves in Switzerland in the Grisons and Valais – dating to 1160 CE in the case of Marmels Castle, Oberhalbstein, Chur (Figure 12.2). These small keeps were no doubt safe on account of their difficult access but were cold and damp refuges.

The first recorded use of a cave as a sanatorium comes from Mammoth Cave, Kentucky, USA. It had been noted as early as 1812 that slaves mining saltpetre in the cave enjoyed excellent health. For many years, visitors had reported feelings of well-being after being exposed to the cave's environment. Some considered its steady temperature, constant humidity, and dry air to be beneficial and restorative. By 1839, Dr. John Croghan had purchased the cave and set to work to make it a tourist attraction. After hearing repeated praise for the quality of air in the cave, and feeling better himself for all the time he spent underground, Dr. Croghan decided to open an underground tuberculosis (TB) hospital. Two small stone buildings with canvas roofs were erected inside the cave (Figure 12.3).

Many of patients who came to Mammoth Cave stayed only for a few months before leaving, while others stayed until their deaths. Their condition was made worse by smoky fires and lard lamps used for lighting. One patient, Oliver Hazard Perry Anderson, wrote:

Figure 12.2 Marmels Castle, Grisons, Switzerland. This small keep was likely to have been constructed around 1100 CE. Source: https://commons.wikimedia.org/w/index.php?curid=45987318.

> 'I left the cave yesterday under the impression that I would be better out than in as my lungs were constantly irritated with smoke and my nose offended by a disagreeable effluvia, the necessary consequence of its being so tenanted without ventilation.'

Dr. Croghan himself died of tuberculosis in 1849, and the cave was sold. The stone cubicles of the TB hospital can still be seen in Mammoth Cave.

Today there are several cave sanatoria in central and eastern Europe. There are cave sanatoria in the Ural Mountains and in Siberia within Russia. In the Czech Republic, the Gombasek Cave is located in the Slovak Karst National Park, in the Slaná River valley. The cave is used for 'speleotherapy' as a sanatorium, focussed on airway diseases. Since 1995, the Gombasek Cave has been inscribed on the UNESCO World Heritage list as a part of the Caves of Aggtelek Karst and Slovak Karst WHA.

Caves are used as burial places in many parts of the world and have been so for tens of thousands of years. It seems likely that our ancestors only used caves as dwelling places episodically, and chose to live in the open in tents or huts. Thus, caves as separate, mysterious places were ideal for the burial of the dead, with associated ceremonies. Quite often the corpse would be exposed outside and allowed to reduce to bones, which would then be coated with ochre and buried with or without grave goods such as tools, food, and ornaments. In Papua New Guinea (PNG), caves have been used for the deposition of skeletons and such sites are regarded as tabu or forbidden places. In both PNG and Sarawak, Malaysia, the dead were placed in ornamented canoe prows which were located in alcoves. During

Figure 12.3   Patients at the TB sanatorium in Mammoth Cave, Kentucky, USA. Source: Library of Congress.

tribal fighting, the vanquished enemy were often forced to suicide by jumping into limestone shafts at spear or arrow point. Thus, there is a pile of shattered bones and skulls at the base of the Nombi shaft in the Chimbu region of PNG.

Cave springs are also important sources of water and food in the form of flying foxes, microbats, and swiftlets. During WWII, many caves in coastal PNG were used as prisons by the invading Japanese troops; in some cases, the prisoners were able to escape via a rear entrance to the cave!

12.3   Caves as Sacred Spaces

Caves are often regarded as ambiguous spaces, offering both protection and shelter but can also trap and imprison people. In many cultures, a location within the earth is identified as female, and caves have been identified as representing the womb of Mother Earth, and are associated with birth and regeneration. In many cultures, people enter caves and become trapped, only being released after some ordeal. Although sacredness may be invested in many other natural forms and objects, such as trees, springs, and mountains, the earliest sacred places in prehistory are in naturally-formed caves, such as those in the Dordogne valley of France. Natural caves have long been a focus of veneration and frequently appear in both mythological and religious stories.

In Greek mythology, the god Zeus was born in a cave on Mount Ida (or Mount Dicte) on the island of Crete. When excavated, this cave was found to be filled with votive offerings. Sacred caves are found throughout Greece, such as the Corycian Cave at Delphi sacred to the nymph Corycia and the god Pan. Rites associated with the goddess Cybele also took place in caves. A sacred cave may also contain a sacred spring which may possess special healing or divinatory properties (for example, the spring and cave at Lourdes in France). Well-known sacred caves are found in India, at Ajanta, Ellora, and Elephanta, all of which have been embellished with carvings and frescoes.

The philosopher Porphyry (234–305 CE) held that before there were temples, all religious rites took place in caves. He argued that the architecture of temples emulated the darkness and single entry of most caves and that the penetration of light into a cave at certain times of the year had ritual significance. This is also noted at New Grange in Ireland and Maeshowe in Orkney, Scotland.

There are many cave temples in SE Asia, both because they are convenient cavities close to towns and because they also have an air of mystery with hidden chambers. There are several well-patronised Taoist and Buddhist temples near the city of Ipoh in northern Malaysia (Figure 12.4). These are well decorated and have modified floors to permit access

Figure 12.4 Taoist shrine in Kek Lok Tong Caves, Ipoh, Malaysia.

(a)

(b)

Figure 12.5 Shinto shrine in Futenma-gu Cave, Okinawa, Japan.

Figure 12.6 Main temple chamber of Batu Caves, near Kuala Lumpur, Malaysia.

and appropriate infrastructure (lighting, parking areas, picnic areas). On the Japanese island of Okinawa, there are several Shinto shrines in the entrance areas of caves.

Near to the Ginowan USA air force base on Okinawa, Japan, Futenma-gu Cave is a Shinto shrine on the middle terrace at 70 m altitude. The cave is some 280 m long with three main chambers. All show some evidence of forming closer to sea level. Three horizontal wall notches are well preserved and suggest former water rest levels. The cave entrance has been modified with a wooden porch (Figure 12.5) while inside the shrines are placed between massive stalagmites.

Many caves in India and Malaysia are used as Hindu temples. The best known is the Batu Caves complex (Figure 12.6) outside Kuala Lumpur (Malaysia). This group of caves were developed in 1891 by Mr. K. Thamboosamy Pillay, a Tamil community leader and businessman. The Batu Caves serve as the focus of the Hindu community's yearly Thaipusam festival. They have become a pilgrimage site not only for Malaysian Hindus but Hindus from countries such as India, Australia, and Singapore.

12.4 Caves as Sources of Raw Materials

The healing properties of caves and their contents have been highly regarded for millennia. Crushed stalactites were used in Chinese traditional medicine as early as the fourth century BCE. They were used as an antacid, to suppress coughs, to stop bleeding, and encourage lactation in nursing mothers. Medieval Europeans used stalactite powder to strengthen bones and treat fevers. Poultices were also made from moonmilk powder to treat eye infections.

Even today, some people believe that selenite, a form of gypsum, eliminates negative energy, relieves stress, and may even reduce the risk of cancer. Epsomite, a natural form of Epsom Salts, has been harvested from caves in central America for treatment of gastrointestinal disorders.

The mining of cave guano deposits for fertiliser is a worldwide phenomenon. Prior to the introduction of artificial or chemical fertilisers, natural or organic fertilisers were widely used from sources such as bird and bat guano. Bird guano was mined on Pacific islands such as Christmas Island (Australia) and Nauru in the Indian Ocean. In Niah Cave, Sarawak, Malaysia, cave swiftlets guano is mined for fertiliser (Figure 12.7). Large accumulations of both bird and bat guano, in many cases 10 m deep, are harvested by local farmers and by small scale entrepreneurs. In the Niah Great Cave, both areas of floor and sections of the wall of the cave are owned by family groups as a traditional right. The infrastructure used for the collection of cave swiftlet nests is permanently in place and must be maintained on a regular time frame. Bamboo scaffolding (Figures 12.8 and 12.9) is tied with rattan and is usually attached to the cave wall. Despite care and maintenance several people are injured or killed in accidents each year.

There is also a large array of phosphatic and ammoniacal minerals derived from bat guano (White 1988). In the USA, nitrate minerals were widely mined for the manufacture of saltpetre during the American Civil War (De Paepe and Hill 1981). In Australia, caves at Wellington, Timor, Moore Creek, and Mount Etna were all extensively mined for horticultural fertilisers well into the twentieth century.

Figure 12.7 Entrance area of Niah Cave, Sarawak, Malaysia. Huts belong to swiftlet nest harvesters and guano collectors.

Figure 12.8 Bamboo scaffolding used to reach swiftlet nests in Niah Cave, Sarawak, Malaysia. Climbers ascend 20 or 30 m to reach nests that are harvested outside of the breeding season.

Figure 12.9 Bamboo poles used to reach swiftlet nests in the roof of Niah Cave, Sarawak, Malaysia.

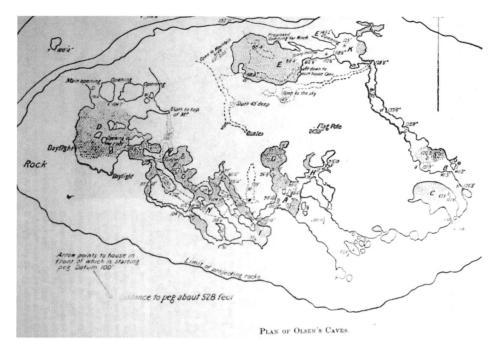

PLAN OF OLSEN'S CAVES.

Figure 12.10 Plan of Olsen's Caves, central Queensland, Australia, drawn in 1903. Source: Queensland Historical Atlas. http://www.qhatlas.com.au/map/plan-olsens-cave-1903.

In 1903, J. Christensen leased the 'cave property' of 80 acres at Mount Etna, central Queensland with the intention to excavate guano from the caves and sell it as a fertiliser. He and H. Buch mapped Olsen's Caves (Figure 12.10).

Cathedral Cave, marked 'F' on the map in Figure 12.10, was a 'beautiful cave' that contained 'the Font' and 'the Pulpit' which were two large stalagmite formations. Christensen observed 'At first sight, the caves appear most irregular in their formation, but a glance at the accompanying map shows them generally to run in the parallel direction from north-west to south-east.'

The Mt Etna Fertiliser Company was established in 1924. A map (Figure 12.11) displays the extent of mining leases taken over the Caves area as early as 1924 and shows Mount Etna in the centre of the map on mining lease R444. The map appeared in the Prospectus of the Mt. Etna Fertiliser company when they issued a call for investment. The company noted that 'From the small beginning the business has greatly increased, other leases being obtained from the Crown, particularly Mt. Etna proper, with very large caves and with enormous quantities of Guano and Limestone.' This limestone resource was destined to be at the centre of the longest-running conservation battle in Australia, from 1966 to 2006 (Lines 2006).

Caves have long been used as water sources. Many Chinese karst springs were important documented water sources in the Shang dynasty (1600–1100 BC), and the Jinci spring was being used for irrigation in 450 BC (Yuan 1991). Many Chinese springs are multiple-use sites exploited for hydroelectricity, irrigation, and as minor tourist caves – for example, Longgong Cave, Guizhou.

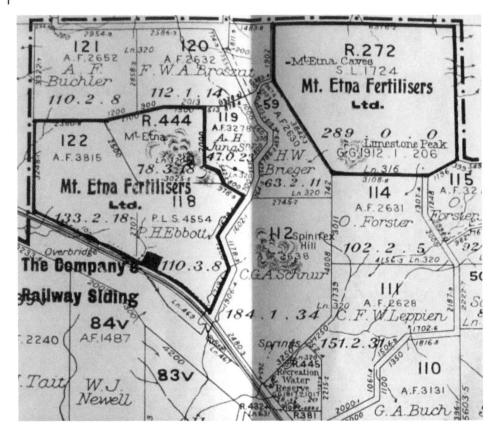

Figure 12.11 Extent of mining leases taken out by the Mount Etna Fertiliser Company (Australia) in 1924. Source: Queensland Historical Atlas http://www.qhatlas.com.au/map/location-plan-mt-etna-fertilisers-1924.

Kentucky bourbon from the Jack Daniels distillery still relies partly on cave spring water (Figure 12.12), though the efficacy of the cave to filter the water may be doubtful. During the Prohibition era 'moonshine' whisky was made in several Tennessee caves, and traces of these activities may be seen today.

Caves have also been widely used for cheese-making for at least two millennia. In the Picos de Europa of northern Spain, cabrales is a matured blue cheese ripened in limestone caves above the town of Arenas de Cabrales. The chilly (8–12 °C) and humid (90%) conditions in the caves facilitate the growth of bluish-green penicillium mould in this precious cheese. The famous Roquefort blue sheep's cheese from southern France has many imitations, but only those cheeses aged in the natural Combalou caves of Roquefort-sur-Soulzon may bear the name. Similarly cheddar cheeses are matured in natural caves of the Cheddar Gorge and at the Wookey Holes Caves in Somerset, England.

Other products are matured in caves. The kimchi much loved by Koreans is matured in Gwangmyeong cave, where families can store their fermented kimchi (Figure 12.13) in large jars on racks in a side passage.

In northern France, champagne is matured by a complex process in caves excavated in the chalk. Taittinger's Gallo-Roman chalk caves date to the fourth century CE and

Figure 12.12 Cave spring at Lynchburg, Tennessee, USA, used for the manufacture of Jack Daniels bourbon.

Figure 12.13 Jars of kimchi (fermented cabbage and chilli) stored for clients in Gwangmyeong Cave, Korea.

Figure 12.14 Champagne maturing in a chalk cave near Rheims, France. Source: Andia / Alamy Stock Photo.

are located approximately 20 m below the surface, preserving constant temperature and humidity (Figure 12.14). They extend for 12 km and were recently named a UNESCO World Heritage site.

Several caves have been used for rope works (Figure 12.15). The near-constant humidity facilitates the twisting of hemp fibres to make hawser laid rope, widely used since medieval times for the rigging of sailing ships. Peak Cavern in Derbyshire was used for this purpose until the mid-twentieth century, and the ropemakers would guide visitors around the cave by candlelight.

Many caves in North America were used as sources of saltpetre (potassium nitrate), an essential ingredient of gunpowder. The black powder used in flintlock rifles contained equal proportions of sulphur, charcoal, and saltpetre. As Kentucky was opened for settlement by Europeans in the 1790s, caves along the Green River were explored and found to contain white, fluffy deposits of saltpetre in the sediments. These deposits were mined and then leached to extract the saltpetre. During the war of 1812 between the United States of America and Great Britain, much of the large quantity of saltpetre needed to fight the war was mined at Mammoth Cave (Crothers et al. 2013) (Figure 12.16). The cave owners relied on a workforce of approximately 70 African–American slaves to mine the mineral and haul it to leaching vats located both at the Rotunda, the large chamber a short distance from the Historic Entrance. Other caves with saltpetre deposits were used in the American Civil War (1860–1865), and some minor battles were fought over access to the resource.

An aircraft storage facility was established in the Grotte de Bédeilhac, Pyrenees, France during the latter stages of WWII. This contained partially dismantled Luftwaffe bombers and it was rumoured that small aircraft were actually flown out (Figure 12.17). In 1972, a French pilot flew a small plane into the cave and managed to land safely.

Figure 12.15 Ropeworks at Peak Cavern, Derbyshire, England. The near-constant humidity facilitated the production of hawser-laid hemp ropes used on the ships of Great Britain's Navy in the 18th and 19th centuries.

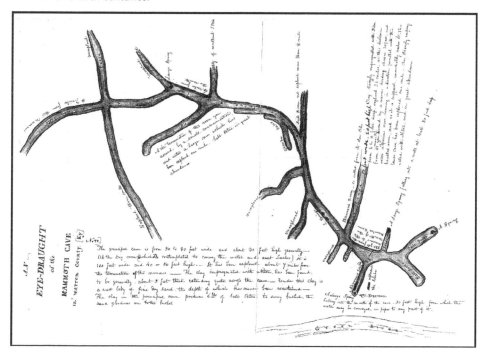

Figure 12.16 Map of the saltpetre deposits in Mammoth Cave, Kentucky, USA. Source: From Brucker (2008) © Elsevier.

(a)

Avion décollant de la grotte de Bédeilhac.

(b)

Figure 12.17 The Grotte de Bédeilhac, Pyrenees, France was used as an aircraft storage facility by the Luftwaffe during WWII. a entrance to the cave b Luftwaffe plane flying out of the cave. Source: Images from http://www.hominides.com/html/lieux/grotte-de-bedeilhac.php and from a French twentieth century postcard by Clement Sans © Hominidés.com.

12.5 Cave Tourism

The growth of cave tourism, from modest beginnings in the late nineteenth century with candle lanterns to today, when fibre-optic lights and electric trains are employed, has expanded both the use and the range of impacts on caves drastically. There are approximately 1300–1500 show (tourist) caves worldwide (Duckeck 2015; Spate and Spate 2013) with some receiving several million visitors each year (Mammoth Cave, USA, and Postojna Cave in Slovenia; Gillieson 2011). These statistics probably underestimate the number of show caves in China, where there may be more than 300 open to the public (Spate and Spate 2013; Zhang and Zhu 1998). Conservatively there may be 170 million visitors to show caves each year, and 200–300 000 people may be employed globally in cave tourism.

Commercial cave tourism dates back to at least 1633, when the Count of Petac opened Vilenica Jama in Slovenia to the paying public (Cigna 2005). During the eighteenth and nineteenth centuries, cave visits became popular with the emerging interest in the natural sciences. In Australia, Major Thomas Mitchell produced some fine lithographs to illustrate his account of the exploration and mapping of inland New South Wales, Australia, (Figure 12.18a). By the nineteenth century, several caves were open to the public – Postojna Jama in Slovenia, the Cheddar Caves in Britain, Mammoth Cave in the USA, and Jenolan Caves in Australia. The first cave photograph was taken in the Blue John Caverns,

(a)   (b)

Figure 12.18 (a) Lithograph of the Wellington caves from Mitchell (1838) (b) photo of Nettle Cave, Jenolan, Australia, taken by Caney & Co in 1883. Source: Chris Howes / Alamy Stock Photo.

Derbyshire, England, by Alfred Brothers in 1865 (Howes 1989). Indeed, some of the earliest cave photographs were taken at Jenolan, Australia by Hart in 1879. A very impressive series of photos were taken by Caney & Co, Mount Victoria in 1883 (Figure 12.18b) and were used as a set of bound postcards. During the early twentieth century, cave images were used by railways to promote the growth of mass tourism; the Paris-Lyon-Méditerranée railway produced a very fine set which today are collector's items.

The modification of caves for tourism requires the alteration of natural passages to improve access, gating of entrances, and installation of stairs and platforms and of lighting of various kinds. Usually, the cave hydrology may be radically altered, and there may be issues with runoff from car parks, introducing hydrocarbons and heavy metals (zinc, cadmium, chromium) as a result. The effects of lighting are well documented (Cigna 1993) and include increased cave temperature, reduced humidity, increased carbon dioxide and the introduction of organic material such as lint and skin cells, fungal spores, bacteria, and protozoa (Gillieson 2011).

Large numbers of visitors in a cave can significantly raise the air temperature. A single person releases heat energy at 80–120 W (Villar et al. 1986), about the same as a single incandescent light bulb. Thus, a party of 50 or 60 people on a cave tour can locally raise temperatures by 1–2 °C. The passage of tourists through Altamira Cave, Spain, raised the air temperature by 2 °C, CO_2 concentration from 400 to 1200 ppm, and decreased relative humidity from 90 to 75% (De Freitas 2010; Villar et al. 1986). The main effect of the reduction in humidity is drying and flaking of flowstone surfaces. According to Cigna (1993), management needs to ensure that these fluctuations lie within the range of natural variation for the cave and that they return to normal levels in a short period of time. Calaforra et al. (2003) provide a good example of determining visitor thresholds in such cases.

Lighting of show caves has tended in the past to over-illuminate the cave passages and chambers as if they were offices or shopping malls. A more enlightened view would light the cave as a cave with deliberate use of darkness and sequencing of illumination on selected cave features. There are two important principles to be borne in mind when designing the lighting for a show cave: access and atmosphere.

Lighting for access should be at the minimum level consistent with the safe movement of all cave visitors. Effective lighting can be used to create safe access through an unfamiliar environment, a zone of familiarity that relaxes the visitors. The use of LED strip lights, 12 V downlights, and other low-energy technology can all achieve this aim. These can be attached to railings or path edges, with necessary inverters or batteries well-hidden below. In general, all fixtures and cabling should be well hidden from visitors but accessible for maintenance without further damage to the cave and its contents. Reduced power consumption has benefits beyond reduction of CO_2 emissions: lower power requirements facilitate the use of local uninterruptible power supply when there is a mains power failure. Less heat is produced as well. There are many technologies available – remote controls, c-bus controlled electrical systems, high lumen per watt output lighting, batteries/inverters, optical fibres, etc. – but they should be used as tools to achieve an end, not as an end in themselves.

The second principle is that of atmosphere. There should be an underlying philosophy to the lighting scheme. A theme should be established which illustrates aspects of cave development or history. The lighting should be sequential, with visitors led from one scene to the next. This avoids the massed illumination of a whole chamber. The manager needs to be very selective about what to light and what not to light. Any light in a dark environment will have a dramatic effect, and sometimes, a very distant light will enhance the illusion of depth and mystery we are trying to foster. Lighting of water features can be very effective. In all of these, the fragility of the cave contents needs to be considered, with some areas being out of bounds for any installation of lighting.

12.6 Cave Dwellings in Turkey

A very distinctive landscape on the Anatolian plateau of Turkey has become world-famous and is a World Heritage site inscribed in 1985. This is the Göreme National Park and the Rock Sites of Cappadocia, where a series of mountain ridges, gorges, and pinnacles have formed by erosion of volcanic tuff. The area is bounded on the south and east by the extinct Erciyes Dağ and Hasan Dağ, both over 300 m high. About 10 million years ago, volcanic eruptions spread ash and welded tuff (ignimbrites) across the area to form level surfaces over 20 000 km², some of which were then infilled with lake deposits (Sarikaya et al. 2015). The dissection of these surfaces along cooling fractures produced gorges and isolated pinnacles, locally termed 'fairy chimneys', and known in North America as hoodoos (Figure 12.19). They are called demoiselles coiffées ('ladies with hairdos') in the French

Figure 12.19 Tuff pinnacles capped by ignimbrite, Pasabagi, Turkey. Source: Photo by Gabriel Crowley.

Figure 12.20 Cave houses excavated in tuff at Zelve, Turkey. Source: Photo by Gabriel Crowley.

Alps. The harder ignimbrites form caps which protect the underlying softer tuff or lake deposits from erosion, forming mushroom-shaped pinnacles. When the caps finally fall off or erode away the pinnacles erode quite quickly.

Subsequently, people have enlarged or excavated cave dwellings in the cliffs or pinnacles. The best-known sites are Karain, Karlık, Yeşilöz, Soğanlı, and the subterranean cities of Kaymaklı and Derinkuyu. The very large number of cave houses, churches, and subterranean cities within the volcanic tuff (Figure 12.20) make it the world's most striking collections of cave dwellings (Gülyaz 2012). Recently mass tourism has led to a great expansion of accommodation in the area, some of it quite poorly designed (Erdogan and Tosun 2009).

As well as the sheer beauty of the landscape and the interesting cave dwellings, the area contains outstanding examples of Byzantine religious art dating back to the fourth century CE. At that time, small monastic communities, acting on the teachings of Basileios the Great, Bishop of Kayseri, began inhabiting cells hewn into the soft rock. Thus, Cappadocian monasticism was already well established by the establishment of the iconoclastic period (725–842 CE) during which decoration was minimalist (Figure 12.21) and largely confined to crosses in tempera paint.

After 842 CE, many new churches were dug in Cappadocia and were richly decorated with brightly coloured figurative paintings (Figure 12.22). These remain in a very good state of preservation.

(a)

(b)

Figure 12.21 Iconoclastic images in a cave church, Göreme Open Air Museum, Turkey. Source: Photo by Gabriel Crowley.

(a)

(b)

Figure 12.22 Figurative images in a cave church, Ilhara Gorge, Turkey. Source: Photo by Gabriel Crowley.

Figure 12.23 The old and the new juxtaposed, Göreme, Turkey. Source: Photo by Gabriel Crowley.

During a period of Islamic invasions, people banded together into underground villages or towns that extend over hundreds of metres in depth and have elaborate stone gates. The best known of these are Kaymakli and Derinkuyu, which served as places of refuge for hundreds of people. Today the area is heavily used by tourists and some vandalism has been reported, as well as some inappropriate buildings (Figure 12.23) being constructed (Tucker and Emge 2010). Long-term bedrock erosion rates have been determined for plateaus, dissected plateau, and fairy chimneys in the Cappadocian landscape using the in-situ cosmogenic 36Cl (Sarikaya et al. 2015). The results show that the caps of chimneys have erosion rates of between 1.26 cm/ky and 3.39 ± 0.36 cm/ky, but once the caps are completely destroyed and removed, the erosion rates of the remaining chimneys increase significantly to about 28.0 ± 9.9 cm/ky. Thus, as the plateau edges retreat new 'fairy chimneys' will form but they are quite rapidly destroyed once the capstones are removed. There has been some earthquake damage to some of the cones and the pillars, but this may be seen as a naturally occurring phenomenon. However, the area is in danger of being 'loved to death' by increasing numbers of tourists.

References

Brucker, Roger. W. (2008). Mapping of Mammoth Cave: How Cartography Fueled Discoveries, with Emphasis on Max Kaemper's 1908 Map. Mammoth Cave Research Symposia, Western Kentucky University Paper 4. https://digitalcommons.wku.edu/mc_reserch_symp/9th_Research_Symposium_2008/Day_one/4

Calaforra, J.M., Fernández-Cortés, A., Sánchez-Martos, F. et al. (2003). Environmental control for determining human impact and permanent visitor capacity in a potential show cave before tourist use. *Environmental Conservation* 30: 160–167.

Cigna, A.A. (1993). Environmental management of tourist caves. *Environmental Geology* 21: 173–180.

Cigna, A.A. (2005). Environmental management of tourist caves. In: *Encyclopedia of Caves* (eds. D.C. Culver and W.B. White), 495–496. Burlington, MA: Academic Press.

Crothers, G.M., Pappas, C.A. and Mittendorf, C.D. (2013). The History and Conservation of Saltpeter Works in Mammoth Cave, Kentucky. Mammoth Cave Research Symposia, Western Kentucky University, Paper 2. http://digitalcommons.wku.edu/mc_reserch_symp/10th_Research_Symposium_2013/Day_one/2

De Freitas, C.R. (2010). The role and importance of cave microclimate in the sustainable use and management of show caves. *Acta Carsologica* 39: 477–489.

De Paepe, D. and Hill, C.A. (1981). Historical geography of United States saltpeter caves. *National Speleological Society Bulletin* 43: 88–93.

Duckeck, J. (2015). Showcaves.com. http://www.showcaves.com/english/index.html (Accessed 26 February 2018).

Earth Homes Now (2018). Cave homes. Grand Terrace, California. http://www.earthhomesnow.com/cave-homes.htm (Accessed 9 March 2018).

Erdogan, N. and Tosun, C. (2009). Environmental performance of tourism accommodations in the protected areas: case of Goreme historical National Park. *International Journal of Hospitality Management* 28: 406–414.

Gillieson, D.S. (2011). Management of Caves. In: *Karst Management* (ed. P. Van Beynen), 141–158. Dordrecht: Springer.

Gülyaz, M.E. (2012). *Göreme National Park and the Rock Sites of Cappadocia*. Republic of Turkey: Ministry of Culture and Tourism Publications.

Howes, C. (1989). *To Photograph Darkness: The History of Underground and Flash Photography*. Gloucester, UK: Alan Sutton Publishing.

Lines, W.J. (2006). *Patriots: Defending Australia's Natural Heritage*. St. Lucia, QLD: University of Queensland Press.

Mitchell, T.L. (1838). *Three Expeditions into the Interior of Australia, with Descriptions of the Recently Explored Region of Australia Felix, and the Present Colony of New South Wales*. London: T. and W. Boone.

Sarikaya, M.A., Çiner, A., and Zreda, M. (2015). Fairy chimney erosion rates on Cappadocia ignimbrites, Turkey: insights from cosmogenic nuclides. *Geomorphology* 234: 182–191.

Spate, A. and Spate, J. (2013). World-wide show cave visitor numbers over the recent past: a preliminary survey. *ACKMA Cave and Karst Management in Australasia* 20: 57–69.

Tucker, H. and Emge, A. (2010). Managing a world heritage site: the case of Cappadocia. *Anatolia* 21: 41–54.

Villar, E., Fernandez, P.L., Gutierrez, I. et al. (1986). Influence of visitors on carbon dioxide concentrations in Altamira cave. *Cave Science* 13: 21–23.

White, W.B. (1988). *Geomorphology and Hydrology of Karst Terrains*. New York: Oxford University Press.

Yuan, D. (1991). *Karst of China*. Beijing: Geological Publishing House.

Zhang, R. and Zhu, X. (1998). Evaluation on exploitation level of show caves – current situation, existing problem and improvement proposal of show caves in China. *Carsologica Sinica* 17: 254–259.

13

Cave Management

The recreational use of caves in a modern sense dates from the early seventeenth century, when the Vilenica Cave in Slovenia was open for visits by paying tourists. By 1816, a number of caves worldwide were being regularly visited (Postojna, Slovenia; Wookey Hole, UK; Mammoth Cave, USA). A rapid expansion in the number of show caves occurred in the second half of the nineteenth century. Today, there are more than 600 show caves open worldwide, with some receiving several million visitors each year (Gillieson 1996; Spate and Spate 2013). According to Zhang and Jin (1996), there are around 800 tourist caves in the world. The estimated number of visits in 197 tourist caves had revealed that more than 25 million people visit them each year. Tourist caves are most attractive in Europe (48% of visits), followed by Asia (36% of visits) whereas North America and other continents have a share of 8% of visits. A rough estimation of income from cave tourism is more than 2.3 billion USD per year, whereas around 100 million people are directly or indirectly affected by the income of cave tourism. Based on data from annual reports for 2016, Postojna Cave, the biggest tourist attraction in Slovenia, was visited by 689 608 tourists, while Škocjan Caves, the only cave system in Slovenia that is under the auspices of UNESCO, had over 145 000 registered visitors. In the Postojna Cave, the economic model prevails, whereas Škocjan Caves are more oriented towards protection of the environment, being a World Heritage site since 1986.

Cigna (2011b) has made estimates (Table 13.1) of global show cave activity:

When we come to consider strategies for cave management, we must first realise that karst and its component caves form a dynamic system which includes landforms, water, gases, life, and energy. This theme has been explored in Chapter 2 of this book. Any management system must, therefore, aim to maintain the flows of mass and energy, the balance of life and its products in the karst environment.

There are some basic premises which should guide cave management:

- Firstly, that caves are a measure of the intensity and persistence of the karst process.
- Secondly, that caves tend to integrate both surface and underground geomorphic processes.

Caves: Processes, Development, and Management, Second Edition. David Shaw Gillieson.
© 2021 John Wiley & Sons Ltd. Published 2021 by John Wiley & Sons Ltd.

Table 13.1   Show caves numbers, visits, and economy.

Number of show caves in the world	>5000
Most important show caves	>800
Total visitors per year	≈170 000 000
Money spent yearly to visit show caves	≈$1.5 billion
People directly employed in show caves	≈200 000
People whose salary comes indirectly from show caves	≈100 000 000

Source: From Cigna (2011b) © 2011 Elsevier.

- Thirdly, that once these products of surface and underground processes enter the cave system, they are likely to be preserved with minimal alteration for tens of millennia, perhaps even millions of years.
- Finally, caves can be regarded as natural museums in which evidence of past climate, geomorphic processes, vegetation, animals, and people will be found by those who are persistent and know how to read the pages of the earth history displayed for them.

These propositions were also made in Chapter 1 and underpin the structure of this book. Visitor management must include components of the protective functions of cave management and should recognise the displayed features of earth history and biology in an individual cave so that they may be appropriately presented and interpreted to excite the imagination of the visitors regardless of their age, ethnicity, and beliefs.

The IUCN Guidelines for Cave and Karst Protection (Watson et al. 1997) provides a general list of the impacts which may result from human activities in caves:

- Alteration of the physical structure of the cave
- Alteration of water chemistry
- Alteration of cave hydrology
- Alteration of air currents and microclimate
- Introduction of artificial light
- Compaction or liquefaction of floors
- Erosion or disturbance to cave sediments or their contents
- Destruction of speleothems
- Destruction of fauna
- Introduction of alien organisms or materials (e.g. concrete, climbing aids), pollutants, nutrients, animal species, algae, and fungi
- Surface impacts, such as erosion, siltation, and vegetation change

It should be said at the outset that any consideration of human impacts in caves should also take into account activities in the catchments overlying the cave. These activities are covered in other chapters. In this chapter, I consider the impacts of visitors to caves through show cave development, and also through recreational caving. I also review some methodologies for assessing the values and vulnerability of caves. Finally, I comment on changing paradigms for protected area management, which have implications for the ways in which caves will be managed in the future.

13.1   Introduction – Caves as Contested Spaces

Humans have used caves for a variety of purposes over tens of millennia (see Chapter 12 for examples). Both our species and closely related species, such as Neanderthals, used the deep zones of caves for ritual activities involving painted and engraved art, sculptures, and stone arrangements. Only the entrance and outer zones of caves were used for occupation and storage of food (cheese, dried meats, wine) and for the small-scale extraction of mineral resources (flint, guano, evaporite minerals). Both Greek and Roman authors refer to the religious use of cave shrines, but these activities involved minimal alteration of the cave environment. Thus, for much of human history, most caves were approached with awe and reverence. This changed in the Renaissance with the emergence of tourism and the harnessing of water resources from cave springs as a precursor to industrialisation.

The change in human attitudes towards caves has had implications for how caves are regarded and used by different interest groups, with the most obvious area of contention being the nexus between cave conservation and cave development for mass tourism. Other areas include the contention between passive recreation or conservation and quarrying; the development of cave springs for water supply and hydroelectricity, the mining of cave minerals; and the use of caves for transport tunnels. For example, the construction of the Three Gorges Dams on the Yangtse River in China flooded large areas of karst containing caves that contained both religious shrines and endemic troglobitic species.

Even in less dramatic contested cave landscapes, tensions arise from the differing expectations of tourists coming from culturally different backgrounds. Doorne (2000) has reviewed the management of the Waitomo Glowworm Cave in New Zealand in terms of visitor perceptions of overcrowding on tours. The Waitomo Caves village, on the North Island of New Zealand, is situated in a landscape comprising numerous caves and underground river systems. The village has a population of around 500 people and a tourist population of around 450 000 international visitors per annum. At the time of Doorne's study (2000), the main nationalities of visitors were estimated as 27% Japanese, 26% Korean, 9% Taiwanese, 8% Australian, and 8% New Zealand. There are clearly cultural differences in perception of overcrowding on tours, with Asian visitors significantly less concerned about large numbers on individual tours. An additional concern is the perception that local visitors are being displaced – a form of recreational exclusion – by the overseas visitors attracted by the aggressive marketing of the New Zealand Tourism Board.

This has created tension between the local Maori landholders, who administer the Ruapuha-Uekaha Hapu Trust, and the lessee Tourism Holdings Ltd. who wish to maximise their profits within the constraint of managing tourist numbers in line with mandated environmental standards set by the NZ Department of Conservation. Doorne's study began as a simple act of local politicking to safeguard access for domestic visitors displaced by visitors with higher tolerances for crowded experiences. However, there is a bigger issue here in terms of who decides on appropriate levels of crowding, and for which group of visitors this should be applied. The de-facto regulator is a 'value-free' market mechanism in the absence of clearly formulated and politically robust alternatives. These alternatives have to take account of the changing climate of cultural relations in New Zealand and the recognition of Maori rights to land and their desire to manage places in a manner consistent with cultural practices.

13.2 Interpretation and Guide Training

The experience of a visitor to a cave is shaped by a number of factors that operate before, during and after the actual visit. These have been summarised by (Hamilton-Smith et al. 1997) as the trip cycle (Figure 13.1).

Of these factors, awareness, anticipation, reception (arrival), and recollection may be more important than the actual cave experience itself, and in the long run, the recollection probably counts most to the individual. So, any monitoring of the visitor experience must be designed to assess these factors. This was done at Jenolan Caves, New South Wales, Australia by a simple process of randomised single questions at point of entry ticket sales and both pre- and post-trip questionnaires. The resulting data was used to inform the development of both infrastructure and the guiding programme at the caves.

Hamilton-Smith (1994) argues that there are five key variables which shape the visitor experience and its outcomes:

- Opportunities and constraints brought by the visitor
- Time–space location
- Physical environment of the cave
- Social environment
- Programme and activities of the visit.

Some basic principles in designing the visitor experience are:

1. Information made available to the public either online or at the site should be accurate, but it should not convey a misleading impression of beauty or solitude which is not authentic.

Figure 13.1 The trip cycle in visitor experience. Source: Adapted from Hamilton-Smith et al. (1997).

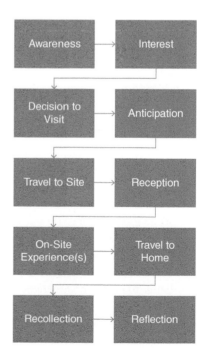

2. Recognise the extent to which this information made available prior to the visit can reduce undesirable behaviour and enhance anticipation.

   The careful use of slogans can also inform visitors and modify their behaviour. For example, at Plitvice in Croatia publicity and information (https://np-plitvicka-jezera .hr/en) emphasises four keywords: '*water, forests, waterfalls, silence*'. The result is that visitors treat the park with remarkable respect, and move through quietly, almost as if they were visiting a cathedral. Similarly, at Škocjan Caves in Slovenia words that are emphasised are '*mysteriousness, wildness and capriciousness of the underground rivers*' (https://www.park-skocjanske-jame.si/en).

3. Ensure that the best possible appearance is given at the entry to the cave area.

4. Every visitor should be welcomed as a real individual, not just as another number in a queue.

   For example, a short walk from the carpark to the cave relaxes visitors and allows appreciation of the scenery. Likewise, a guide greeting visitors and providing orientation creates a good initial impression.

5. Each tour should be tailored for an appropriate number of people and last for an appropriate time.

   The actual tour size will obviously depend on the nature of the cave, any environmental constraints, such as carbon dioxide levels, and the ability of the visitors to see and hear the guide at all times. Scheduling tours to allow some extra time at the end for questions is a good idea if workloads permit this.

6. Given the fundamental importance of a good guide, each tour should aim to develop a relationship or rapport between the guide and the visitors.

   Having multiple guides on the same tour, for example is best avoided unless they have well-defined roles, for instance providing interpretation on different aspects of the tour, such as cave biology or geology. Disruptions such as construction or repair work should be avoided, while having other parties close enough to disturb each other is also bad practice.

7. Visitors should be accurately informed as to what they can expect from each tour.

   A preliminary briefing at the cave entrance (with the emphasis on 'brief') is essential. This should cover the duration and exertion required, what to expect in terms of temperature and lighting, and what the tour will cover. Think about how good it is to have a restaurant menu, which not only gives a name to each dish but which tells you in a line or two what the ingredients are and how it is cooked.

8. Avoid mixing unduly diverse kinds of people on a tour if this will have a negative impact on the experience.

   As an example, it might not be a good idea to have a school group on the same tour as elderly people, nor mix individuals with an organised tour group. Allowing photography on a tour is a vexing question as it can be very disruptive. Flash units should not be allowed, and there are health issues associated with this source of light. It might be better to have scheduled photographic tours for groups or individuals who have made prior bookings.

9. Every effort should be made to identify the specific needs and interests of all visitors and to provide for them.

We should recognise the individuality of visitors and provide appropriate opportunities for them – specialist tours that range from interpreted cave science to history, ghost tours for entertainment, or tours that cater for specific language groups. Given the large numbers of international visitors to most cave areas, it is a good idea to have multilingual guides or at least provide a range of interpretive material in print form or as audio guides.

10. Every guide should be expected to develop their own repertoire of tours.

    The best tour will be one where the guide is enthusiastic about what they have to deliver – and that is most likely if it is their own tour programme, not something put together by someone else and delivered in a parrot-like fashion. They should be able to define their own objectives, decide on the content, and the most appropriate means of delivery. Less may be more. Allowing for quiet contemplation of the beauty of a chamber in a cave may be much more effective than continuous delivery. This leads to the next point.

11. Recognise that good guiding does not just consist of talking to the tour group.

    Excessive talking by the guide can be quite irritating and can degrade the whole experience. It may be better to have periods of total silence and time for questions or to elicit comments from the visitors. Above all the visitors must feel that the guide acknowledges them and speaks to them personally, making eye contact. The guide must be audible but not overbearing or condescending. Finally, at the conclusion of the tour, there is some common courtesy – thanking people for joining the tour, and wishing them a safe and pleasant journey and many happy memories.

At Naracoorte Caves, South Australia, guides have come up with some novel approaches to their craft (Bourne and Plowman 2007):

- See several caves with one guide so that the same information is not being repeated.
- A photographic tour that focuses on giving people time to take photographs rather than commentary.
- Adults only tours of Victoria Fossil Cave (as young children become bored and restless).
- Family-focused fossil tours.
- Storytelling in Alexandra Cave for children, rather than a tour.
- Having an 'open cave' at Tantanoola Cave, where people can roam around for as long as they want, rather than attend a structured tour.
- Children re-enacting the journey of children from the past as they explore Blanche Cave.

In developing a guide for an individual cave site, a number of topics could be identified, such as:

- History
- Water
- Geology
- Speleothems
- Biology
- Bats
- Management

These topics could be further developed into themes. For water, some themes might be:

- Water is the artist at work in this cave.
- Water is the life of the cave.
- Water links all the other parts of cave development.

Once a cave site had some themes developed these could be used both in tours and in promotional material on the internet, and could easily form the basis for an education programme. Although developing these themes presents some challenges for guides and managers, they can result in more meaningful cave experiences for visitors and a more fulfilled workforce.

To conclude, while cave tours generally follow a similar pattern in many countries, there are alternatives available. Firstly, there is a range of possibilities besides the guided tour. Secondly, the use of thematic interpretation means that a variety of tour messages can be developed. The use of interpretive themes means that guides will be talking about different things, and a guide working in just one cave can have a range of interpretive presentations. There's no reason at all for visitors to feel as if 'you've seen one cave, you''ve seen them all' (Bourne and Plowman 2007).

13.3   Cave Lighting

Electric lighting releases both light and heat inside the cave, so the energy balance of a cave should not be modified beyond its range of natural variations. Any lights should be of high efficiency and produce more light than heat. For this reason, there has been a shift to light-emitting diode (LED) lighting in recent years. An individual cave should be divided into zones so that lighting can be controlled by the guide and switched on or off as the party moves through the cave. Where possible a non-interruptible power supply should be provided to avoid problems for the visitors in the event of a failure of an external power supply.

Lighting should have an emission spectrum with the lowest possible contribution to the absorption spectrum of chlorophyll (in a range from 440 to 650 nm) to minimise the growth of *lampenflora*. The growth can also be prevented by keeping a safe distance between the individual light and any rock or speleothem surfaces where plants might establish. This will greatly reduce the energy reaching any surface where the plants may live. The safe distance between the lamp and the cave surface depends on the intensity of the lamp, but in general, a distance of 1 m should be safe. Special care should also be paid to avoid heating speleothems and any paintings that may exist in the cave. Finally, lighting should be installed to illuminate only the portions of the cave that are currently occupied by visitors.

The lighting of show caves has, in the past, tended to over-illuminate the cave passages and chambers as if they were offices or shopping malls. A more enlightened view would light the cave as a cave with deliberate use of darkness and sequencing of illumination on selected cave features. There are two important principles to be borne in mind when designing the lighting for a show cave: access and atmosphere.

Lighting for access should be at the minimum level consistent with the safe movement of all cave visitors. Effective lighting can be used to create safe access through an unfamiliar environment, a zone of familiarity that relaxes the visitors. The use of LED strip lights, 12 V downlights, and other low-energy technology can all achieve this aim. These can be attached to railings or path edges, with necessary inverters or batteries well-hidden below. In general, all fixtures and cabling should be well hidden from visitors but accessible for maintenance without further damage to the cave and its contents. Reduced power consumption has benefits beyond the reduction of CO_2 emissions: lower power requirements facilitate the use of local uninterruptible power supply when there is a mains power failure. Less heat is produced as well. There are many technologies available – remote controls, c-bus controlled electrical systems, high lumen per watt output lighting, batteries/inverters, optical fibres, etc. – but they should be used as tools to achieve an end, not as an end in themselves.

The second principle is that of the atmosphere. There should be an underlying philosophy to the lighting scheme. A theme should be established which illustrates aspects of cave development or history. The lighting should be sequential, with visitors led from one scene to the next. This avoids the massed illumination of a whole chamber. The manager needs to be very selective about what to light and what not to light. Any light in a dark environment will have a dramatic effect, and sometimes, a very distant light will enhance the illusion of depth and mystery we are trying to foster. The lighting of water features can be very effective. In all of these, the fragility of the cave contents needs to be considered, with some areas being out of bounds for any installation of lighting.

A final principle is that of creating a performance, with analogy to an orchestral performance (Kell 2002). All too often, a show cave tour proceeds from the entrance to the rear of the cave over half an hour or so, then there is a 'bolt and run' back to the entrance. Clever use of lighting and c-bus sequencing allows for a different experience on the return journey to the entrance at a leisurely pace. This provides a more satisfying experience and may allow for a different theme to be explored.

13.4 Some Engineering Issues in Caves

Caves are heritage places, both in terms of natural and cultural heritage. So, it is important that any infrastructure or alteration to the cave be carried out in line with accepted materials and methods for heritage places. Unfortunately, there are many caves in which past practices have led to degradation of the cave, usually inadvertently. The purpose of this section is to review some basic principles for modifications to the cave environment, principally for infrastructure in show caves, but also provide some guidance on suitable and unsuitable materials to use in the cave environment.

- Repair things, but do not replace them. If a speleothem is broken by all means repair it but don't replace it with another.
- Any additions or alterations should be readily reversible. Any new infrastructure should be designed so it can be removed without causing any damage to the cave.

- Any new development should be clearly distinguished from the original. There is a clear parallel here with the way in which archaeologists backfill excavations, using clearly different materials and labels to avoid confusion in the future.
- Avoid imitation of natural features. For example, use durable lampshades rather than cemented rock basins that poorly imitate natural features.
- Aim for design excellence in any work that is carried out, within the limits of cost and logistics. Any alterations or additions should be compatible with the cave and aesthetically pleasing.
- Accept that there is natural weathering and ageing of surfaces in any cave. This can be a balancing act in preserving natural surfaces but periodically cleaning surfaces which have accumulated dust, lint, and skin cells from visitors. The latter can have profound negative impacts on cave biology, so some good judgement is called for. The precautionary principle will apply here.
- Respect previous modifications if they have historical or cultural significance. Inappropriate alterations should be removed if this can be done without serious damage.
- Discontinue any unsound practices. It is now evident that significant pollution in caves can result from electrolytic action between dissimilar metals and from the decomposition of wood used for steps and railings.

Many metals are subject to electrolytic processes or corrosion, especially in the very humid environment found in caves. For corrosion to occur, the process requires an electrolyte in contact with the metal, and a potential difference either within the metal, between two different metals, or between a metal and a non-metal (Chang 1995). The whole then acts as a galvanic cell, where electrons are lost by the metal at the anode, producing the characteristic pitting and deterioration of corroded metal. The cathode reaction usually involves the consumption of oxygen, so the cathodic area is usually that exposed to air. The electrolyte, which may be simply a moisture film, allows migration of ions and completes the circuit.

Metal pipes or stair treads may be painted to prevent corrosion, however, if the painted surface is abraded severe, corrosion can occur due to the high potentials being generated by a very large anode area (the metal under the paint) and a small cathode (the exposed bit). Galvanising is a more effective means of protection because the zinc in the galvanising material is more easily oxidised (has a lower potential) than iron or steel, so continues to be preferentially oxidised even if the metal underneath is exposed by a scratch. However, there are significant problems due to the leaching of zinc into cave waters (Jameson and Alexander 1995). Aluminium pipes form a protective oxide coating quite readily which will protect the underlying metal from corrosion, but alkaline water conditions will erode the coating – which is why most aluminium boat hulls are painted with epoxy coatings.

Today most cave managers use stainless steel for cave infrastructure, especially for handrails. The aesthetics and functionality of dimpled stainless-steel handrails can be appreciated in the Waitomo Glowworm Cave and in Aranui Cave at Waitomo, New Zealand. However, they are expensive to purchase, can be difficult and thus expensive to fabricate, and some varieties may be readily altered by welding or grinding to produce 'non-stainless steel' fragments which will oxidise to produce intractable rusty stains on

Figure 13.2 Walkways and other infrastructure are made of stainless steel in Hwangseongul Cave, South Korea. This large cave is visited by more than one million people each year.

flowstones. In Korea, stainless steel is cheaper to obtain and has been very widely used for all cave infrastructure (Figure 13.2). Despite the initial costs, stainless steel is virtually maintenance-free for many years if good quality steel is obtained.

Timber has been used in many caves, but it is generally unsuitable as it readily breaks down under biological attack from slime moulds and fungi to release organic substances whose ecological impact is largely unknown. Rotten wood supplies large amounts of food modifying the equilibrium and trophic structure of the cave life. Organic material may also stain speleothems, especially flowstones. There may be a case for using hardwoods such as jarrah, teak, and tropical hardwoods, such as Belian. Softwoods should never be used as they degrade quickly, and CCA treated timber (copper chrome arsenate) should never be used in a cave though it may be suitable on the surface. If any timber is used for formwork, scaffolding, and similar temporary purposes, it should not be worked in the cave. It should be removed on completion of the job and care should be taken to remove any sawdust, scraps, or splinters resulting from working the timber.

Over the last decade, inert plastic materials have become widely available and preliminary research suggests that they may be stable in cave environments. These products are often derived from recycled plastics and can be moulded and coloured to produce handrails, steps, platform matting, and other infrastructure. In combination with stainless steel, these are probably the materials of choice for cave managers.

In the past, concrete was very widely used for steps and walkways in caves. It is certainly durable, but has a number of drawbacks including its weight, the mess created when it is mixed and poured, and the difficulty of removing it once it is in place. There is also some evidence that leachates from concrete may have adverse biologic impacts. Concrete should never be poured onto raw rock, flowstone, or other natural floor materials. Low-density concrete can be made using perlite, pumice, or volcanic scoria, and this has some advantages in terms of reduced weight but retaining adequate strength for walkways.

13.5 Impacts of Visitors and Infrastructure on Show Caves

The development of caves for mass tourism requires physical alteration of natural passages, installation of lighting, pathways, platforms, and associated infrastructure. Under such conditions, cave biology is the first to suffer, and heavily used areas are generally depauperate of cave fauna (Culver and Pipan 2009). The cave hydrology may also be altered, and there may be pollution from car park and road runoff, greywater, and sewage in extreme cases. The organic pollution of Hidden River (Horse) Cave, Kentucky, USA was both catastrophic and long-lasting. The cave was commercialised in 1916, and at that time, its underground river had a rich fauna including fish and crayfish. Contamination by sewage and creamery waste led to the extinction of the cave fauna and closure of the cave by 1943. Since the 1980s, the pollution has been stopped, and the cave is now recovering as fauna recolonises from more pristine upstream sections of the cave. It has now reopened to visitors (Lewis 1996).

Cave entrance modification is widespread in show caves but can also occur in wild caves. This action can have a profound effect on terrestrial fauna. Infilling or gating can restrict or stop the movement of animals, especially bats. Enlarging an entrance or creating a new artificial entrance can alter air flows, changing the microclimate. At Mammoth Cave, Kentucky, the Historic Entrance was blocked to stop cold winter air from entering (Elliott 2000); bats abandoned the cave as a result. Cave entrance gates can be designed to allow free passage of bats (Figure 13.3), though some species will avoid any gate and alternative solutions like fencing must be found.

Large numbers of visitors in a cave can significantly raise the air temperature. A single person releases heat energy at 80–120 W (Villar et al. 1986), about the same as a single incandescent light bulb. Thus, a party of 50 or 60 people on a cave tour can locally raise temperatures by 1–2 °C. The passage of tourists through Altamira Cave, Spain, raised the air temperature by 2 °C, CO_2 concentration from 400 to 1200 ppm and decreased relative humidity from 90 to 75% (De Freitas and Littlejohn 1987). The main effect of the reduction in humidity is drying and flaking of flowstone surfaces. According to Cigna (1993), management needs to ensure that these fluctuations lie within the range of natural variation for the cave and that they return to normal levels in a short period of time (Figure 13.4). Calaforra et al. (2003) provide a good example of determining visitor thresholds in such cases (Figure 13.5).

Increases in CO_2 concentration due to visitor respiration can range from 1500 to 2000 ppm, at which point people start to be distressed. At such concentrations, speleothems will start to dissolve. A threshold for this corrosion is reached at 2400 ppm in the Glowworm Cave, New Zealand, while at Jenolan Caves, the threshold is at 2700–2800 ppm Managing carbon dioxide levels requires effective monitoring, limiting the party size and

Figure 13.3 Construction of the Tumbling Creek Cave chute gate (Missouri, USA), March 2004. The open chute allows pregnant Grey Bats to use the entrance freely, but excludes human intruders. Source: From Elliott and Aley (2006) © Springer Nature.

frequency of tours, and modifying doors to improve air circulation. Studies at Jenolan Caves have shown that well-ventilated sites reach background levels in a few hours, while poorly ventilated sites can take days (James 2004).

The continuous lighting of cave features provides an opportunity for the establishment and growth of green plants (blue–green and filamentous algae, mosses, and ferns) in a concentric zone around the light. This is known as lampenflora (Figure 13.6) and has been the subject of considerable research (Aley 2004). Reduction in light intensity below the threshold for photosynthesis, movement-activated switches, and C-Bus technology controlling lights can all help to eliminate this problem. C-Bus is a microprocessor-based control and management system for cave lighting. It is used to control lighting and other electrical services such as audio-visual devices, pumps, motors, etc. A twisted pair cable network operates at 36 V and allows installation in wet places where conventional mains power would be dangerous. A lighting or audio-visual sequence can be pre-programmed or can be interrupted or changed at any location by a ranger or tour guide. The technology has now been employed in many caves in Australasia, Europe, and North America.

The growth of this 'maladie verte' at the Lascaux prehistoric art site was one factor leading to the latter's closure to the public. Low-wattage lamps, 'cool-lights', periodic ultraviolet irradiation and timed lights can all help to minimise this problem. Cleaning using

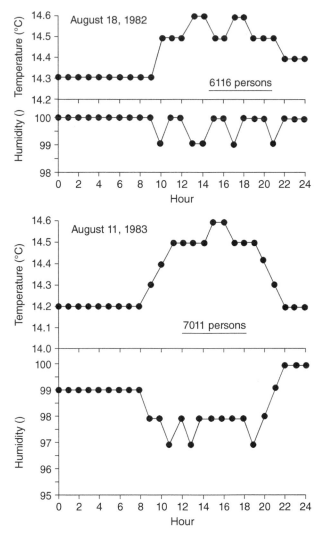

Figure 13.4 Sample 24-hour records of air temperature and relative humidity in Ancona Hall, Grotta Grande del Vento, Italy, during the peak of the tourist seasons in 1982 and 1983. Source: From Cigna (1993) © Springer Nature.

high-pressure jets and/or strong oxidising agents (calcium hypochlorite) may be necessary in extreme cases (Slagmolen and Slagmolen 1989). In a Hungarian cave, Rajczy and Buczko (1989) noted 39 species colonising newly installed lights, with smooth substrate texture being an important factor inhibiting establishment. Significant lampenflora was established on 20% of lamps within a year.

Within caves, pathways and stairs may concentrate the flow of drip water away from flowstones or stream channels, causing desiccation of cave formations and localised sedimentation. Leachates from concrete can also form deposits mimicking natural speleothems. Construction debris, including wire clippings, can leach toxins into cave

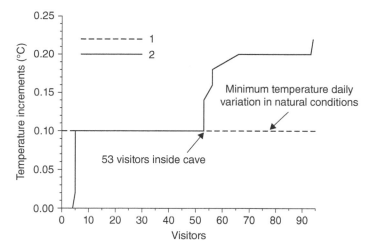

Figure 13.5 Number of visitors producing an increase in temperature of more than 0.1 °C. 1: maximum mean daily variation of temperature under natural conditions; 2: mean variation caused by visitors. Source: From Calaforra et al. (2003).

Figure 13.6 Lampenflora growth on flowstones at Wombeyan Caves, New South Wales, Australia. Source: Photo by Andy Spate.

streams. Galvanised steel railings and steps have been shown to leach various metallic ions, including cadmium, which can harm freshwater fauna (Buecher 1995). Today, there is increasing use of stable materials such as stainless steel, plastic decking made from recycled PET bottles, plastic rails, and similar products (Figure 13.7). In Australia and in Europe, there is now widespread use of recycled materials to produce inert resin products for guide rails, decking, and other outdoor infrastructure. The materials derive from

Figure 13.7 Re-lighting of the Donna Cave, Chillagoe, Queensland, Australia. Main lighting sequence is controlled by c-bus programming and can be modified by a guide. Tracklights are low energy use LEDs. Note use of inert materials – plastics and stainless steel – for all stairs and railings.

post-consumer plastics, principally HDPTE, and wood waste. These recycled products do not leach any chemicals, require low energy for production and have almost zero carbon emissions. See www.advancedplasticrecycling.com.au/index.php for further details.

In many caves, speleothems and pathways are cleaned on a regular basis due to the deposition of dust, hair, and lint from visitors and of algae and fungi. High-pressure water jets are commonly used, often with external mains water supply. Steam cleaning and use of surfactants have also been used. All of these methods have some impact on the surface being cleaned. Spate and Moses (1994) studied the impact of cleaning at Jenolan Caves. They found that repeated cleaning with high-pressure jets damaged crystal facets and recommended other strategies, such as protective clothing for visitors and mesh entrance walkways to limit dust tracking.

These impacts highlight the need for effective ongoing monitoring of the cave atmosphere, water quality, and particulate deposition. A set of biophysical indicators can be defined, and their state reviewed periodically. Coupled with this is the need for a set of social indicators that address the issue of the level of satisfaction of both visitors and staff (Davidson and Black 2007; Hamilton-Smith 2004). These should form part of an ongoing process of adaptive management (Figure 13.8) that takes account of monitoring results and is informed by the latest research.

Every year around 20 million people visit tourist caves, with Mammoth Cave, alone receiving over 2 million. There are some 650 tourist caves with lighting systems worldwide,

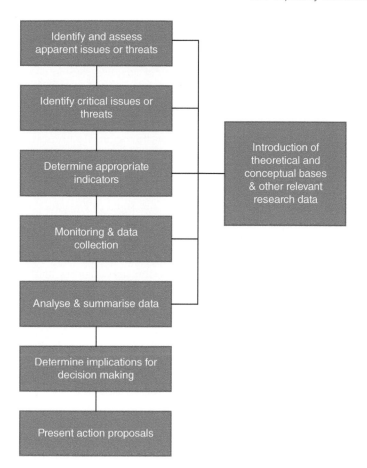

Figure 13.8 Outline of Visitor Impact Management (VIM) process implemented at Jenolan Caves, New South Wales, Australia. Source: From Jenolan Caves Reserve Trust (1995).

not counting caves used for 'wild' cave tours, where visitors carry their own lights. The infrastructure of tourism has a major impact on the cave system. In Carlsbad Caverns, New Mexico, there are more than 1400 lights, 65 km of electrical wiring, 6 km of walking paths, 70 signs, and 56 drains! Surface impacts resulting from the construction of car parking areas, walking tracks, kiosks, toilets, hotels and motels, and interpretive centres may be added to the direct underground impact.

There are few well-documented studies of the incidental effects of tourist cave operations. Most tourist caves have a fine layer of dust on the speleothems (Jablonsky 1990, 1992). This is made up of lint from clothing, dead skin cells, fungal spores, insects, and inorganic dust. At Carlsbad Caverns over 50 kg of lint has been removed manually in the last five years, while at Jenolan Caves, over 0.2 kg.m^{-2} of lint has been removed from heavily soiled areas. This material has the capability to act as a source of pathogenic bacteria and may develop an unpleasant odour upon decomposition. It also serves as an extra energy source for cave biota. Visitors' vehicles may contribute a source of dust and other pollutants. James et al. (1990) investigated lead levels in cave spiders' webs at the Grand Arch, Jenolan Caves, and compared levels with nearby non-impacted caves: they found that, although lead levels in the webs were high, they were also high at some other sites, and attributed the high values to dust trapping by webs. The dust caused the collapse of webs and may have increased spider mortality. Also at Jenolan, Kiernan (1989) reported minor sewage seepages into the caves from old pipes.

Probably the most extreme example is Hidden River Cave, Kentucky. The underground river served as the water supply for the town of Horse Cave from the 1880s to the 1930s, and there was a booming tourist industry based on the caves of the region. Groundwater pollution on account of domestic sewage from two towns, creamery wastes, and industrial wastes destroyed the water source and caused the cave to close in 1943. Blind cave fish in Hidden River Cave became locally extinct, and the only aquatic organisms surviving in the near oxygen-free water were bloodworms and sewage bacteria. Since 1989, reduction in sewage as a result of altered municipal practices has led to improved water quality. Troglophilic crayfish, troglobitic isopods, and amphipods have returned to the cave stream. Recently the blind fish have been sighted again in the main Hidden River Cave, having recolonised the site from upstream refugia. The cave is now the headquarters of the American Cave Conservation Association, and the recovery of the site is a key feature of their interpretive programme.

The range of factors producing environmental impacts on caves and karst has been identified by Williams (1993, Figure 13.9). Changes to cave hydrology and atmosphere, cryptogam growth, speleothems, and cave biota are all possible. There is a great potential for hydrologic change within caves from the construction of pathways, entrance structures, car parks, and toilets. Above a cave, the surfacing of the land with concrete or bitumen renders it nearly impermeable, in contrast to the high natural permeability of karst. Thus, the feedwater for stalactites may be drastically reduced or eliminated. Drains may alter flow patterns and may deliver additional percolation water to certain areas of a cave, causing changes in speleothem deposition. One way to minimise these effects is to use gravel-surfaced car parks or to include infiltration strips and cross-drains in the car park design. Similarly, pathways may need to be hardened for foot traffic, but this should be permeable (gravel, raised walkways, pavers) rather than concrete or bitumen. Toilet facilities may leak into karst fissures or conduits. There are many tourist sites where sewage reticulation or septic tank systems have leaked or overflowed into caves. Today there is a growing trend to use either pump-out toilet systems, where wastes are dispersed as sprays or sludges away from the karst, or composting toilets (e.g., Clivis Moltrum or Dowmus), where residues are dehydrated and may be subsequently used for fertiliser.

13.6 Radon Risk in Caves

Radioactivity in the environment is a cause of increasing concern to the public, and the gas Radon−222 has been highlighted as a potential hazard in houses, mines, and caves. ^{222}Ra is released by the radioactive decay of uranium salts weathered from volcanic or plutonic rocks or shales, and it may accumulate in sediments or in areas with poor air circulation. It has a half-life of four days. If this radioactive substance, a daughter of Uranium−238, is ingested or inhaled, the alpha and beta radiation associated with it or its daughters may

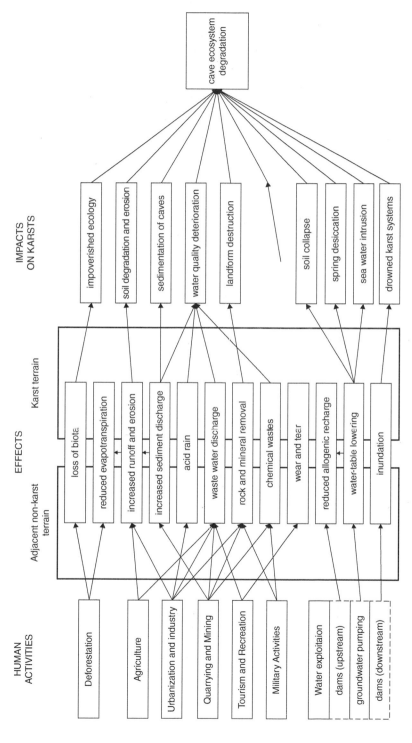

Figure 13.9 The range of effects and consequent impacts of human activities on caves and karst. Source: From Williams (1993).

cause serious cell damage, leading to an increased risk of cancer. Radon gas may be absorbed onto particles of dust or onto water droplets, and thus may gain entry to the body. While coarse particles are filtered by nasal hairs and fine particles are exhaled, intermediate-sized particles >10 μm are likely to lodge in the lungs.

The basic unit of radioactivity is the Becquerel (Bq), which is one disintegration per second, and activities are often expressed as Bq.m^{-3} for air or Bq.kg^{-1} for sediments. Health guidelines are usually expressed in milliSieverts (mSv), which is a measure of the energy absorbed by the body, or as working levels (WL), which are a measure of the radiation dose received related to recommended maximum exposure times; thus we have working-level hours (WLH), the dose received by someone exposed to a dose rate of 1 WL for one hour. As a rule, the conversions between these units are as follows:

$$1WL = 0.0735mSv = 3.8kBq.m^{-3} \tag{13.1}$$

Thus, a person working for 170 hours in an environment of 1 WL would receive a dose of 12.5 mSv, which is nearly the maximum annual dose for a non-classified radiation worker in the UK (15 mSv). United Kingdom ionising regulations apply above a level of 0.03 WL, while the threshold for government action for domestic radon levels is 0.05 WL. Several cave sites in Britain exceed these thresholds (Table 13.2), especially in areas like the Peak District, where there is mineralisation associated with the caves. In Giant's Hole, Derbyshire, UK, an activity of 40 WL means that the annual permissible dose will be reached in one six-hour

Table 13.2 British cave radon survey results, 1989–1992.

Region	Average[a]	Maximum[a]	Minimum[a]	No. of detectors	No. of caves
Peak District	8716	46 080	9	161	12
South Wales	2466	19 968	127	168	9
Forest of Dean	1820	4663	182	10	3
Mendips	n/a	19 000	12 000	88	1
North Pennines	1079	27 136	27	347	18
Portland	454	1024	86	18	5
Potential radiation dose[b]	*Based on averages*	*Based on maximums*			
Peak District	0.36	1.88			
North Wales	0.38	1.16			
South Wales	0.10	0.82			
Forest of Dean	0.07	0.19			
Mendips	0.34	0.44			
North Pennines	0.04	1.11			
Portland	0.02	0.04			

a) *Values in Bq m^{-3} radon gas concentration.*
b) *Potential doses from a 4-h caving trip, in mSv. A non-classified radiation worker can receive 15 mSv per year while at work under UK legislation.*
Source: From Gunn et al. 1991, Hyland and Gunn 1992, Reaich and Kerr 1991.

Table 13.3 Radon measurements in some Australian caves.

Region	Range (WL)	Mean (WL)	No. of observations
New South Wales			
Jenolan	0.05–0.53	0.33	15
Wombeyan	0.03–0.42	0.14	4
Abercrombie	0.02–0.03	0.03	2
Yarrangobilly	0.23–0.90	0.43	12
Victoria			
Buchan	0.01–0.08	0.06	18
Western Australia	0.02–0.16	0.05	6
Nullarbor (WA & SA)	0.02–0.16	0.05	6

Source: From Lyons (1992).

trip. This does not mean that there is any risk of acute radiation sickness, rather that there is an increased risk of lung cancer in the future as a result of this level of exposure. Caves in Australia (Table 13.3; Lyons 1992) have somewhat lower levels, but there is high variability in the results, and more monitoring needs to be carried out before definitive statements about radon exposure risk can be made (Lyons 1992).

The concentration of radon in cave air will depend on the amount of radon reaching the air and its subsequent dilution. The net amount reaching the air will be the instantaneous sum from all the materials in the cave – rock, sediments, speleothems, and water. Dilution factors include the air movement and cave ventilation, the surface area to cave volume ratio, and the loss of radon by plating out on exposed surfaces. Rock usually has low emanation rates, while speleothems have negligible emanation rates. Dry, well-sorted sandy sediments will allow ready diffusion of radon and will thus contribute proportionately more radon to the cave atmosphere. Radon concentrations may be higher in areas where there is a higher concentration of uranium salts in rocks or sediments. Caves with deep sediments, especially those which are porous and dry, occurring in smaller or blind passages may have higher concentrations, especially if there is poor air circulation. Caves with extensive speleothems, rock surfaces, and water, and with good air circulation, may be less problematic. There may be diurnal or seasonal variations in radon concentration where pressure or temperature gradients drive cave air circulation.

Cave sites where high radon concentrations are suspected should be monitored to take account of these seasonal variations, as well as spatial variation within the cave. Staff who spend significant time underground, such as cave guides and electricians, need to have their personal exposure monitored according to prevailing radiation health standards using accepted techniques and instrumentation. Where personal exposure levels are high, it may be necessary to modify work practices to limit within-cave time or introduce multi-skilling. In extreme cases, it may be necessary to close the cave to visitors seasonally or permanently. Clearly, the radon problem is one which affects both tourist caves and recreational cavers, and it is one where levels need to be monitored, work practices amended, and cave ventilation installed if necessary. There has been a good deal of inappropriate modification to cave airflows and spread of misinformation as a result of the 'radon panic' (Ahlstrand and Fry 1978; Aley 1994; Hyland and Gunn 1992; Yarborough 1976).

13.7 Cave Cleaning and its Impacts

In many tourist caves, formations are cleaned on a regular basis because of the accumulations of dust, lint, inwashed sediments, fungus, and algae (lampenflora). A number of approaches have been tried, with high-pressure water jets being the most common method employed (Bonwick and Ellis 1985). In some cases, scrubbing, use of surfactants, and steam cleaning (Newbould 1976) have also been tried. All these methods can be expected to have some impact on the speleothem surfaces being cleaned.

The impact of cleaning has been assessed by Spate and Moses (1994) for one of the Jenolan Caves. Examination of the unwashed surfaces under ultraviolet light revealed many fibres derived from fabrics washed or bleached with optical brighteners. Scanning electron microscope (SEM) observations of surfaces before and after cleaning revealed a mean surface lowering of 0.018 mm per wash, with damage to crystal facets being pronounced. On unwashed surfaces, individual crystals were obscured by amorphous clay layers thought to be derived from atmospheric dust. Hair and lint were embedded in the clay and in active calcite growths. A single wash at pressures less than 5500 kPa was sufficient to remove these foreign objects, but a second wash was found to expose calcite crystals and damage their facets. Clearly, cave cleaning has an undesirable impact; operators should try to limit the number and frequency of washes, and use the minimum number of nozzle passes over a calcite surface. Alternatives to limit particulate entry in tourist caves include plastic mesh walkways at the cave entrance and protective clothing.

The use of strong bleaching chemicals for the reduction or removal of lampenflora has been investigated by Cigna (2011a). The two most commonly used chemicals are sodium hypochlorite (household bleach at 5% by vol) and hydrogen peroxide (at 15% by vol). Sodium hypochlorite releases chlorine into the cave environment and although an effective cleaning agent this is a poison for cave life, though it may disperse quickly. Hydrogen peroxide releases water and carbon dioxide so is less toxic, but it is still a very strong oxidising agent. Both have an effect on calcite surfaces, with the latter having increased solution rates. Therefore, thorough washing of the surfaces after cleaning is recommended. Annual cleaning is probably the most appropriate frequency.

13.8 Impacts of Recreational Caving on Caves

The aesthetic and scientific values of wild caves (caves not modified for tourist development) are being degraded as a result of increased recreational use (Spate and Hamilton-Smith 1993). Aley (1976) stated that the carrying capacity of a cave is effectively zero. That is, the capacity of a damaged cave system to regenerate in anything like human generational timescales is very limited or non-existent. This is because natural rehabilitation processes in caves operate very slowly in the comparatively low-energy cave

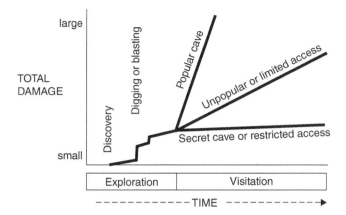

Figure 13.10 Damage vs. time during the exploration and visitation phases of a hypothetical cave. Total damage will vary widely depending on popularity, knowledge, limitations on access, experience of visitors, etc. Source: From Ganter (1989).

environment. It is therefore important to attempt to quantify both the extent and severity of actual or potential impacts to establish appropriate management regimes for wild caves and to determine the acceptable limits of environmental change (Wilde and Williams 1988). Coupled with this is the need to establish some monitoring to provide baseline data, to assess the direction and rate of change, and to identify the causes of change. It is unusual to have absolute baseline data as many caves have been entered before; we therefore often start with an existing level of damage and must mitigate further damage. Various strategies exist to achieve this (Figure 13.10), for example, by limiting access and reducing the number and frequency of visitors, permit systems, and trail marking (Ganter 1989).

Recreational caving results in a variety of impacts on the physical cave environment (Bunting and Balks 2001; Gillieson 1996). In the first instance, there is direct contact between the caver and the floor, walls, and ceiling of the cave. Cavers moving through a cave cause direct impacts to the physical cave environment, such as disturbances to cave sediments and cave breakdown deposits, erosion of cave rock surfaces, damage caused by bolting and rigging, modification of cave entrances and passages, speleothem breakage, and disturbances to fossil deposits (Gillieson 1996). In addition, there are impacts associated with sediment transfer to previously clean areas of the cave, carbide dumping, carbide staining on cave walls and ceilings, and the introduction of energy sources from mud and food residues, and sometimes deposition of faeces and urine.

The impacts listed above may not occur in every cave or in every part of one cave system. Not all parts of caves are equally vulnerable to disturbance, and the challenge for management is to correctly evaluate the relative vulnerability of cave passages. One approach that is valuable takes account of the relative energy status of the cave passage (New Zealand Department of Conservation 1999: pp. 10–11) in evaluating potential visitor impacts.

High-energy cave passages are subject to high-energy events, such as flooding or rockfall, on a regular basis. These cause the regular modification of the passage and may involve rapid sedimentation as well. Speleothem formation is rare because they are quickly scoured away or broken off unless they are in elevated or secluded alcoves. The effects of visitors on such passages will be minimal, and they may cope with high visitor numbers. Examples include the large river caves of Southern China, Sarawak, and Vietnam.

In medium-energy cave passages, forces such as running water, air currents, or animal activities operate at a somewhat lower magnitude. These caves often contain the most abundant active speleothem formations, reflecting an inflow of abundant saturated water. Cave pools may have abundant life with a periodic influx of organic matter providing an energy source (Tercafs 2001). The effects of visitors may be more evident than in high-energy caves, although they may be masked by occasional flooding and sediment reworking. Many tourist caves fall into this category.

For low-energy cave passages, a major event may be a single falling droplet of water. Speleothems in a low-energy cave are characterised by small and delicate formations resulting from the slow rate of crystal growth. Sediment deposits may have been in place for tens or hundreds of millennia. The influx of energy may be measured on decadal timescales. The presence of visitors in a low-energy cave may have a serious cumulative effect on the cave environment, as the amount of energy released by them in even a short visit may be more than what the cave has experienced in hundreds of years.

In most cases, individual caves are likely to contain components of all three different types of energy levels. Many caves are medium or low-energy environments, with essentially little input of energy on a human timescale. The entry of a single caver can change the energy balance by affecting the heat, light, and nutrients therein. One factor that is only now being realised is the potential introduction of microflora and microfauna by cavers (Cunningham et al. 1995). The effects of visitors to caves are generally cumulative and quite possibly synergistic.

Damage to bone deposits in caves has been well covered by Griffiths and Ramsey (2005). Significant paleontological/archaeological assemblages are less likely to occur in high-energy caves or cave passages since regular flooding or rockfalls are liable to redistribute or cover them. Such assemblages are far more likely to be found in medium or low-energy passages. The same stable environmental conditions which would tend to facilitate the preservation of paleontological assemblages and their contexts are those most vulnerable to disturbance. Excavations in such passages may entail significant changes to the energy regime, with corresponding impacts on the underground environment. In contrast to disturbances to surface sites, traces or effects of human activities in medium or low-energy underground environments may persist for hundreds or even thousands of years – consider what is believed to be a Cro-Magnon footprint discovered on the surface of a sediment deposit in Chauvet Cave, France (Tattersall and Schwartz 2001).

Documented attempts at quantifying recreational impacts in caves have been limited (Bodenhamer 1995; Bunting and Balks 2001; Gillieson 1996) and may indicate difficulties in quantifying and, subsequently, documenting recreational impacts in the cave environment. Cave photomonitoring is one method that can be used to assess and monitor the effects of recreational use of caves. This involves the collection of precise photographs of selected points within the cave, taken on a regular basis, appropriate to the management cycle (Uhl 1981). The main problem with photomonitoring is the difficulty in accurately and efficiently replicating the image view (Bunting and Balks 2001). Many of the problems listed above can be avoided or minimised by the adoption of a minimal impact code for caving (Table 13.4; Webb 1995).

Table 13.4 Minimal impact code for caving.

1.	Remember, every caving trip has an impact. Is this trip into this cave necessary? If it is just for recreation, is there another cave that is less vulnerable to damage that can be visited? Make this assessment depending on the purpose of your visit, the size and experience of the proposed party, and if the trip is likely to damage the cave
2.	Where possible the party leader should have visited the cave previously and hence should be aware of sensitive features of the cave, the best anchor points, and generally reduce the need for unnecessary exploration
3.	Cave slowly. You will see and enjoy more, and there will be less chance of damage to the cave and to yourself. This especially applies when you are tired and exiting a cave
4.	If there are beginners on a trip, make sure that they are close to an experienced caver, so that the experienced caver can help then when required, e.g. in difficult sections. Ensure that the party caves at the pace of the slowest caver
5.	Keep your party size small – four is a good party size
6.	Cave as a team – help each other through the cave. Don't split up unless the impact is reduced by doing so
7.	Constantly watch your head placement and that of your party members. Let them know before they are likely to do any damage
8.	Keep caving packs as small as possible and don't use them in sensitive caves or extensions
9	Ensure that party members don't wander about the cave unnecessarily
10.	Stay on all marked or obvious paths. If no paths are marked or none is obvious – define one!
11.	Learn to recognise cave deposits or features that may be damaged by walking or crawling on them. Examples are: Drip Holes, Stream Sediments, Paleo soils, Soil Cones, Crusts, Flowstone, Cave Pearls, Asphodilites, Bone materials, Potential Archaeological sites, Cave Fauna, Coffee & Cream, Tree Roots
12.	Take care in the placement of hands and feet throughout a cave
13.	Wash your caving overalls and boots regularly so that the spread of bacteria and fungi are minimised
14.	If a site is obviously being degraded, examine the site carefully to determine if an alternative route is possible. Any alternative route must not cause the same or greater degradation than the currently used route. If an alternative is available, suggest the alternative route to the appropriate management authority and report the degradation
15.	Carry in-cave marking materials while caving and restore any missing markers. Tape off sensitive areas you believe are being damaged and report the damage to the appropriate management authority
16.	If it is necessary to walk on flowstone in a cave, remove any muddied boots and or clothing before proceeding or don't proceed! Sometimes it is better to assess the situation and return at a later date with the appropriate equipment
17.	Treat the cave biota with respect, watch out for them, and avoid damaging them and their 'traps' webs, etc. Also, avoid directly lighting cave biota if possible
18.	If bone material is found on existing or proposed trails, it should be moved off the track to a safer location if at all possible. Collection should only be undertaken with appropriate permission

Table 13.4 (Continued)

19.	If you eat food in a cave, ensure that small food fragments are not dropped as this may impact the cave biota. One way is to carry a plastic bag to eat over and catch the food fragments. This can then be folded up and removed from the cave
20.	Ensure that all foreign matter is removed from caves. This includes human waste. If long trips are to be made into a cave, ensure that containers for the removal of liquid and solid waste are included on the trip inventory
21.	When rigging caves with artificial anchors, e.g. traces, tapes, rope etc., ensure that minimal damage occurs to the anchor site by protecting the site. For example, protect frequently used anchors, e.g. trees, with carpets, packs, cloth, etc. Bolts should only be used where natural anchors are inappropriate
22.	Cave softly!

Source: Adopted by the Australian Speleological Federation (Webb 1995).

All visitors to caves have an impact on the cave itself and on its contained biology. We have all been guilty, at some time in our caving careers, of wittingly or unwittingly causing some degradation to a particular cave. Unfortunately, the documentation of the impacts of recreational caving is restricted to a few observations, often gained fortuitously as part of another study. Stitt (1976) has outlined the range of internal and external impacts on caves, while Everson et al. (1987) have surveyed recreational impacts on Missouri caves. Specific types of wilderness cave impacts are described by Gamble (1981) and Kiernan (1989). Direct observations of the impact of caving on fauna are restricted to Tercafs (1988) for Belgium, and Carlson (1993), Crawford and Senger (1988), and McCracken (1989) for the United States of America. This is an area that desperately needs good systematic research, given the fragility of cave ecosystems outlined in the previous chapter. It should be pointed out that many cave scientists also cause significant impacts on the cave environment in the course of their research. These may include the excessive breakage of formations, excessive collection of biotas, excavation of pits subsequently left unfilled, permanent marking of study sites or survey stations with inappropriate media (paint, permanent tags, flagging tape), and leaving monitoring infrastructure in the cave. All cave scientists have been guilty of this, to a greater or lesser degree, during their careers. It is now unacceptable practice under most natural or protected area plans of management.

A typical cave is a low-energy environment with essential little input of energy over human activity periods. The entry of a single caver will change the energy regime, albeit slightly, in terms of heat, light and, possibly, nutrients. This impact has little effect on the rock itself unless visitor numbers are large, as we have seen for tourist caves. Cave resources are essentially non-renewable, and these impacts are cumulative, possibly synergistic (Stitt 1976), and irreversible, even when technology and time are brought to bear on a specific problem. For example, a dug entrance could be reblocked, but there will have been changes to the cave atmosphere and possibly its hydrology which may have disadvantaged or eliminated cave fauna. Compaction of sediments may have synergistic effects in that more water is channelled through the cave, sediment may be eroded, infilling of cave pools occurs, cave biota may be smothered by poor aeration, and slumping of sediment banks blocks passages. The impact of a party of 10 may be more than twice the impact of a party of five (Stitt 1976). In my youth, I was privileged to be one of five people to enter a virgin cave with flat silt floors rich in organic material 10 to 15 cm deep. The cave had a rich invertebrate fauna which was collected by an entomologist on the second trip of 12 people. After six trips about 100 people had seen the cave, the floor was compacted, water flow had been channelised, and the porous silts had been compacted over large areas. The biota was drastically reduced in numbers and species richness in less than a year. Ironically the cave is now under the waters of an irrigation supply dam.

Some specific effects of recreational cavers on caves are:

- Carbide dumping and marking of walls
- Compaction of sediments and its effects on hydrology and fauna
- Erosion of rock surfaces (ladder and rope grooves, direct lowering by foot traffic)
- Introduction of energy sources from mud on clothes and food residues
- Introduction of faeces and urine leading to water pollution
- Entrance and passage enlargement by traffic or digging
- Cave vandalism and graffiti.

Many caves which are frequently visited, and which are not under any form of access control, have suffered from some form of vandalism. This ranges from graffiti to breakage of speleothems and mechanical enlargement of passages. Interestingly, ancient or vintage graffiti become heritage items – for example, eighteenth century signatures in Baradla (Hungary) and Mammoth Caves (Kentucky, USA); engraved Sung or Ming dynasty lettering in Lung Yin Tung Cave, Guilin, China; and late Victorian era signatures on speleothems beyond the inverted syphon of Murray Cave, Cooleman Plain, Australia. Modern graffitists use paint spray cans; their more durable works can be removed only with strong detergents or solvents.

There is a range of ways of limiting the impacts of visitors on both tourist and wild caves:

- Hardening the environment to reduce the impact, by installing tracks, paths, and marked routes.
- Decreasing the demand for cave experiences by restricting the flow of information about caves in the media.
- Increasing the supply of caves by new discoveries.
- Exporting the demand to other countries richer in caves.
- Restricting access to caves by gating and/or permits.
- Reducing impacts of visitors through education and development of minimal impact codes.

Of all these options, the last is most likely to be productive in the long term and should involve all identifiable cave users. The development of individual cave management plans by speleologists is one way to ensure responsible conduct (Glasser and Barber 1995) and allows for community consultation in the planning process.

13.9 Cave Rescue

The exploration of caves is not without some risk and, following an accident, the rescue of an injured person from a cave poses problems quite different from those encountered on a surface rescue. Caves are dark beyond the entrance zone, and there may be hazards due to water, low temperatures, elevated carbon dioxide levels, rockfalls, and constricted passages. Cave rescue is, therefore, a specialised situation requiring a high level of training and competence in the rescue team (McCurley and Mortimer 2017).

Cave rescue teams can thus be called out for a wide variety of reasons:

- Cavers may become lost or may be late in returning to the surface.
- A caver may suffer a fall of one form or another. These can range from a twisted ankle, a slip on wet rocks or falling a great distance.
- A caver may become exhausted and need help getting back to the surface.
- Heavy rain can result in water levels rising, flooding a passage, and causing it to become impassable.
- Rockfalls may cause cavers to be entrapped.
- Cavers can become physically trapped in small passages.

An analysis of cave accidents requiring rescue in the USA was carried out by Stella-Watts et al. (2012). A total of 1356 victims were identified, of which 83% were male and 17% were female. Ages ranged from two to 69 years old, with an average of 27 years. The greatest number of events occurred in summer months, peaking in July. The most common incident leading to traumatic injury was a caver fall (74%), also contributing to 30% of caver fatalities. Lower extremities were most commonly injured (29%), followed by the upper extremities and head (21% and 15%, respectively). Fractures comprised 41% of injuries, followed by lacerations (13%), bruise, haematoma, and abrasions (12%), and sprains and strains (7%). Most of these injuries will involve stretcher immobilisation and lengthy rescues (Figure 13.11).

A rescue in Ellison's Cave in Georgia, USA, illustrates some of the challenges of a rescue underground. Two groups entered separate pit entrances for a crossover trip; each group left its rigging in place for the other group to ascend. Ellison's Cave has two of the deepest underground pits in the United States, at 179 and 134 m. While in the bottom of the cave, one of the explorers fell and sustained a compound fracture of the femur. His comrades stabilised it as best they could and sent for help. It took two hours for a fast caver who knew the route to get up the 179 m pit and out to where he could call for help.

The response involved cave rescue teams from three different states. Because it was a holiday weekend, many team members were unavailable, so cavers from the region who had been trained in rescue procedures were called in to supplement the agency response. Initial responders carried supplies to stabilise the patient further and prepare for transport. Blood was airlifted to the site and transfused at the bottom of the pit. Once packaged in a rescue stretcher, the patient was moved to the base of the 179 m pit. Separate rigging teams had made a counterbalance haul system using two cavers as the counterweight. Once the rescuers had carried the patient out of the pit and the cave, the patient was flown to a hospital.

Although most cave evacuations require stretcher transport, each individual situation should be evaluated carefully to determine whether a stretcher is really necessary. A properly stabilised 'walking wounded' caver can be helped out of a cave in a short time and with little manpower. If that same individual is placed in a stretcher the numbers of rescuers and hours to the hospital increase exponentially. Two incidents from Lechuguilla Cave illustrate this. A caver several kilometres into the cave had a high tibia–fibula break from an 80 kg rock falling on her leg. She required a stretcher evacuation that took 176 rescuers a total

(a)   (b)

Figure 13.11 Practice vertical rescue in Jenolan Caves, New South Wales, Australia. Source: Andrew Baker/NSW Cave Rescue Squad. © Rod Burton/NSW Cave Rescue Squad.

of 68 hours to accomplish and cost nearly a quarter of a million dollars. See https://www .youtube.com/watch?v=J7I7bXcSWK8. Details of a protracted, 12-day European rescue are given in Schneider et al. (2016).

Two years later, another caver broke his ankle further into the cave. Rather than call out a rescue, his teammates splinted him and supported him as he crawled out under his own power. Twenty-five hours later, with the help of 10 people and no real expense, he was out of the cave. The difference was not the stoutness of the cavers but the anatomic location of the injury. For one, a time-intensive stretcher carry was needed, but for the other, self-rescue was possible with relatively little help. A decision to stretcher-carry a patient has significant consequences. Any measures that can support the self-rescue of a patient are a benefit to the patient, rescuers, and cave (Hooker and Shalit 2000).

Virtually all cave rescue teams have training seminars and practice rescues; there is an active exchange of experiences and equipment design at an international level; and details of individual cases are published allowing the compilation of statistics on the diverse caving accidents that require rescue.

Cave diving is, without doubt, one of the more hazardous recreational activities (Exley 1979). Cave divers visit environments with inherent hazards that include flowing water and low or no visibility, a confusing number of possible exits, restrictions, entanglement in a guideline, and a physical overhead barrier preventing divers from ascending to the surface. In the UK, most cave divers started as cavers and then took up diving, whereas, in the USA, most cave divers started as open water divers (Farr 1991). Most open-water diving skills apply to cave diving, and there are additional skills specific to the environment, and to the chosen equipment. These are good buoyancy control, trim, and finning techniques helping to preserve visibility in areas with silt floors, and the ability to navigate in total darkness using the guideline to find the way out. Emergency skills for dealing with gas supply problems are complicated by the possibility of the emergency occurring in a confined space and low visibility or total darkness.

Buzzacott et al. (2009) examined fatality records for American cave-diving fatalities (n = 368) occurring between 1969 and 2007. The number of deaths per year peaked in the mid-1970s and has diminished since. Drowning was the most frequent cause of death, most often after running out of gas (compressed air or nitrox), which usually followed getting lost or starting the dive with insufficient gas. Compared with untrained divers, trained divers tended to be older, dived at deeper depths and further inside caves, carried more dive cylinders and more often dived alone. Untrained divers were more likely to have dived without a guideline, without an appropriate number of lights, and/or without adequate gas for the planned dive. In both cases, the common causes leading to death were a failure to use a continuous guideline to the surface and bringing inadequate dive cylinders for the intended dive. They recommend every cave diver give additional consideration to the use of a guideline and turn around after using one-third of the available gas in their dive cylinders. Based on the statistics, they recommend that gas management rules should receive the greatest emphasis in cave diving courses.

Rescues of cave divers have been reasonably well documented, though often concerned with body retrieval rather than a live patient. In 2010, a team attempted a rescue in the Dragonnière Gaud Cave in the Ardèche region of France, which was ultimately unsuccessful. In 2011, British cave divers assisted in the recovery of the body of a Polish cave diver from a cave in Kiltartan, Ireland. The bodies of two Finnish divers were recovered from about 135 m deep within Jordbrugrotta, Norway in 2014. In June and July 2018, a widely publicised cave rescue successfully extricated members of a junior football team trapped in Tham Luang Nang Non cave in Thailand. Twelve members of the team, aged 11 to 16, and their 25-year-old assistant coach entered the cave on 23 June after football practice. Shortly afterwards, heavy rains partially flooded the cave, trapping the group inside. On 2 July British cave divers found the group alive on an elevated rock about four kilometres from the cave mouth. Rescue organisers discussed various options for extracting the group, including whether to teach them basic diving skills to enable their early rescue, wait until a new entrance was found or drilled, or wait for the floodwaters to subside at the end of the monsoon season months later.

On the morning of 8 July officials instructed the media and all non-essential personnel around the cave entrance to clear the area as a rescue operation was imminent, due to the threat of monsoon rains later in the week, which were expected to flood the cave until October. For the first part of the extraction, 18 rescue divers consisting of 13 international cave divers and five Thai Navy SEALs were sent into the cave to retrieve the boys, with one diver to accompany each boy on the dive out. The international cave diving team was led by British divers Vernon Unsworth, John Volanthen, Richard Stanton, Jason Mallinson, and Chris Jewell, assisted by two Australian cave divers, Dr. Richard Harris, a physician specialising in anaesthesia, and Craig Challen. Their portion of the journey would stretch over 1 km going through submerged routes while being supported by 90 Thai and foreign divers at various points performing medical check-ups, resupplying air-tanks for the main divers and other emergency roles.

The boys were each dressed in a wetsuit, buoyancy jacket, harness, and a positive pressure full face mask. Dr. Harris administered anaesthetic to the boys before the journey, rendering them unconscious, to prevent them from panicking on the journey and risking the lives of their rescuers. The boys were manoeuvred out by the swimming divers who held onto their back or chest, with each boy on either the right or left side of the diver, depending on the guideline; in very narrow spots divers pushed the boys from behind. The boys arrived at 45-minute intervals, and each journey took three hours. All of the boys were rescued, and the cave divers had exited by 10 July. The boys have made a complete recovery, and their rescuers have gained the highest civilian awards for bravery in their respective countries. Several books have been published on the rescue, and a feature film is in production.

13.10 Cave Inventories and Alternative Management Concepts

Caves have values that relate to their aesthetics, their geology and geomorphology, their biology, their archaeology and palaeontology, and their history, to name but a few. Cave inventories provide a means of documenting all of the values of an individual cave (Lewis 1996). It also allows for comparison between caves and therefore underpins management classification and/or zoning. Detailed inventories can also provide a valuable snapshot of baseline conditions in caves, against which subsequent degradation of cave values may be measured. Many recreational cavers produce basic inventories as part of mapping and documentation activities. These provide very valuable, but sometimes unacknowledged, sources of data for management agencies. Maps produced by cavers are especially valuable, as a wealth of detail is usually recorded. In caves with exceptional values, a more detailed multidisciplinary approach may be needed. This can include geomorphology and ecology, archaeology and palaeontology, groundwater hydrology and ecology, microbiology, tourism, and cultural heritage.

Caves should be carefully evaluated on an individual basis for the significance of each category of values and the vulnerability of the cave to disturbance. The significance of each category of values can vary from cave to cave. One cave might have high recreational and hydrological values but demonstrate low biological, archaeological, and geological values. Another might only be significant in terms of its historical associations.

Kiernan (1988) points out:

> *There is a need for cave managers to recognise that each cave has a limiting factor on usage (i.e., the value most at risk). Each site needs to be managed on the basis of the particular limiting factor for that site.*

The limiting factor for some caves may be biological, mineralogical, or any of the other values listed above. Management authorities need to establish what that limiting factor is, prior to making any decisions about a cave's use, either by researchers or the general public.

This provides an alternative view to more conventional recreational planning concepts and tools such as Limits to Acceptable Change, Recreation Opportunity Spectrum, and Visitor Impact Management. All of these are well entrenched in the outdoor recreation planning literature. They are all based on the premise that the system in question has some capacity to regenerate over a relatively short timescale. This is simply inappropriate for most caves. They are also based on the premise that visitor needs should be more important than natural or cultural heritage considerations

In Western Canada, parks are preparing to adopt cave management guidelines using a three-tier classification system to manage access (Horne and Senior Park Warden 2005). Other countries have similar systems of cave classification in place or under development.

- Class 1 caves are 'access by application' only – they have the highest resource value and are not for recreation. Each visit must add significant knowledge or give nett benefit to the cave.
- Class 2 caves are 'access by permit' – recreational use is allowed, and there are some management concerns. Entry for education/orientation is possible under the permit.
- Class 3 caves are 'unrestricted public access' – there are few or no management concerns, and no permit is required. In order to determine which class a cave will fall into, three sets of factors are considered: (a) cave resources, (b) surface resources, and (c) accident and rescue potential.

It is probably true for caves that whatever we have today is more than we will have tomorrow. Although new caves are found every year, the growth in leisure time and demand for outdoor recreation outstrip the increase in the resource under pressure. Most caves and their contents are relict features, often formed under climatically more favourable conditions than today. This is especially true of caves in seasonally humid or arid climates, but it also applies to those in more equable climates. Thus, to all intents the regeneration of speleothems, for example, may occur on timescales far beyond those of our species.

'*The carrying capacity of a cave is zero*' (Aley 1976). The concept of carrying capacity was first enunciated in agriculture, for which it was defined in terms of the number of animals that might be placed on a given area on a sustainable basis (Dasmann 1964). The concept was transferred to natural area management by scientists trained in biology and was widely used for up to two decades (Stankey et al. 1990). The idea rests on the assumption that there is a fixed limit to the use that an area can withstand, and by extension, there is also a fixed population which can be supported by that area. This is simply not true. Factors such as the timing and spatial pattern of use, changes in climate and vegetation, and extreme events, such as fires and floods, result in a changing resource base. Different organisms have different ecological vulnerability and resilience. The original concept of carrying capacity took no account of socio-economic issues and did not recognise the non-reversibility of many ecological changes. In particular, many cave animals are habitat specialists and may be vulnerable to minor environmental changes (light, heat, humidity); their populations may not recover from an imposed stress. Thus the concept of carrying capacity is now recognised as a beautiful idea that was murdered by a gang of ruthless facts (Burch 1984).

The issue is now seen as one of identifying the quality of recreational experience which is appropriate to a particular cave environment and determining the environmental conditions consistent with that use. Thus, the question has moved away from just simple numbers of people in caves to the social and environmental conditions which should prevail. Several complementary management tools have emerged. All of them involve the translation of qualitative management goals to quantitative management objectives using environmental indicators and standards.

The first, the Recreation Opportunity Spectrum (ROS), identifies the range of recreation opportunities at a site, the basis for identifying these, and the physical resources needed. This approach requires that caves be classified in terms of their potential uses, likely impacts and resulting environmental conditions. These can range from undisturbed natural environments used for low-density recreation (wilderness caves) through to highly modified environments suitable for high-density use (tourist caves).

The second, the Limits of Acceptable Change (LAC) (Stankey et al. 1985), is concerned with defining those environmental conditions and maintaining them. LAC involves selecting key indicators, setting standards of achievement for each indicator, monitoring to allow comparison to that standard, and modifying recreational use and management strategies in the light of non-conformity to that standard. The first need in the application of LAC is to determine target features or organisms that can be used as indicators. A physical indicator might be the level of breakage or soiling of speleothems in a heavily used passage or compaction of cave sediments. A biological indicator might be the maintenance of a meta-population of a particular obligate cave species, such as glow-worms or syncarid shrimps. Clearly, the choice of an indicator must take account of its representativeness and the feasibility of monitoring it. The need for environmental monitoring implies considerations of replication, frequency, and cost. Frequent monitoring of one key parameter is preferable to occasional monitoring of many. It is generally better to monitor an indicator which is simple and cheap to measure at many sites than one so complex and so expensive that it can be afforded only at one or two sites. For example, measuring the evaporation of fixed volumes of water from many open Petri dishes in a cave may give a better picture of desiccation problems than a single thermo-hydrograph near the entrance.

Use of the LAC concept implies a need for visitor monitoring. This may be done simply by visitor logbooks, questionnaires, cheap infra-red beam counters, and fixed photographic reference points. Such monitoring systems must take into account the possibility of vandalism, the intrusiveness of the monitoring apparatus, and, for logbooks, the compliance of visitors (often distressingly low). In many cases, a programme of visitor education may repay the effort involved many times over and may be incorporated into site interpretation.

These two concepts are combined in Visitor Impact Management (VIM; Jenolan Caves Reserve Trust 1995). The key elements of the VIM process are:

- Establish precise goals for management and define management units.
- Identify key indicators which can be used to monitor change.
- Define desired conditions for these indicators.
- Design a monitoring programme using these indicators.
- Establish management responses to monitoring results.

Indicators can be drawn from both the environmental and social sciences. For example, a regular visitor survey could be conducted to determine levels of satisfaction/dissatisfaction with the management of the cave, and the VIM modified accordingly. Populations of keystone cave invertebrates could be monitored seasonally to determine if visitor impacts were causing significant population decline. Or cave carbon dioxide levels could be monitored continuously to see what level was attained after a tour, and how quickly the cave atmosphere returned to normal (the relaxation time). Each site will require a different set of indicators and guidelines, and management must be flexible to take account of this.

Finally, cave managers may have to ask if existing community standards are relevant to caves. In many countries, conservation legislation does not specifically cater to caves and karst. It may be necessary to embark on the long and often difficult process of legislative reform. If so, then support from cave user groups (speleologists, youth groups, local tourism boards, scientific societies) may be crucial to eventual success. Prior education programmes for the public may greatly enhance the level of support that is obtained.

13.11   Rehabilitation and Restoration of Caves

In the nineteenth century, the development of show caves often involved guiding visitors along narrow passages. There was a risk that people would inadvertently damage delicate speleothems, such as helictites, or soil them by handling them or brushing against them.

> *'The formations on walls are extremely delicate, some of it is white and some like yellow coral. The roof has been slightly defaced by certain nineteenth century cads. In various places the "mark of the beast" in lampblack has been produced by holding candles near to the ceiling and moving them about gradually. The sooty Heiroglyphics remain unto this day as an evidence of vanity and folly. The floor which was once like alabaster, is now soiled by the tramping of feet'.* (Cook 1889)

In a number of caves in Australia (Jenolan, Buchan), wire netting was used to separate visitors from the speleothems (Holland 1992). Over time, this method spread to a number of sites in Australia and New Zealand. Fortunately, the idea did not catch on in the larger European show caves, though individual speleothems might be protected in this way.

In 1886, electrical lighting was introduced at Jenolan, New South Wales to replace damaging magnesium wire and candles as a means of lighting. On July 22, 1880, E. C. Cracknell temporarily illuminated the Margherita Cave, Jenolan with electric light. This was the first recorded use of electricity for lighting any cave in the world and came only two years after the electric lighting of the Thames Embankment, London (Havard 1934). At that time there was no attempt to hide the electrical wires; mains wires were strung from insulators attached to the cave walls and light bulbs were suspended in a similar manner. Later, electrical wires were pinned to walls and speleothems and special grooves or trenches were cut to take the wire cables. Poorly planned cave lighting inherited from earlier times is still in place today and can be observed throughout Australian and New Zealand caves.

Over time the netting deteriorates and results in the formation of iron and zinc compounds which cause unpleasant staining of speleothems, contamination of pools, and slippery areas on the pathways. The breakdown of old electrical wires sheathed in lead produces lead carbonate (cerussite), an unpleasant white compound, and copper wires react to form the blue and green basic copper carbonate minerals, azurite, and malachite.

Wire netting collects large accumulations of lint from the air and from visitors' clothing as they brush past. Eventually this drops onto paths and when wetted forms a pulpy slime; whether wet or dry it will spread around the cave. Defective netting and wire rope handrails can splinter when held or brushed against. The sharp pieces of wire also tear clothing.

The removal of the netting and electrical wires has to be done with great care as they fragment into small pieces which must then be vacuumed up and the area washed. In some cases, the wires in grooves or trenches can be encased in stable plastic resin and the surface restored and colour matched. Supporting posts can be tapped periodically to loosen them and then lifted out. Sawing or oxyacetylene cutting liberates fine metal dust and heat which is injurious to the cave.

Areas in and around rimstone or gour pools must be treated by hand, a slow laborious process. Where concrete has intruded onto calcite surfaces, chiselling by hand can be used and any mud tracked into the area removed with brushes and trowels and then rinsed several times with cave water. The process is not too different from archaeological excavation.

In many cases, show caves are extensively modified to facilitate visitor access. The World Heritage site of Mammoth Cave illustrates some of the alterations and problems that result from them (Thurgate and Olson 2002). The original entrance passage, known as Houchins Narrows, has been modified several times over the past 180 years. The passage has had its floor lowered to double its original height, while the cave floor has been smoothed and altered to make way for paths. Masonry walls and gates that restricted airflow were installed to provide security and to moderate the cave temperature during winter. These changes made access easier but modified air flows, reducing airflows in some areas, and allowing cold sinking air into the cave during winter.

The cave hosted large colonies of hibernating bats but these abandoned the caves, the problem compounded by visitor disturbance and unsuitable gating. Woodrats that once lived in the loose rocks on the floor of the entrance passage have also abandoned the cave. Increased moisture due to condensation freezes between bedding planes causing small rock falls. Fallout of fine gypsum from the roof in some sections of the cave has also increased. The increased moisture has also promoted fungal growth on wooden artefacts in the cave, up to 2 km from the Historic Entrance. Since 1989, cold winter air entering through the Historic Entrance has mixed with warmer moist air flowing from Gothic Avenue. A cloud forms at the ceiling level and water collects on all surfaces in contact with the cloud. This condensation drips onto the War of 1812 saltpetre leaching vats below, and has promoted active fungal growth.

Work is now underway at Mammoth Cave to restore air flows and climate to their original conditions (Toomey et al. 2009). Around 10 million bats, mostly Grey and Indiana Bats, hibernated in areas that are now used as main tour routes. These species prefer a very narrow temperature range and this provides the staff with a target for cave climate restoration work. Other measures to make the cave more attractive to bats include:

- Restriction of access to the cave by visitors at the critical times of dusk and dawn.
- Reconfiguration of entrance lighting to be less intrusive.
- Rerouting of cave tours to leave some areas free of disturbance.

The use of plexiglass baffles on the entrance gate allows manipulation of winter cave temperatures to approach the range required by the bats. The baffles have also improved the rockfall incidence and the fungal growth.

In common with many show caves, paths in Mammoth Cave were made from cave sediment, which was mined from within the cave. In the drier winter, dust was raised by tours and combined with lint from visitors. In 1960s and 1970s, water was applied to the paths to control dust and calcium chloride and ammonium nitrate were added to retain moisture. The effect of these chemicals on the cave environment is unknown, but it is likely that there were major impacts on water quality and cave invertebrates.

Sections of the historic paths near the River Styx included raised wooden boardwalks and platforms. These were colonised by fungal communities and the additional food supply would have upset the ecological balance of invertebrate populations. The wood was also treated with creosote, which leached into the underground stream. A lighting system from the 1960s began to corrode and polluted cave streams. These pollutants have probably impacted on the aquatic ecosystems which support the endangered Kentucky Cave Shrimp. To prevent further declines in water quality, the wooden pilings, decking, and old lighting system have been manually dismantled and carried out of the cave. This has been a massive operation and has relied on volunteer efforts coordinated by the National Speleological Society. Over 300 m of boardwalk have been dismantled and removed. More modern materials, such as concrete, recycled plastic decking, and posts, are being used in Mammoth Cave, while lint curbs have been added to capture low-level dust and lint. Finally, the surfaces of many formations are being hand cleaned with brushes to remove lint and dust.

13.12   Cave Classification and Management

As seen in Section 13.10, cave classification schemes are a vital ingredient in any rational cave management planning. Cave classification needs to consider an individual cave site in relation to the immediate area that surrounds it, to the rest of the park or reserve in which it is located, and to its national and global context. The criteria used to evaluate the significance of an individual cave site (Worboys et al. 1982) include:

- Geological considerations – for example, specific features that relate to structure, stratigraphy, palaeontology, or mineralogy.
- Geomorphological considerations, including features that illustrate genetic or chronological relationships, or particularly fine examples of cave morphology.
- Hydrological considerations, such as the presence of major underground streams or lakes, unusual networks involving breaches of surface divides, or key elements in understanding the conduit network.
- Biological considerations relating to species richness, the presence of rare and endangered species, unusual trophic structures, or key bat maternity sites.
- Archaeological and cultural considerations, such as the presence of deep, well stratified deposits, the cave's role in the evolution of a regional prehistory, examples of historic cave use such as mining or water management, or its spiritual significance to indigenous people.

- Geographical considerations of remoteness and wilderness values, proximity to park infrastructure such as roads and camping grounds, recreational opportunities, and accessibility from major population centres.

A more detailed discussion of cave and karst significance is provided by Davey (1977).

Once the significance of an individual cave has been assessed (and significance is relative to a particular community viewpoint or management philosophy) then that cave can be placed in one of a number of use categories. Two examples of systems are provided here. One is that used by the National Parks and Wildlife Service of New South Wales (Worboys et al. 1982):

Group 1 – *Closed caves.*
Those caves in which access is not permitted because of danger from instability or foul air, or caves awaiting classification.
Group 2 – *Scientific reference caves.*
Those caves which are the best representatives of particular attributes of geology, geomorphology, biology, or archaeology. The management aim will be to preserve such caves in their natural state so that reference sets of caves and cave life are available in perpetuity. Access will be strictly limited by scientific research permit to experienced cavers. Each permit will indicate the nature and value of the research and its likely impacts on the cave. Party size and experience of users will be specified and impacts should be monitored.
Group 3 – *Limited access caves.*
These caves have such a quality in their physical or biological attributes that they warrant special protection, or have a high level of difficulty which limits cave exploration to very experienced cavers. The principal management aims are to preserve their high quality and/or maintain a high safety standard. Gating is desirable where possible, standards of party size and experience will be strictly maintained, and detailed speleological research is to be encouraged.
Group 4 – *Speleological access caves.*
In this group are caves where the physical or biological attributes do not warrant special protection and the degree of difficulty is suitable for cavers and novices with some training. The principal management aims are to preserve the cave's natural features and provide recreational opportunities for cavers where this does not conflict with its natural values. Gating is usually desirable where possible, party size may be limited, and each party must have an experienced or accredited leader.
Group 5 – *Adventure caves.*
These are caves with little or no inherent value other than their morphology, suitable for exploration by inexperienced but properly equipped groups such as organised youth groups, tourist groups led by experienced guides, or novice speleologists.
Group 6 – *Public access caves.*
These include caves in open areas which have been developed as public inspection caves or are suitable for that use, and require no special equipment or clothing unless specified by the ranger. Public access would usually be either with a ranger or as a self-guided tour.
Within large and complex caves, further zoning may be necessary, such as that employed by the United States National Parks Service in Mammoth Cave:

Zone A – *Closed passages.*
   Closed by gate for scientific or safety reasons.
Zone B – *Scientific reference.*
   Part of a cave which may be an excellent example of the geomorphological, geological, biological, and/or speleothem attributes of a cave. May be used as a reference against which to evaluate effects of visitor use on the rest of the cave. Usually gated, with access on the basis of a scientific permit.
Zone C – *Limited access.*
   Relates to passages or chambers whose physical or biological attributes warrant special protection, or where the degree of difficulty limits exploration to experienced cavers only. Where possible, access is limited by a strategically placed gate.
Zone D – *Natural passage.*
   Unimproved passages which may be traversed only by properly equipped and experienced speleologists.
Zone E – *Partially developed passages (no electric lighting),*
   Partially or previously developed passages which are now abandoned. Paths range from good to somewhat primitive. Such passages provide a 'wild' cave experience for untrained visitors with hand-held lights.
Zone F – *Fully developed passage (electrically lit).*
   All passages provided with electric lighting aesthetically arranged and developed with paths, steps, signs, etc. Either self-guided or guided by ranger staff; a fee is charged and the size and frequency of visitor groups is regulated.
Zone G – *Intensive use area.*
   Assembly points, visitor facilities and lifts located where possible in areas of the cave with low aesthetic and/or scientific value.

These schemes may serve as the basis for cave classifications elsewhere, bearing in mind that not all zones may be present and that local conditions may render some classes inappropriate. Often such systems of zoning of land use can be interpreted to visitors in such a way that the concepts of rational, ecologically based management are made clear. This is a good way to ensure compliance with such schemes and to increase visitor satisfaction.

13.13   Policy Approaches to Cave and Karst Protection

Over recent years, many cave sites worldwide have been listed as World Heritage, either in their own right or as part of larger nominations. There are now 59 cave and karst properties inscribed on the World Heritage register, with a further 33 listed as being worthy of nomination (Williams 2008). A National Geographic survey of 415 World Heritage sites, undertaken by over 400 independent, suitably qualified people, found that the standard of management across these sites had declined significantly since 2004. Presumably, World Heritage sites should be the best managed, as the sites with outstanding universal values to be presented to the world. Identified causes for this decline were mass tourism, inappropriate tourism products. and rampant commercial and industrial development. There is a strong conflict between tourism and heritage management on one hand and a focus on economic return on the other. These issues also apply to other caves being managed by state or federal agencies in many countries.

There are a number of common issues which are raised by cave managers. Funding is usually dispensed on an annual basis with little chance of carryover into the next financial year. Long-term plans are difficult if not impossible to implement under this model, and so many activities, especially monitoring, are compromised. Funding is more available for tourism development than for scientific studies aimed at managing the resource. Staff training to build capacity and enable them to deliver professional outcome has suffered in quality and availability. Cave guide exchange schemes, a valuable source of cross-fertilisation of ideas, have been virtually abandoned. Many organisations are preoccupied with occupational health and safety issues to the exclusion of all else. Clear lines of communication between agencies are not maintained and thus a holistic approach to management is made more difficult.

For many governments, the environment is now low on the political agenda and is being overshadowed by industries such as logging and mining. There is a recent proposal to allow mining in the National Parks of New Zealand – including its karst areas. Geodiversity conservation is also very low on the political agenda. Environmental concerns and conservation planning are dominated by biodiversity issues and fail to recognise the importance of geological heritage in integrated landscape management. There is a new paradigm emerging of public versus contracted management or private ownership of cave and karst areas. While in many cases outsourcing of park management can inject new ideas and funding, safeguards need to be in place to ensure that nature conservation values are not compromised in favour of accelerated tourism development.

On a more positive note, there has been a quantum shift globally in the underlying philosophy of protected area management. Previous management regimes about protection were exclusionary and restrictive and took little regard of public opinion. We are now moving to more enlightened management regimes, where good relations with park neighbours are seen as critical, and parks are run using principles of adaptive management (Table 13.5; Phillips 2003). The challenge for cave managers will be to embrace the new paradigms while conserving what are essentially non-renewable resources.

13.14 Management of the Gunung Mulu World Heritage Area, Sarawak, Malaysia

The Gunung Mulu karst of northern Sarawak hosts some of the longest caves in Southeast Asia. Mulu is located about 100 km east of Miri, a coastal city in northern Sarawak. The small town of Mulu is reached by daily air services from Miri and can also be reached by boats up the Baram and Tutoh rivers. Gunung Mulu National Park (Figure 13.12) has an area of 90 000 ha and most visitors concentrate on the southernmost karst, accessible from the park headquarters adjacent to the town. Over 90% of the park remains unvisited and is in pristine condition (Gill 1999). The climate of Gunung Mulu is equatorial perhumid, with a diurnal temperature range from around 22 °C to 32 °C. Humidity is high, usually around 90% in the early morning and decreasing to 60% in the afternoon. Annual rainfall is also high, from 4400 to 6800 mm (Gillieson and Clark 2010).

Table 13.5 Contrasting Paradigms for Protected Areas, from Phillips (2003).

As it was: protected areas were...	*As it is becoming: protected areas are...*
Planned and managed against people	Run with, for, and in some cases by local people
Run by a central government	Run by many partners
Set aside for conservation	Run also with social and economic objectives
Paid for by taxpayer	Paid for from many sources
Managed by scientists and natural resource experts	Managed by multi-skilled individuals
Managed without regard to the local community	Managed to help meet the needs of local people
Developed separately	Planned as part of national, regional, and international systems
Managed as 'islands'	Developed as 'networks' (strictly protected areas, buffered and linked by green corridors)
Established mainly for scenic protection	Often set up for scientific, economic, and cultural reasons
Managed mainly for visitors and tourists	Managed with local people more in mind
Managed reactively within a short timescale	Managed adaptively in long term perspective
About protection	Also, about restoration and rehabilitation
Viewed primarily as a national asset	Viewed also as a community asset
Viewed only as a national concern	Viewed also as an international concern
Managed in a technocratic way	Managed with political considerations

The karst mountains exhibit classical tropical karst features, pockmarked with dolines, closed depressions, valleys, and caves. In the southern hills, 'The Garden of Eden' is a fine example of a tiankeng, being over 1 km in diameter. Hidden Valley on the east side of Gunung Api is a deeply incised closed valley with a misfit stream sinking in its bed. The Melinau, Melinau Paku, and Medalam Rivers have truncated the limestones, forming deeply incised gorges with towering 300 m high cliffs and remnants of high-level caves. Some of the world's finest examples of pinnacle karst (Figure 13.13) can be found on Gunung Api and Gunung Benarat Figure 13.14.

The caves of Mulu are currently the longest in Southeast Asia and more will undoubtedly be found. The caves are predominately phreatic in origin but exhibit vadose, phreatic, and paragenetic profiles with some of the world's finest examples of phreatic pendants. The cave sediment deposits are still in place at all altitude levels, affording the opportunity for important scientific investigations of climatic change. Reversal of the earth's magnetic field, recorded in the sediments at 1.8 million years before present, indicates that the caves are at least 2 million years old, possibly as much as 3 million years. Notches in the walls of the caves at various levels (Figure 13.15) can be correlated with interglacial periods. The caves and associated karsts are, therefore, of outstanding significance in recording the major changes in the earth's history, and have thus joined the rich array of karst sites inscribed on the World Heritage List.

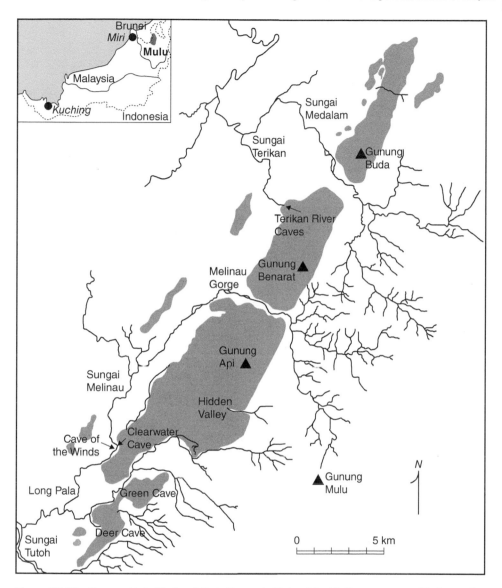

Figure 13.12 Gunung Mulu National Park (Sarawak, Malaysia) with limestone massifs (Inset of location in Borneo).

Gunung Mulu National Park (NP) was the subject of a Royal Geographical Society (UK) expedition in 1978 and the society subsequently wrote a management plan for the park in 1982. This was followed by a new management plan covering the period 1992–1995, and subsequent evaluations led to the World Heritage nomination (Gill 1999). The park was inscribed on the World Heritage List in November 2000 and has subsequently become one of the most iconic national parks in Southeast Asia, and a model for sustainable

(a)

(b)

Figure 13.13 a The Pinnacles on Gunung Api b Upstream entrance of Deer Cave, Gunung Mulu National Park, Sarawak, Malaysia.

Figure 13.14 Rugged limestone terrain on the edge of Hidden Valley, Gunung Mulu National Park, Sarawak, Malaysia.

development that has been emulated elsewhere. Mulu clearly met four criteria for World Heritage status:

- Be an outstanding example of the world's geological history (caves and cave deposits).
- Be and outstanding representative example of on-going evolutionary processes.
- Be of exceptional beauty.
- Contain significant natural habitat for in-situ conservation of biological diversity and the protection of threatened species.

The caves of Gunung Mulu NP have been systematically explored by annual expeditions since 1980. Clearwater Cave contains 227 km of mapped passages on several levels with a major stream passage. Survey data produced after the 2018 expedition indicates a total of 330 km of mapped cave passages in the area (Figure 13.16).

The Sarawak Forestry Corporation (SFC) is responsible for the management of all Protected Areas of Sarawak, including Gunung Mulu National Park. SFC is responsible for the management of the Wilderness Zone which is more than 90% of the Park. Borsarmulu Park Management Sdn Bhd (BPM) has been appointed by the Sarawak Government as the managing agent for the Tourism Zone (less than 10% of the Park) with the responsibility

(a)   (b)

Figure 13.15 a Stream incut in Clearwater Cave and b the Shower speleothem in Deer Cave, Gunung Mulu National Park, Sarawak, Malaysia.

to implement the recommendations of the Integrated Development and Management Plan (Manidis Roberts Consultants 2000). Borsarmulu Sdn Bhd is required to implement and support a programme of developmental, operational, and maintenance activities under the supervision of the Sarawak Government. The Integrated Development and Management Plan is complemented by a five-year plan which is reviewed annually (IUCN World Heritage Outlook 2017).

In national park management of this type, there must be an effective policy where tourism ventures on park must give a substantial return to the protected areas they depend and profit from. This must be not confined to just annual permit fees, but making a real contribution to local employment, education, science, conservation, and management actions. As a result of this outsourcing of management of the Tourism Zone there have been significant improvements in the standard of accommodation at Park HQ, complemented by the Marriott Resort down river which provides luxury accommodation and is also operated by Borsarmulu Sdn Bhd. The Park boardwalks and cave infrastructure (lighting, paths, platforms) have also been improved to international standards. A guide training programme has been implemented for a number of years and the local school is also supported by Borsarmulu. Borsarmulu Sdn Bhd employs local people as far as possible and staff consist of Berawan, Penan, Kelabit, Iban, Kayan, Kenya, Lun Bawang, Saban, and Tering people. Ninety seven percent of the workforce at Mulu Park is of local origin (Gerstner et al. 2018) and researchers are encouraged to involve the guiding staff in the ongoing research. This

(a)

(b)

Figure 13.16 a Sarawak Chamber in Gua Nasib Bagus Cave and b Speleothems in Wild Boar Cave, Gunung Mulu National Park, Sarawak, Malaysia.

has led to park staff attending overseas conferences and at least one member of staff carrying out research for a higher degree overseas.

The range of activities available to visitors ranges from guided cave visits (total of five caves including Deer Cave and Clearwater Cave), adventure cave tours (five caves graded for difficulty), nature walks, a canopy walk and trekking (Gunung Mulu summit, the Pinnacles and Headhunter Trail). Most, if not all visitors witness the daily evening flight of millions of bats from Deer Cave (Figure 13.17). Current annual visitor numbers are around 60 000 with most visitors staying two to three nights. There are daily flights from Miri, Kuching, and Kota Kinabalu.

One of the key requirements of ongoing World Heritage status is the need to provide correct, scientifically accurate information to visitors and facilitate research. Coupled with this is the need to empower the local community and provide significant employment opportunities in this remote area. According to the principles of World Heritage, the management plan, and the agreement with the State, it was evident that local people should be trained as guides and interpreters (Gerstner et al. 2018). The locals already have a sense for the forest and amazing skills, but language disadvantages and a lack of education as far as the sciences are concerned have prevailed. The park management have instituted a training scheme for train new guides or to refresh existing guides. This course is presented in modules covering the karst and cave aspects as well as the forest, totally protected areas (TPAs) and associated ordinances, biodiversity, and geodiversity. There is a special section dealing with the handling of clients and presenting a tour in both a show cave and adventure setting. A course in basic accredited First Aid is normally a part of the training.

According to the IUCN World Heritage Outlook (2017), ongoing survey and research in Gunung Mulu National Park is regularly enhancing the known World Heritage values defined at the time of nomination in 1999. The condition of all cited World Heritage values is good and the condition remains stable. The location of the park and limiting of tourism to less than 10% of the area provides a relatively stable environment. However, some forward vision to anticipate change is necessary. Internal changes could be the result of changes in park tourism, with pressure to develop new visitor experiences and cater for increasing numbers of visitors. External changes could be the result of renewed forestry activities, oil palm plantation development, mineral exploration, and mining and hydroelectricity development. Roads up to the park boundary for whatever purpose represent a significant potential threat. Thus, forest clearance for the establishment of a major oil palm plantation near Batu Bungan in early 2019 is a significant concern. This emphasises the need for ongoing monitoring of several issues:

- There is a need for ongoing management of community relations, especially those who feel aggrieved as a result of government decisions relating to the park which affect land rights and compensation.
- Monitoring of the impacts of tourism on the caves.
- Monitoring of the impacts of traditional hunting and gathering.
- Monitoring of the quality of interpretation and guiding in the show caves and on guided surface activities.
- Monitoring of the major bat and cave swiftlet colonies in the park, especially in relation to their feeding areas which are outside the park.

Gunung Mulu National Park is one of the best protected and appropriately managed protected areas in Southeast Asia, and has served as a model for the development of other parks in Sarawak. However ongoing monitoring and enforcement is needed to ensure that inappropriate development both within and outside the park is not allowed to occur. That will be the challenge for park management over the next few decades.

(a)

(b)

Figure 13.17 a Entrance of Deer Cave and b a sunset flight of wrinkled-lipped bats from Deer Cave, Gunung Mulu National Park, Sarawak, Malaysia.

References

Ahlstrand, G.M. and Fry, P.L. (1978). Alpha radiation project at Carlsbad caverns: two years and still counting. In: *Proceedings of the First Cave Management Symposium* (eds. R. Zuber, J. Chester and S. Gilbert). Albuquerque: Adobe Press.

Aley, T. (1976). Caves, cows and carrying capacity. In: *Proceedings of the National Cave Management Symposium*, 70–71. Albuquerque: Speleobooks.

Aley, T. (1994). Some thoughts on environmental management as related to cave use. *Journal of the Australasian Cave and Karst Management Association* 17: 4–10.

Aley, T. (2004). Tourist caves: algae and lampenflora. In: *The Encyclopedia of Caves and Karst Science* (ed. J. Gunn), 733–734. New York: Fitzroy Dearborn.

Bodenhamer, H.G. (1995). Monitoring human caused changes with visitor impact mapping. In: *Proceedings of the National Cave Management Symposium*, 28–37. Bloomington, Indiana.

Bonwick, J. and Ellis, R. (1985). New caves for old: cleaning, restoration and redevelopment of show caves. *Cave Management in Australasia* 6: 134–153.

Bourne, S. and Plowman, C. (2007). Interpretation workshop: if you've seen one cave, have you seen them all? In: *Proceedings of the 17th Australasian Conference on Cave and Karst Management*. Buchan, Victoria.

Buecher, R.H. (1995). Footprints, routes, and trails: methods for managing pathways in the cave environment. In: *Proceedings of the National Cave Management Symposium*, 47–50. Bloomington, Indiana.

Bunting, B. and Balks, M. (2001). A quantitative method for assessing the impacts of recreational cave use on the physical environment of wild caves. *Australasian Cave and Karst Management Association Journal* 44: 10–18.

Burch, W.R. (1984). Much ado about nothing – some reflections on the wider and wilder implications of social carrying capacity. *Leisure Sciences* 6: 487–496.

Buzzacott, P.L., Zeigler, E., Denoble, P., and Vann, R. (2009). American cave diving fatalities 1969-2007. *International Journal of Aquatic Research And Education* 3: 7.

Calaforra, J.M., Fernández-Cortés, A., Sánchez-Martos, F. et al. (2003). Environmental control for determining human impact and permanent visitor capacity in a potential show cave before tourist use. *Environmental Conservation* 30: 160–167.

Carlson, K.R. (1993). The effects of cave visitation on terrestrial cave arthropods. In: *Proceedings of the National Cave Management Symposium*, 338–345. Bowling Green, Kentucky.

Chang, R. (1995). *Chemistry*. McGraw-Hill.

Cigna, A.A. (1993). Environmental management of tourist caves. *Environmental Geology* 21: 173–180.

Cigna, A.A. (2011a). The problem of Lampenflora in show caves. *Journal of the Australasian Cave and Karst Management Association* 82: 16–19.

Cigna, A.A. (2011b). Show caves. In: *Encyclopedia of Caves* (eds. D.C. Culver and W.B. White), 690–697. Amsterdam, The Netherlands: Elsevier/Academic Press.

Cook, S. (1889). *The Jenolan Caves: An Excursion in Australian Wonderland*. London: Eyre & Spottiswoode, Her Majesty's Printers.

Crawford, R.L. and Senger, C.M. (1988). Human impacts on populations of a cave Dipluran (Campodeidae). In: *Proceedings of the Entomological Society of Washington State*, 827–830.

Culver, D.C. and Pipan, T. (2009). *The Biology of Caves and Other Subterranean Habitats*. Oxford: Oxford University Press.

Cunningham, K.I., Northup, D.E., and Pollastro, R.M. (1995). Bacteria, fungi and biokarst in Lechuguilla cave, Carlsbad caverns National Park, New Mexico. *Environmental Geology* 25: 2–8.

Dasmann, R.F. (1964). *Wildlife Management*. New York: Wiley.

Davey, A.G. (1977). Evaluation criteria for the cave and karst heritage of Australia. *Helictite* 15: 1–41.

Davidson, P. and Black, R. (2007). Voices from the profession: principles of successful guided cave interpretation. *Journal of Interpretation Research* 12: 25–44.

De Freitas, C.R. and Littlejohn, R.N. (1987). Cave climate: assessment of heat and moisture exchange. *Journal of Climatology* 7: 553–569.

Doorne, S. (2000). Caves, cultures and crowds: carrying capacity meets consumer sovereignty. *Journal of Sustainable Tourism* 8: 116–130.

Elliott, W.R. (2000). Conservation of the north American cave and karst biota. In: *Subterranean Ecosystems* (eds. H. Wilkens, D.C. Culver and W.F. Humphreys), 665–690. Amsterdam, The Netherlands: Elsevier Press.

Elliott, W.R. and Aley, T. (2006). Karst conservation in the Ozarks: forty years at Tumbling Creek Cave. In: *Proceedings of the 2005 National Cave and Karst Management Symposium*, 204–214. Albany, New York.

Everson, A.R., Chilman, K.C., White, C., and Foster, D. (1987). 1987. Recreational use of seven wild caves in Missouri. In: *Proceedings of the National Speleological Society* (ed. J.M. Wilson), 11–21. Huntsville, Alabama: Cave Management Symposium, 1987.

Exley, S. (1979). *Basic Cave Diving: A Blueprint for Survival* 38pp. National Speleological Society: National Speleological Society Cave Diving Section.

Farr, M. (1991). *The Darkness Beckons*. London: Diadem Books.

Gamble, F.M. (1981). Disturbance of underground wilderness in karst caves. *International Journal of Environmental Studies* 18: 33–39.

Ganter, J.H. (1989). Cave exploration, cave conservation: some thoughts on compatibility. *National Speleological Society News* 47: 249–253.

Gerstner, H., McArthur, E., and Clark, B. (2018). Feeding the furnace of information. In: *Proceedings of the 22nd Australasian Conference on Cave and Karst Management*, 6–10. Margaret River, WA: Australasian Cave and Karst Management Association.

Gill, D. (1999). *Nomination of the Gunung Mulu National Park, Sarawak, Malaysia for World Heritage Listing* 216 pp. Report to UNESCO World Heritage Committee/Kuching: Sarawak Forestry Department.

Gillieson, D.S. (1996). *Caves: Processes, Development and Management*. Oxford: Blackwell.

Gillieson, D.S. and Clark, B. (2010). Mulu: the World's Most spectacular tropical karst. In: *Geomorphological Landscapes of the World* (ed. P. Migon), 311–320. Springer.

Glasser, N.F. and Barber, G. (1995). Cave conservation plans: the role of English nature. *Cave and Karst Science* 21: 33–36.

Griffiths, P. and Ramsey, C. (2005). Best management practices for palaeontological and archaeological cave resources. *Journal of the Australasian Cave and Karst Management Association* 58: 27–31.

Gunn, J., Fletcher, S., and Prime, D. (1991). Research on radon in British limestone caves and mines, 1970-1990. *Cave Science* 18: 63–65.

Hamilton-Smith, E. (1994). Outdoor recreation and social benefits. In: *New Viewpoints in Australian Outdoor Recreation Research and Planning* (ed. D.C. Mercer), 79–87. Melbourne: Hepper Marriott & Associates.

Hamilton-Smith, E. (2004). Tourist caves. In: *The Encyclopedia of Caves and Karst Science* (ed. J. Gunn), 1554–1560. New York: Fitzroy Dearborn.

Hamilton-Smith, E., McBeath, R., and Vavryn, D. (1997). 1997. Best Practice in Visitor Management. In: *Proceedings of the 12th ACKMA Conference*, 85–96. New Zealand: Waitomo.

Havard, W.L. (1934). The romance of Jenolan caves. *Journal and Proceedings of the Royal Australian Historical Society* 20: 20–65.

Hazlinsky, T. (1985). *International Colloquium on Lamp Flora* (10–13 October 1984) Budapest, Hungarian Speleological Society.

Holland, E. (1992). Away with wires! *Journal of the Australasian Cave and Karst Management Association* 10: 22–24.

Hooker, K. and Shalit, M. (2000). Subterranean medicine: an inquiry into underground medical treatment protocols in cave rescue situations in national parks in the United States. *Wilderness & Environmental Medicine* 11: 17–20.

Horne, G. and Senior Park Warden (2005). Cave management guidelines for western mountain national parks of Canada. In: *Proceedings of the National Cave and Karst Management Symposium, September*, 53–61. Albany.

Hyland, R. and Gunn, J. (1992). Caving risks. *New Scientist* 135: 183.

IUCN World Heritage Outlook (2017). Gunung mulu national park – 2017 conservation assessment. International Union for Conservation of Nature and Natural Resources.

Jablonsky, P. (1990). Lint is not limited to belly buttons alone. *National Speleological Society News* 48: 117–119.

Jablonsky, P. (1992). Implications of lint in caves. *National Speleological Society News* 50: 99–100.

James, J.M. (2004). Tourist caves: air quality. In: *The Encyclopedia of Caves and Karst Science* (ed. J. Gunn), 1561–1563. New York: Fitzroy Dearborn.

James, J.M., Gray, M., and Newhouse, D. (1990). A preliminary study of lead in cave spiders' webs. *Helictite* 28: 37–40.

Jameson, R.A. and Alexander, E.C. Jr., (1995). Zinc Leaching from Galvanized Steel in Mystery Cave, Minnesota. In: *Proceedings of the National Cave Management Symposium*, 178–186. Indianapolis: Indiana Karst Conservancy Inc.

Jenolan Caves Reserve Trust (1995). *Determining an Environmental and Social Carrying Capacity for Jenolan Caves Reserve, Applying a Visitor Impact Management System*. Bathurst: Jenolan Caves Reserve Trust.

Kell, N. (2002). Re-lighting Newdegate cave. *Journal of the Australasian Cave and Karst Management Association* 47: 40–43.

Kiernan, K. (1988). *The Management of Soluble Rock Landscapes: An Australian Perspective* 61 pp. Sydney: Sydney Speleological Society.

Kiernan, K. (1989). Karst management issues at the Jenolan tourist resort, NSW, Australia. In: *Resource Management in Limestone Landscapes: International Perspectives* (eds. D. Gillieson and D. Ingle Smith), 111–131. Department of Geography & Oceanography, University College, Australian Defence Force Academy.

Lewis, J.J. (1996). Bioinventory as a management tool. In: *Proceedings of the 1995 Cave Management Symposium*, 228–236. Indianapolis: Indiana Karst Conservancy.

Lyons, R.G. (1992). Radon hazard in caves: a monitoring and management strategy. *Helictite* 30: 33–40.

Manidis Roberts Consultants (2000). Gunung Mulu National Park Integrated Development and Management Plan. Sydney.

McCracken, G.F. (1989). Cave conservation: special problems of bats. *National Speleological Society Bulletin* 51: 49–51.

McCurley, L. and Mortimer, R.B. (2017). Caving and cave rescue. In: *Auerbach's Wilderness Medicine*, vol. 63, 1403–1412.

New Zealand Department of Conservation (1999). *General policy and guidelines for cave and karst management in areas managed by the Department of Conservation*. Wellington, New Zealand: Department of Conservation.

Newbould, R. (1976). Steam cleaning of the Orient Cave, Jenolan. Cave Management in Australia. In: *Proceedings of the First Australian Conference on Cave Tourism (10–13 July 1973)*. Broadway: Australian Speleological Federation.

Phillips, A. (2003). Turning ideas on their head: the new paradigm for protected areas. *The George Wright Forum* 20: 8–32.

Rajczy, M. and Buczko, K. (1989). The development of the vegetation in lamplit areas of the cave Szelo-Hegyi-Barlang, Budapest, Hungary. In: *Proceedings of the 8th International Congress of Speleology*, 514–516. Kentucky: Department of Geography and Geology, Western Kentucky University.

Reaich, N.M. and Kerr, W.J. (1991). Background alpha radioactivity in cuckoo Cleeves cavern, Mendips. *Cave Science* 18: 83–84.

Schneider, T.M., Bregani, R., Krammer, J. et al. (2016). Medical and logistical challenges of trauma care in a 12-day cave rescue: a case report. *Injury* 47: 280–283.

Slagmolen, C. and Slagmolen, A. (1989). La maladie verte. *Regards* 6: 25–31.

Spate, A. and Hamilton-Smith, E. (1993). Caver's impacts – some theoretical and applied considerations. In: *Proceedings of the Ninth Australasian Cave Tourism and Management Conference*, 20, 30. Margaret River, WA.

Spate, A. and Moses, C. (1994). Impacts of high pressure cleaning: a case study at Jenolan. *Cave Management in Australasia* 10: 45–48.

Spate, A. and Spate, J. (2013). World-wide show cave visitor numbers over the recent past: a preliminary survey. *ACKMA Cave and Karst Management in Australasia*. Conference Proceedings 20: 57–69.

Stankey, G., McCool, S., and Stokes, G. (1990). Managing for appropriate wilderness conditions: the carrying capacity issue. In: *Wilderness Management*, 2e (eds. J. Hendee, G.H. Stankey and R.C. Lucas). Golden, CO: North American Press.

Stankey, G.H., Cole, D.N., Lucas, R.C., Petersen, M.E., and Frissell, S.S. (1985). *The limits of acceptable change (LAC) system for wilderness planning*. U.S.D.A. Forest Service General Technical Report 37, Intermountain Forest and Range Experiment Station, Ogden, Utah, USA.

Stella-Watts, A.C., Holstege, C.P., Lee, J.K., and Charlton, N.P. (2012). The epidemiology of caving injuries in the United States. *Wilderness & Environmental Medicine* 23: 215–222.

Stitt, R. (1976). Human impact on caves. In: *Proceedings of the National Cave Management Symposium*, 36–43. Albuquerque: Speleobooks.

Tattersall, I. and Schwartz, J.H. (2001). *Extinct Humans*. Boulder, CO: Westview Press.

Tercafs, R. (1988). Optimal Management of Karst Sites with cave Fauna protection. *Environmental Conservation* 15: 149–158.

Tercafs, R. (2001). *The Protection of the Subterranean Environment. Conservation Principles and Management Tools*. Luxembourg: Production Services Publishers.

Thurgate, M. and Olson, R. (2002). Cave management at Mammoth Cave. *U. S. A. Journal of the Australasian Cave and Karst Management Association* 49: 33–35.

Toomey, R.S.I., Olson, R.A., Kovar, S. et al. (2009). Relighting Mammoth Cave's new entrance: improving visitor experience, reducing exotic plant growth, and easing maintenance. In: *Proceedings of the 15th International Congress of Speleology* (ed. W.B. White), 1223–1228. Kerrville, Texas: Greyhound Press.

Uhl, P.J. (1981). Photomonitoring as a management tool. In: *Proceedings of the 8th International Congress of Speleology*, 476–478. Bowling Green, Kentucky: Department of Geography and Geology, Western Kentucky University.

Villar, E., Fernandez, P.L., Gutierrez, I. et al. (1986). Influence of visitors on carbon dioxide concentrations in Altamira cave. *Cave Science* 13: 21–23.

Watson, J., Hamilton-Smith, E., Gillieson, D., and Kiernan, K. (1997). *Guidelines for Cave and Karst Protected Areas* 53 pp. Gland: World Conservation Union (IUCN).

Webb, R. (1995). Minimal impact caving code. *Journal of the Australasian Cave and Karst Management Association* 19: 11–13.

Wilde, K.A. and Williams, P.W. (1988). Environmental monitoring of karst and caves. In: *Proceedings of a Symposium On Environmental Monitoring In New Zealand, With Emphasis On Protected Natural Areas*, 82–90. Otago University, Albany, New Zealand: New Zealand Department of Conservation.

Williams, P. W. 1993. Environmental change and human impact on karst terrains: an introduction. *In:* Williams, P. W. (ed.) *Karst Terrains: Environmental Changes and Human Impact*. Catena Supplement 25: 1-19.

Williams, P.W. (2008). *World Heritage Caves and Karst: A Thematic Study*. Gland: IUCN.

Worboys, G., Davey, A., and Stiff, C. (1982). Report on cave classification. *Cave Management in Australia* 4: 11–18.

Yarborough, K. (1976). Investigation of radiation produced by radon and thoron in natural caves administered by the National Park Service. In: *Proceedings of the National Cave Management Symposium*, 703–713. Arkansas.

Zhang, S. and Jin, Y. (1996). Tourism resources on karst and caves in China. In: *Proceedings of the International Show Caves Association 2nd Congress. Malaga, Spain*, 111–119.

14

Catchment Management in Karst

14.1 Introduction

Throughout this book, the connectivity of caves and their overlying karst has been stressed. Conservation of caves is a very short-sighted preoccupation unless it is accompanied by energetic attention to conservation of karst as a whole. Nearly all of the karst solution process is moderated by factors operating on the surface of the karst, in the epikarst, and in the subcutaneous zone. Surface vegetation regulates the flow of water into the epikarst through interception, the control of litter and roots on soil infiltration, and the biogenic production of carbon dioxide in the root zone. The metabolic uptake of water by plants, especially trees, may regulate the quantity of water available to feed speleothems. Trees, in particular, are like large carbon dioxide pumps, releasing 20–25% of the atmospheric gas uptake through root respiration (Aley 1994). Thus, clearfelling of karst, or major changes consequent on plantation establishment, may radically change the flow and quality of water in the karst. Soil erosion in excess of the natural rates may infill streamsinks, dolines, or caves, and may totally smother cave life. Changes to surface drainage resulting from contour banking, irrigation, or river regulation may interrupt or drastically reduce the supply of karst water. The release of fertilisers, herbicides, and insecticides from agricultural activities may compromise cave ecosystems beyond their capacity to recover. Water is the primary mechanism for the transferral of surface actions to become subsurface impacts.

14.2 Basic Concepts in Karst Management

Karst management must be holistic in its approach and should aim to maintain the quality and quantity of water and air movement through the subterranean environment, as well as the surface. Managers of karst areas should recognise that these landscapes are complex three-dimensional integrated natural systems comprised of rock, water, soil, vegetation, and atmosphere elements (Figure 14.1). In all cases, the intention should be to maintain natural flows and cycles of air and water through the landscape in balance with prevailing climatic and biotic regimes. All users of karst terrains should recognise that, in karst, surface actions may be rapidly translated into impacts underground and elsewhere.

Figure 14.1 Karst catchments, such as in this polygonal karst area at Waitomo, New Zealand, will often extend beyond the limestone boundary and are especially vulnerable to altered land use and water pollution. Within this catchment, land uses include grazing, forestry, and quarrying.

Pre-eminent among karst processes is the cascade of carbon dioxide from low levels in the external atmosphere through greatly enhanced levels in the soil atmosphere to reduced levels in cave passages. Elevated soil carbon dioxide levels depend on plant root respiration, microbial activity, and a healthy soil invertebrate fauna. This cascade must be maintained for the effective operation of karst solution processes.

Thus, the integrity of any karst system is dependent upon a specific relationship between water and land; this water is often drawn from a very wide catchment area, and any alteration in the hydrologic system will threaten the karst and those caves which have a continuing relationship to the water levels. However, many caves, abandoned by the original formative waters as groundwater levels have been lowered, will be relatively dry, relatively static in character, and essentially non-renewable. Their contents – formations, sediments, and bones – are especially vulnerable, and may need special management provisions.

Climate change has occurred over the geological timescales within which karst systems have evolved. Human intervention has the potential to alter climate in ways that may radically affect natural karst processes. Management prescriptions must be flexible, must recognise this possibility and must maximise the resilience of the system. The effects of high magnitude-low frequency events such as floods, fires, and earthquakes must be perceived in management strategies at regional, local, and site-specific scales.

14.3 Defining Karst Catchments

The catchment of a karst drainage system is usually much larger than just the area of limestone outcrop and the obvious non-karstic contributing catchment. However, defining the contributing catchment of a cave may be difficult, if not impossible, in some cases. A minimalist approach would be to define a catchment as an area of limestone outcrop. This is often convenient for agencies wishing to restrict the application of environmental protection legislation to a given situation, such as a quarry or landfill. This neglects the possibility that the limestone is continuous though not outcropping in a given terrain, or that surrounding non-karstic rocks are contributing significant quantities of water by surface or subsurface flow. In many cases, a thick mantle of colluvium lies over the limestone and directly feeds cave systems. This is especially true in areas which were formerly glaciated or which have been subject to repeated mass movements over geologic time. The complexity of karst drainage systems has been illustrated for the Mammoth Cave area in Kentucky, USA (Chapter 3). The elucidation of this drainage network was the result of over 20 years' investigation and hundreds of dye-tracing experiments. It is rare for managers to have this level of detail. Usually, decisions must be made on the basis of a few dye-tracing experiments, aided by local geological knowledge and intuition. Karst managers must, therefore, learn to expect the unexpected!

Subterranean breaches of surface drainage divides are often the norm rather than the exception, and the exact conditions for the activation of conduits may depend on storm events or antecedent rainfall. In some cases, palaeokarst conduits may be reactivated. Thus, the definition of a karst catchment is imprecise and must have a dynamic boundary to take account of extreme events. This is best achieved by constructing buffer zones around limestone massifs, in which any change to land use must be preceded by investigations of the drainage network and its dynamics using repeated dye-tracing experiments.

For karst areas, the concept of total catchment management becomes more vital than in many other lithologies. This involves the coordinated management and utilisation of physical resources of land, water, and vegetation within the boundaries of a catchment to ensure sustainable use and to minimise land degradation. Proper environmental management on karst terrains rests on a base of public acceptance that clear linkages exist between surface and underground systems, and that these linkages are of fundamental importance to karst system function.

Pollutants readily enter karst drainage systems and are rapidly transmitted in cave conduits. The range of likely pollutants and their relative importance are given in Table 14.1. In Britain, some 148 licensed landfill sites are located on limestone (Chapman 1993, p. 197). Many of these take industrial waste, usually because of their remote location. Leachates from these sites may travel to contaminate underground watercourses and springs over several kilometres. In Ireland, up to $100\,000\,\mathrm{m}^3$ of slurries from feedlot operations are discharged into limestone aquifers via sinkholes, quarries, and caves. This nutrient boost to underground waters may cause dramatic changes in cave ecosystems and severely affect water quality for people. In Slovenia, typhoid–paratyphoid epidemics as a result of sewage pollution of limestone groundwater affected populations in the classical karst between 1925 and 1975. Chapman (1993, pp. 198–199) provides further graphic examples of sewage pollution in British karst areas.

Table 14.1 Sources of water pollution in caves.

Source	Oxygen demand	Nitrogen, phosphates	Chlorides	Heavy metals	Hydrocarbons, organic complexes	Bacteria, viruses
Domestic and municipal wastes						
Septic tanks	●●●	●●				●●●
Outhouses or privies	●●●	●●				●●●
Sewage lines	●●	●				●●
Landfills	●●●	●	●	●●●	●●	●●
Dumps in sinkholes	●	●●	●	●●	●●	●●●
Agricultural wastes						
Feedlots	●●●	●●●				●●●
Fertiliser leaching		●●●				
Insecticides and herbicides					●●●	
Construction and mining						
Salting of roads			●●●			
Mine tailings				●●●		
Car park runoff			●●●	●	●●	
Oil fields			●●●		●●●	
Industrial						
Petroleum storage					●●●	
Chemical dumps	●			●●●	●●●	
Chemical wastes			●	●●●	●●●	

Note: Number of dots Indicates very approximate severity of pollution threat.
Source: Modified from White (1988. p. 389).

Documentation of these probably commenced with the 1845 cholera epidemic but continues to the present day. The comparatively rapid transmission of groundwater flows in karst provides little opportunity for natural filtering or other purifying effects, and so problems, such as disease transmission, may arise much more readily than in any other terrain. Again, the source of the pollution may be located far outside of the karst area itself but can still have devastating impacts. Management should aim to maintain the natural transfer rates of fluids, including gases, through the integrated network of cracks, fissures, and caves in the karst. The nature of materials introduced must be carefully considered to avoid adverse impacts on air and water quality.

Limestone and marble are quarried worldwide and used for cement manufacture, as high-grade building stone, for agricultural lime, and for abrasives. Most resource conflict over limestone mining revolves around visual and water pollution, as well as a loss of recreational amenity and conservation values (Figure 14.2). In Britain, some important caves have been destroyed by quarrying, and this industry remains one of the most potent threats

(a)

(b)

Figure 14.2 Mount Etna, a limestone hill in central Queensland, Australia with a large bat colony: (a), prior to mining in; (b), after blasting and quarrying commenced in 1968. Source: Photos by Paul Caffyn.

to caves in that country. In the Guilin tower karst of southern China, there are numerous small quarries operated for cement manufacture and for industrial fluxes. These, coupled with devegetation as a result of acid rain from coal burning, have scarred many of the towers around the city. Ironically, much of this cement is used to expand the infrastructure of tourism – hotels and shops – for people who have come to see the famous karst landscape.

Limestone bodies with high relief are ideal for mining and are often the most cavernous, and there is often conflict and compromise when there is a high community expectation of continued access to this resource as well as a strong conservation movement. Recognising that the extraction of rocks, soil, vegetation, and water will clearly interrupt the processes that produce and maintain karst, such uses must be carefully planned and executed to minimise environmental impact. Extractive industries may be incompatible with the preservation of natural and cultural heritage.

Meiman (1991) has related changes in land-use practices in Kentucky to water quality decline in Mammoth Cave National Park. Flood pulses from heavily used agricultural land produced significant rises in turbidity, chloride, faecal coliform levels, and triazine herbicides. Hardwick and Gunn (1993) have summarised land use impacts on British caves. Of the 79 caves identified, 68% were affected by agricultural pollution: 35 sites were affected by rubbish tipping, 16 by farm tipping, and 14 by infilling of caves.

Dolines and open shafts in karst have long been seen as 'good places' to dispose of rubbish. The 'out of sight, out of mind' principle applies here. Such places have commonly been used for the dumping of animal carcasses, and this poses a considerable health risk. At Earls Cave, Mount Gambier, South Australia, more than 5000 sheep carcasses were dumped in the cave entrance. This connects directly to the karst groundwater, which is used for town water supply. Blood and unwanted body parts from slaughterhouses on Kaua'i Island, Hawaii, USA, have been disposed of in caves in the past, altering the cave ecology in favour of exotic carrion feeders and scavengers. In the Chimbu region of New Guinea, the vanquished in a tribal fight were compelled to suicide at arrow point by jumping down the 60 m shaft of Nombe. The water from this site drains to major springs used for villages near the town of Chuave.

14.4 Vegetation and Caves

The thin mantle of soil on most limestone areas has a great significance for karst processes, as is seen in Chapter 2. First, there are a number of interactions between limestone soils and vegetation. The free vertical drainage of most limestone soils creates special conditions for evapotranspiration, gas exchange, and root penetration. Large eucalypts act as water pumps, taking up $250–270 \, l \, day^{-1}$ for each tree of river red gum (*Eucalyptus camaldulensis*). Tree roots can penetrate to depths of 30–50 m in search of water, especially so in the seasonally humid climates of northern Australia. Litter retention is of great importance for nutrient cycling, with the most release by decomposition occurring within six months of litterfall.

Secondly, the vegetation structure (especially projected foliage cover) is important for interception of rainwater, soil water infiltration, and temperature control in soil and subcutaneous zones. This directly affects both the quantity and quality of water available as feed water for speleothem growth in underlying caves. The penetration, of tree roots, and the release of complex organic acids and phenolic compounds by them, aid the enlargement of bedrock fissures in karst and ensures the high degree of secondary porosity characteristic of limestone terrains.

Native plants growing on limestone are frequently displaced by exotic species, creating management problems. Most plant species can be classified according to their life strategies – their life-forms, resistance to disturbance, rates of seed production, and viability. Highly competitive species such as blackberries grow rapidly, resist disturbance, bear copious viable seed, and may inhibit the establishment and growth of other species near them by poisoning the soil (allelopathy). Stress-tolerating plants, such as figs and eucalypts, have adaptations to ensure adequate water supply (deep roots, high root-shoot ratios), to resist fire (resistant bark, lignotubers, and vegetative reproduction), and to exploit low levels of nutrients (root fungi associations, high root hydrogen ion production). Ruderal species, those that habitually colonise disturbed habitats such as roadsides, have tuberous energy stores, very effective seed dispersal, and easily pollinated flowers. Dandelions and St John's Wort are good examples of ruderal plants. In general, stress-tolerating plants are more numerous on limestone soils, with some ruderals present, but highly competitive species are rare. In part, this explains why many karst areas have been colonised by competitive exotic species which grow naturally in other limestone areas. Those native species which are stress-tolerators are very susceptible to displacement by exotic species when the disturbance regime is altered radically. Their reproductive strategies may be adapted to infrequent disturbance and they may be unable to complete their reproductive cycle if repeatedly stressed. This may happen as a result of increased fire frequency, after earth-moving operations, or after accelerated erosion.

In the case of fire, the transference of prescriptions derived from one area to another may not produce the same response in vegetation. A change in the vegetation may, therefore, have major consequences for karst hydrology and the limestone solution process. This has been noted for the replacement of native woodland by exotic pines with higher water use and a tendency to acidify the soil, resulting in interruption of growth, and even erosion of speleothems in caves below the introduced forest. In many karst areas, old growth native forests have been cleared and replaced with monospecific plantation forests, often coniferous. These plantations have higher basal area and often higher water demand per hectare than the forests they replace (Costin et al. 1984). Thus, there may be a reduction in the flow of percolation water to the karst system, as well as some sediment transfer associated with felling and roading. At Naracoorte Caves, South Australia, exotic pine plantations over caves were cleared in 1990; already speleothems deprived of feed water for decades have reactivated and extended.

At Yarrangobilly, in the Snowy Mountains of Australia, eucalypt forests were cleared in the 1930s for exotic pine plantations. Caves underlying the pine forest had high root biomasses visible and were relatively dry. Since clearance of the pine and partial regeneration of native vegetation, some cave formations have reactivated. Soil loss in the area has been minimal, with most sediment transfer occurring in spring (National Parks Wildlife Service 1983).

14.5   Accelerated Soil Loss in Karst

There are numerous examples of accelerated soil erosion on karst areas worldwide (Gams et al. 1993; Gillieson 1989; Jiang 2006; Kiernan 1988; Urich 1991; Urushibara-Yoshino 1993; Yuan 1993). Limestone soils tend to be shallow and stony, with low to moderate nutrient holding capacity because of excessive leaching as a result of free drainage. There is thus a strong tendency for devegetated or heavily used limestone soils to erode down to bedrock surfaces quite rapidly. This soil stripping (Figure 14.3) can be seen in the classic karst of the Burren, Ireland; the Dinaric karst in Slovenia and Croatia; the Guizhou polygonal karst of China; and the karst of Vancouver Island, British Columbia, Canada.

The process started some 2000 years BP in Greece and continues today in areas such as the Bohol cone karst, the Philippines. Eroded soil material is rapidly transferred underground to block passages, divert or impound cave streams, or smother cave life (Kranjc 1979). Soil erosion control is, therefore, a high priority for karst managers, and much depends on the effectiveness of revegetation. There is usually accelerated soil loss and tree decline associated with forestry operations on karst. On Vancouver Island, forests clear-felled since 1900 have regained only 17% of the original timber volume after 75 years, and soil depth loss ranges from a mean of 25% 5 years after logging to 60% after 10 years (Harding and Ford 1993). Clear guidelines for forestry operations on karst need to be developed and adhered to. Soil management in karst must aim to minimise erosive loss and alteration of soil properties

Figure 14.3   Rocky desertification in the karst of Anshun region, Guizhou Province, China. Springs are fed into channels to permit irrigated rice production.

such as aeration, aggregate stability, organic matter content, and a healthy soil biota. Pivotal to the prevention of erosion and maintenance of critical soil properties is the presence of a stable vegetation cover.

Erosion prediction and control on limestone soils follows the same provisions as most other soil types with some important exceptions. Water erosion by sheet and rill processes may be predicted at a point using the Universal Soil Loss Equation (USLE). This has been calibrated for some soils from long-term erosion plot studies (Rosewell and Edwards 1988). Annual soil loss in tonnes/ha (A) is a function of the rainfall intensity or erosivity (R), the soil erodibility (K), the slope angle and length (L, S), vegetation cover (C), and cultivation practices (P):

$$A = f(R \times K \times L \times S \times C \times P) \tag{14.1}$$

Since soil erosion is a function of ground slope and slope length, any factors locally increasing the slope angle will result in more erosion. On non-karstic terrain, the slope angle and length are functions of the slope distance to the nearest drainage line. On limestone, the intimate connections between the surface and the underground cave conduit network lead to locally steeper hydraulic gradients which enhance erosive processes. There is also greater potential for the vertical abstraction of material down joints and fissures by sinkhole collapse, gullying, or soil stripping (Figure 14.4). The preservation of adequate vegetation cover and of soil structure (reducing erodibility) assumes greater importance on limestone than elsewhere in mitigating this effect.

Figure 14.4 Stripped limestone pavements with in washed soil in grikes, Inishmore, Ireland.

Caesium-137 has been used extensively as a tool to measure rates and patterns of soil redistribution. The fallout of this radionuclide is derived from atmospheric testing of atomic weapons during and after the 1950s, with the 1958 concentrations of ^{137}Cs in soil being still detectable today within Australia. After reaching the soil surface, ^{137}Cs has been shown to be irreversibly bound to soil particles, especially clays. Its fate and redistribution within the landscape thereafter occur only by soil transport processes (Campbell et al. 1986). Its properties as a unique and discrete label of surface soil make it ideal as a tracer to test comparative rates of erosion, in which measurements of the areal concentration of ^{137}Cs in erosion or deposition sites are compared with the known total or 'reference' fallout at that location (Murray et al. 1990). Increases in this isotope compared with the 'reference' value indicate areas of soil accretion, and areas of decrease represent zones of erosion. These proportional differences may be converted to quantitative loss by a variety of models (Gillieson et al. 1994; Walling and Quine 1990). The measurements allow us to make estimates of erosion integrated over a long time and thus free of the biases of short-term sampling, which may miss major events in the landscape.

The literature of soil science is extensive and often complex. Under the widely used USDA Soil Taxonomy, a very large number (>70) of soil properties must be measured before the soil can be classified and its behaviour predicted. This is also true for soil classifications based on soil formation or genesis, such as Great Soil Groups. A simpler factual key for the classification of soils (Northcote 1979) may be used in which a few key properties are measured in the field. These are the soil profile form, the texture of the visible layers, their pH and colour. From these much of the behaviour of soils can be inferred:

- The profile form tells us about soil drainage, the rooting ability of trees, and aeration.
- Soil texture tells us about the water-holding ability and nutrient retention.
- pH tells us about nutrient availability and the acidity of soil water.
- Colour tells us about organic matter content, clay minerals, and precipitates.

To these should be added total organic matter content, soil water infiltration capacity, and penetrometer resistance, a measure of compaction. These are all useful attributes of limestone soils which can be measured in the field and provide a good basis from which to predict soil behaviour, especially susceptibility to erosion and suitability for plant growth. They should be part of the bask toolkit for karst managers.

14.6 Agricultural Impacts

14.6.1 Rocky Desertification

Forest clearance in the extensive karsts of Europe and Asia has been undertaken to obtain land for agriculture from Mesolithic and Neolithic times to the present. The process of extreme land degradation known as rocky desertification, thus has some antiquity (Williams 1993). Limestone soils are clearly very prone to rapid erosion leading to widespread exposure of the epikarst. One region where this is particularly severe is in southern China, in Guizhou, the Guangxi Zhuang Autonomous Region, and in Yunnan Province (Huang et al. 2008; Xiong et al. 2009). Land is being transformed by vegetation loss and soil erosion into exposed rock slopes at a rate of 25 000 km² per year. About half of the region is affected by soil erosion and the accelerated sedimentation affects both rivers and reservoirs (Wang et al. 2004). Silt transport rates for rivers in this region subject to deforestation and rocky desertification are 208–1980 t/km² per year (Yuan 1993).

Similar accelerated soil erosion has occurred in many Mediterranean karst regions since prehistoric times (Calò and Parise 2006; Gams et al. 1993). There has also been significant soil loss into subsoil cavities in the karst of Northern Italy (Sauro 1993) and in the classical karst areas of Slovenia and Croatia (Mihevc et al. 2016). The karst landscapes of Lebanon have thin soils which are prone to erosion due to deforestation, burning, and overgrazing (Bou Kheir et al. 2008). Gillieson and Thurgate (1999), reviewing karst and agriculture in Australia, comment that there is a crucial link between landscape stability and vegetation cover in Australian karsts.

14.6.2 Infilling of Dolines

Blockage of cave and doline entrances are a common occurrence in agricultural lands on karst (Drew 1996). For property owners, these openings and depressions represent a potential hazard to stock and reduce the value of grazing land. Regrettably, in many instances, this has led to the deliberate infilling of these features with agricultural and domestic waste (Figure 14.5). There are numerous examples of refuse sites being located within karst depressions, and causing point source pollution. At Wee Jasper, in New South Wales, Australia, the entrance to Dip Cave has served as a rubbish dump (Figure 14.6) for the local community (Jennings 1983), while the Heywood municipal tip in western Victoria, Australia, is located on, and completely infills Dyers Cave (Davey and White 1986). At Borenore Caves, also in New South Wales, dolines have been used as rubbish dumps and an old grike has been used as a sheep dip, the latter probably releasing arsenic into the groundwater (Holland 1991). Farm refuse has also been dumped in numerous sinkholes in the Mole Creek area, Tasmania, Australia, including drums of the chemical 2,4-D, prompting the site to be named 2,4-D Cave by local cavers (Kiernan 1989).

Figure 14.5 Car bodies dumped in a doline, Naracoorte, South Australia.

Figure 14.6 Domestic rubbish is often dumped in cave entrances, and provides both an energy source and potential pollutants for cave organisms. Dip Cave, Wee Jasper, New South Wales, Australia. Source: Photo by Stefan Eberhard.

14.6.3 Altered Drainage

Deforestation of karst areas may result in an increase in flood frequency and severity. This is usually due to accelerated soil erosion, blocking karst conduits, and creating backflooding of enclosed valleys. Removal of deciduous oak-maple forest cover in Kentucky (Crawford 1984) and cultivation of tobacco and corn in the 1930s resulted in disastrous valley floods known locally as 'valley tides'. The combination of increased runoff and blocking of streamsinks resulted in the flooding of the valleys upstream of river sinks (Dougherty 1981). Forest clearance in Mole Creek, Tasmania, caused cave conduits to become choked by sediments, resulting in flooding of pastures during the winter months (Gillieson and Thurgate 1999). At Naracoorte, in South Australia, thick mats of pine roots from a 90-year-old plantation had intercepted cave air space, and many decorations had dried. Logging of these pine trees resulted in regeneration of speleothems and refilling of calcite rimstone formations (Kiernan 1988). Fire associated with hazard reduction in agricultural areas can cause sealing of the epikarst, reducing infiltration capacity, so that surface waters may pond, and runoff may carry high sediment loads, eventually impacting on groundwater quality (Holland 1994). Some agricultural changes may cause a decrease rather than an increase in flooding. Karst poljes, which provide areas of flat, fertile land within rocky Mediterranean karst landscapes, have also been subject to drainage schemes by various means and with varying degrees of success (Figures 14.7 and 14.8). In recent decades, polje water regulation schemes have been multi-purpose, for hydroelectric power generation and water supply as well as agriculture, and have involved major tunnel construction (Milanovic 2002).

Figure 14.7 Brsnica polje in Slovenia with a seasonally flooded floor; in summer used for agriculture.

14.6.4 Groundwater Lowering

Irrigation of agricultural land may cause a hydrological impact at the source of the irrigation water or at the irrigation site. In karst aquifers, the lowering of the water table due to irrigation water withdrawals may be very uneven and difficult to predict. The irrigated land may then provide a new source of recharge to the karst aquifer, with potential problems of salinisation noted below. Problems may also arise with damming for irrigation purposes, for example, Gillieson and Thurgate (1999) record damage to the Texas cave system in Australia due to the building of an irrigation dam. Land drainage operations carried out with the aim of decreasing soil moisture levels to improve agricultural productivity may also have significant hydrological impacts. In the Carboniferous limestone lowland areas of western Ireland, channelisation to relieve flooding of agricultural land has resulted in a lowering of water tables and a change to the flooding regime of ecologically important seasonal karst lakes or *turloughs* (Drew and Coxon 1988).

The most spectacular example of agriculture impacting on surface drainage at a large scale can be found in the syngenetic karst areas of south-eastern South Australia (Boulton et al. 2003). This area was first settled in the 1840s; however, settlement was typically scattered into isolated pockets on high ground. Natural drainage is impeded by extensive karst dune ridges running parallel to the coast, the water table occurs close to the surface (Figure 14.9), and rainfall is relatively high.

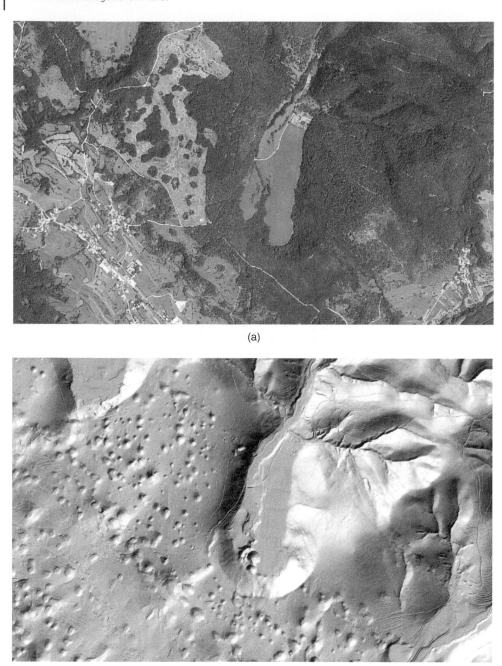

(a)

(b)

Figure 14.8 Aerial photograph (a) and LIDAR image (b) of Brsnica polje in Slovenia. Source: From Atlas of the Environment, Environmental Agency of the Republic of Slovenia (EARS).

Figure 14.9 The Sisters cenotes near Mount Gambier, South Australia. Many of these cenotes have had their watertables lowered by 2–3 m in recent years.

This resulted in between half to three-quarters of the land surface in any given area being seasonally waterlogged or flooded. In 1863, the South Australian government, eager to prevent the loss of potentially valuable lands, began a programme of drainage to open up large areas of land for grazing and agriculture. Drainage has uncovered 53% of formerly flood-prone lands and has resulted in the alteration of surface flows, allowing agricultural improvement to the detriment of the wetland ecosystems that once were so widespread. Along the southern coast, approximately 1800 ha of swamps, characterised by lines of spring lakes and surrounded by peat fens, were drained (Figure 14.10), burnt, and ploughed for soldier settlement schemes following the Second World War. Water levels have been lowered in several cenotes (Goulden Hole, Ewens Ponds, Picaninnie Ponds) by at least 1.5–2 m below original survey levels.

14.6.5 Fertiliser and Herbicides

Globally inorganic fertilisers are widely used, and that use has impacted on groundwater quality, especially in nitrate concentrations. A range of organic material is also spread on agricultural land, including livestock manure as a slurry, silage effluent, farmyard runoff, human wastes (sewage effluent, sewage sludge), and agro-industrial wastes (dairy effluent, blood, offal) (Berryhill 1989; Coxon 1999). The water quality issues that result from these practices are widespread but become more acute in limestone landscapes due to the high aquifer permeability and thin soil cover. In many cases, contamination may not be restricted to more mobile anions such as nitrate and chloride. Compounds associated with

Figure 14.10 Drain being excavated by hand in South East karst district c. 1947. Source: Photograph From State Library of South Australia.

point pollution sources such as phosphorus, potassium, ammonium, and also faecal bacteria may gain entry to vulnerable karst aquifers. In addition, both inorganic and pesticide contaminants may be adsorbed to colloidal soil particles and may be washed down to karst groundwater through enlarged bedrock fissures.

14.6.6 Pesticides

The commonly used pesticides vary in their toxicity, mobility, and persistence in the natural environment. They may be carcinogenic at very low concentrations, and the thin soils of karst landscapes and high permeability make contamination more likely than in other landscapes. The herbicide Atrazine has been widely used in maize cultivation and also as a general broad-spectrum weed killer. It has been the most widely reported pesticide in karst groundwater over the last two decades: two karst catchments in West Virginia, USA, where atrazine and its metabolite desethylatrazine were detected in more than 50% of samples (Pasquarell and Boyer 1996); the Green River Basin in Kentucky, where it was detected in 100% of karst spring water samples (Crain 2002); the Illinois, USA, sinkhole plain, where concentrations ranged from <0.01 to $34\,\mu g\,l^{-1}$ (Panno and Kelly 2004); and two karst springs in Northern Alabama, USA (Kingsbury 2008). In the USA, the legal limit for atrazine in potable water is $3\,\mu g\,l^{-1}$ as a yearly average (with higher concentrations permissible in the shorter term), while in the European Union, the limit is $0.1\,\mu g\,l^{-1}$. Atrazine has been recorded from several European karst aquifers including a Jurassic karst aquifer in Germany and a weakly karstified Cretaceous Chalk aquifer in France, where it was present in

83% of samples at concentrations up to 5.3 µg l^{-1} (Baran et al. 2008). Atrazine was present in Slovenian groundwater for several years after it was banned and also present in the French Chalk aquifer mentioned above, where it was recorded three years after the ban in 2003 (Baran et al. 2008).

Other herbicides recorded in karst groundwaters include other triazines such as simazine; alachlor, a widely used herbicide in maize and peanut cultivation (Dalton and Frick 2008; Panno and Kelly 2004); and fluometuron, used in cotton cultivation (Dalton and Frick 2008; Kingsbury 2008). These pesticides may also be moved adsorbed onto colloidal particles, and thus concentrations may increase with flood peaks in karst catchments. There are several metabolites of common pesticides which are also transported in karst aquifers. For example, lengthy storage of atrazine in soil results in the loss of its metabolite desethylatrazine (DEA) from the soil to karst groundwater. In many cases, concentrations of other metabolites exceed those of the parent pesticide; Dalton and Frick (2008) reported that degradates of alachlor and metoalachlor were generally found at higher concentrations than their parent compounds.

Antibiotics are commonly used in intensive feedlot operations to boost growth rates in cattle. The disposal of manure slurries or spreading on karst landscapes may also contaminate water supplies with antibiotics, giving rise to concern about the spread of antibiotic resistance. Dolliver and Gupta (2008) documented a study of leaching and runoff losses of antibiotics from the application of hog and beef manures in a karst area of Wisconsin, USA. Chlortetracycline was detected in runoff while monensin and tylosin were detected in both runoff and leachate.

14.6.7 Microbial Contamination of Groundwater

The pollution of karst groundwater by pathogenic microorganisms is a widespread problem. It is exacerbated by the rapid flow of water in karst conduits, allowing pollutants to travel hundreds of metres in a day to emerge at springs and in wells. The microorganisms involved include bacteria, viruses, and protozoan parasites, such as giardia. High concentrations of faecal coliforms and faecal streptococci due to animal and human faecal contamination have been reported in many karst areas (Figure 14.11).

In Ireland, faecal bacterial contamination of groundwater supplies from karst aquifers has been reported since the 1980s (Coxon 2011) and continues to the present. Celico et al. (2004) recorded bacterial contamination of karst springs in Southern Italy resulting from manure spreading, with the highest bacterial numbers found when intense rainfall produced concentrated infiltration of runoff in a swallow hole. Faecal bacterial contamination is also well documented from karst aquifers in North America. Boyer and Pasquarell (1999) compared bacterial contamination in cave streams in a beef cattle area and a dairying area and found that faecal coliforms and faecal streptococci were present in much greater numbers in the dairying area. Kozar and Mathes (2001) found that 32% of wells in a karst limestone and dolomite aquifer in West Virginia contained *E. coli*, with contamination rates being higher where there was nearby agricultural activity.

A disturbing strain of bacterium produces a verotoxin-producing *E. coli* (VTEC; *E. coli* 0157) and causes dysentery and haemolytic uremic syndrome, which destroys red blood cells and damages the kidneys. The most serious recorded case took place in Walkerton,

Figure 14.11 The entrance of Earls Cave, Mount Gambier, South Australia, was the site of dumping for more than 5000 sheep carcases. Casual dumping still occurs, polluting the karst aquifer which serves many farms and towns. Source: Photo by Andy Spate.

Canada in May 2000, when 2300 people became ill, and seven people died. Cattle faeces entered the groundwater after heavy rain and *E coli 0157:H7* with *Campylobacter jejuni* were rapidly transported to domestic water sources. Dye tracing experiments demonstrated flow velocities 80 times faster than predicted in the initial modelling, which treated the aquifer as an equivalent porous medium (Worthington et al. 2002).

Protozoan parasites, particularly *Cryptosporidium parvum*, can be easily transported in karst groundwater and are resistant to conventional water treatment using chlorination. *Cryptosporidium* causes acute gastroenteritis and can be a potentially fatal disease in children and immune-compromised individuals. The first outbreak of cryptosporidiosis linked with karst groundwater occurred in Braun Station, San Antonio, Texas, USA, in 1984, when more than 200 people became ill due to contamination in the Edwards aquifer (D'Antonio et al. 1985). *Cryptosporidium* has been detected in many karst groundwater supplies in the last decade, including springs in West Virginia, a karst spring in Switzerland, and the karst springs which provide the water supply for the town of Ennis, in Ireland (Page et al. 2006. p. 64).

Regular water sampling of karst aquifers on a weekly or monthly basis may not pick up microbial contamination. Temporal variation in faecal coliform numbers is large, and peak

numbers often coincide with peaks inflow and turbidity (Pronk et al. 2006, 2007). Soil moisture levels may control temporal variations in faecal coliform numbers, as bacteria stored
in the soil are released groundwater recharge occurs (Pasquarell and Boyer 1995). As even
brief exposure to pathogenic bacteria in groundwater could have serious consequences, the
need to predict temporal variations is acute.

14.6.8 Golf Courses on Karst

As we have seen, karst drainage systems are highly susceptible to contamination by runoff
and also by the adsorption on sediments of chemicals, micro-organisms, low solubility
herbicides and pesticides not usually transmitted by drainage water (Watson et al. 1997).
Golf courses on karst landscapes (Figure 14.12) are especially prone to groundwater impacts
leading to adverse environmental outcomes (Balogh and Walker 1992, p. 63). They constitute an intensive form of agriculture, and they are ubiquitous, even in climates unsuitable
for natural grass growth.

Substantial inputs of fertilisers and water to maintain turf systems on golf courses have
led to the realisation that golf courses are a major contributor to non-point source water pollution. In particular, elevated concentrations of nitrate–N, ammonium–N, and phosphate–P
have been recorded after storm-generated surface runoff from golf courses (King et al. 2007).
In the Yucatan Peninsula of Mexico, significant pollution of the groundwater in a porous
karst landscape similar to Kangaroo Island, South Australia, has occurred (Metcalfe et al.
2011), leading to eutrophication. Soil cover through the whole area is very thin and patchy,
with high-porosity limestone bedrock commonly exposed. Chlorophenoxy herbicides used

Figure 14.12 Stone Forest (Shilin) Golf Couse near Kunming, China. This development has
occurred on a World Heritage site. Source: CPA Media Pte Ltd / Alamy Stock Photo. © Alamy Images.

Figure 14.13 Collapse sinkhole formation in a golf course on karst, Branson, Missouri, USA. Source: https://www.news-leader.com/story/news/local/ozarks/2015/05/22/large-sinkhole-opens-top-rock-golf-course/27799219 © Dr David Weary.

on golf courses were detected in a number of sinkhole lakes (Figure 14.13) and caves, as well as sewage effluent, PAHs (polycyclic aromatic hydrocarbons, usually benzene derivatives) and other chemicals derived from road runoff.

In several areas, herbicide applications to golf courses have been identified as a major cause of groundwater pollution. Studies in Japan (Suzuki et al. 1998), concluded that golf courses have a *'high pollution potential for pesticides relative to agricultural areas'*. In the karst areas of Florida, USA, the herbicide monosodium methanearsenate (MSMA) is widely used as a herbicide on golf courses. A study of arsenic levels in lakes in the limestone areas (Pichler et al. 2008) concluded that arsenic levels in lakes on golf courses were significantly higher than levels in non-golf course lakes and that the potential for groundwater pollution was very high. This poses problems for drinking water supply as arsenic levels exceeded the USA standard of $10\,\mu\mathrm{g}\,\mathrm{l}^{-1}$ by a factor of more than 10 times.

14.7 Fire Management in Karst

One of the more vexing questions about karst catchment management is concerned with revegetation and fire control. Fire management on limestone areas is a contentious subject, especially when severe wildfires have caused loss of life or property. In traditional societies, fire is widely used as a vegetation clearance tool (Gillieson et al. 1986; Head 1989;

O'Neill et al. 1993; Urich 1991). Most karsts have a low natural fire frequency on account of the shielding effects of limestone outcrops, reduced ground cover, and often a more mesic canopy with rainforest elements in the flora. In the impounded karsts of eastern Australia, natural fire frequencies are poorly documented, but the fire interval may be 35–50 years or greater (Holland 1994; Williams et al. 1994). Under these conditions, relict vegetation types may survive – for example, the vine thickets of North Queensland, Australia, (Figure 14.14) or the monospecific *Acacia* scrubs on the Bendethera karst, New South Wales. In Britain, there are rare fern species which are now restricted largely to limestone outcrops (Goldie 1993). In these karsts, sediment transport occurs only immediately after fires, with minimal soil erosion in the intervening periods.

Hazard reduction burning is widely used by land managers but may have major deleterious effects on karst areas. In Australia, many authorities aim to burn individual areas on a five- to seven-year cycle. This increased frequency reduces the fuel load but promotes more fire-tolerant vegetation, principally shrubs which are often more flammable than the understorey of grasses and herbs that is replaced. Thus, there is potential for changes to the hydrology of the epikarst. Although there may be a management prescription to avoid burning limestone outcrops, there may be an inadvertent escape of fires into sensitive areas on account of weather changes. There is a possibility of increased cave sedimentation (Stanton et al. 1992). Careful zoning of fire management, aided by precise mapping of past fire

Figure 14.14 Spalling of limestone slope after fire at Chillagoe karst, north Queensland, Australia. The vine scrub vegetation is being invaded by exotic grasses which increase the fuel load.

boundaries with buffers around karst areas, may help to reduce these impacts. Elucidation of fire histories using sedimentary charcoal in caves is a promising avenue for research (Holland 1994).

14.8 Conservation Issues in Karst

Karst waters can be viewed as types of wild rivers where the drainage network is not as obvious as in surface streams, and there is complexity in hydrological linkages and in flow regimes. In many mountain areas, the highest parts of karst catchments are still forested and inaccessible. In such areas, both water quantity and quality are maintained along with the integrity of ecosystems. It is estimated that one-quarter of the world's population gain their water supplies from karst, either from discrete springs or from karst groundwater. The maintenance of water quality in karst can be viewed as a common good which is becoming increasingly important in those areas where rural populations are increasing rapidly and the settlement of karst is well established. In other areas, such as China and the Philippines, recent settlement of karst terrain is creating both opportunities and constraints for sustainable management of karst resources, especially water and soil. Establishment and maintenance of karst protected areas can contribute to the protection of both the quality and quantity of groundwater resources for human use. Catchment protection is necessary both on the karst and on contributing non-karst areas. Activities within caves may have detrimental effects on regional groundwater quality.

The rugged nature and physical isolation of most karst landscapes ensure that they act as refugia for rare and endangered species, such as leaf-eating monkeys in Vietnam and cave-adapted animals such as salamanders and blind fish in North America. Karst landscapes have high importance in the maintenance of *biodiversity*, and there are still opportunities to protect the habitats of these organisms. In many ways, karsts are buffered against climate change, and their biota is less vulnerable in climates characterised by high natural variability. There is often a high degree of endemism in karst biota. Many karst areas preserve relict populations of organisms whose distributions were much greater during past colder or wetter climatic regimes.

Because of the importance of karst areas as biological refugia, further fragmentation by road building and similar activities should be avoided. If they are unavoidable, then corridors for animal dispersal should be maintained as a high priority. Within caves, both terrestrial and aquatic fauna are best protected by the preservation of air and water quality. Accelerated stream siltation and compaction of sediments by visitors may be detrimental to cave fauna. The infrastructure of tourist caves (paths, steps, lights) should be designed to avoid decomposition and the release of either toxic substances or additional energy sources into the cave environment.

Caves have long been used for shelter, for religious purposes, for hunting, and as objects of veneration and awe. For indigenous people living with traditional lifeways, access to cave resources is but one component of a complex of land use and rights, which is central to their being. For the Tifalmin people of the Star Mountains, New Guinea, caves are the resting places of their ancestral female deity Afekan, and constructed pits in Selminum Tem Cave are of great significance (Gillieson 1980). Cave entrances and streamsinks also mark the boundaries of hunting territories. On the Nullarbor Plain, southern Australia, the flint mines of Koonalda Cave have been used by more than 600 generations of First Nations peoples, while other caves mark dreaming sites on a mythologically linked songline from the West Kimberleys, in Western Australia, to Eucla on the Great Australian Bight. In many areas, traditional owners are now in partnership with government agencies concerned with the management and interpretation of caves; for example, the Berawan and Penan people of Gunong Mulu National Park, Sarawak, Malaysia, are employed as cave guides, boatmen, and in daily site management (Gill 1993). Aboriginal cooperatives in the West Kimberley karst of Australia are running ecotourism ventures involving cave visits, while there are several leaseback arrangements where National Park agencies lease the land from traditional owners who sit on the board and are employed as rangers (Young 1992). Their involvement adds a whole new dimension to interpretive programmes. This form of community involvement in cave management deserves to be more widely employed, even in non-traditional societies.

14.9 Assessing Vulnerability in Karst Management

14.9.1 Karst Disturbance Index

The Karst Disturbance Index, first developed by van Beynen and Townsend (2005), is a method for evaluating human impacts on karst landscapes. It employs five categories of environmental indices: geomorphology, hydrology, atmosphere, biota, and cultural, from which levels or ranges of disturbance can be defined. Efforts to protect karst are rated under the cultural category. In principle, the indicators in each category have to be inexpensive to obtain, easily reproducible, and responsive to changes in environmental condition. Data sources have included field surveys, geographic information system (GIS) data layers, topographic maps, aerial photography, and expert opinions from local cavers and State officials. Scoring of indicators can be either semi-quantitative (ranked data, categorised areas, or percentage cover) or qualitative (type of settlement, type of cave development). Full details of this scheme are given in Table 14.2.

If an indicator is not relevant to the area concerned it can be discarded. The total disturbance index is calculated by summing all obtained scores and dividing by the summed maximum score, to produce a fraction. Total disturbance can be measured by tallying individual indicator scores and dividing this total by the highest possible score. A value between 0 and 1 will be obtained, corresponding to a karst disturbance level: 0.0–0.19 (pristine), 0.20–0.39 (little disturbance), 0.40–0.59 (disturbed), 0.60–0.79 (highly disturbed), and 0.80–1.0 (severely disturbed). The advantage of an index is that stakeholders can examine each indicator and see how it was derived, while the overall state of the karst environment is reduced to an easily comparable category for environmental managers and policymakers. The disadvantage of this index method is a certain level of subjectivity and the difficulty of comparing indices between different karst regions.

Table 14.2 Karst disturbance index main categories.

Category	Attribute	Scale	Indicator
Geomorphology	Surface landforms	Meso to Macro	Quarrying/mining
			Human-induced hydrologic change
			Infilling of cave and dolines
			Dumping in dolines
	Soils	Micro to Macro	Erosion
			Compaction due to livestock or humans
	Subsurface karst	Micro	Flooding
			Speleothem vandalism
			Mineral/sediment removal
			Floor sediment compaction
Atmosphere	Air quality		Desiccation
			Human induced condensation corrosion
Hydrology	Water quality (surface practices)	Micro to Meso	Pesticides and herbicides
			Industrial spills or dumping
	Water quality (springs)	All scales	Concentrations of harmful chemicals
	Water quantity	All scales	Changes in the water table
			Leakage of hydrocarbons from underground storage tanks
			Changes in cave drip rates
Biota	Vegetation disturbance	All scales	Vegetation removal
	Subsurface biota (groundwater)	Micro	Species richness decline
			Population density
	Subsurface biota (caves)	Micro	Species richness decline
			Population density
Cultural	Human artefacts	All scales	Removal or destruction of historical artefacts
	Stewardship of karst region	All scales	Regulatory protection
			Enforcement of regulations
			Public education
	Building infrastructure		Building of roads
			Building over karst features
			Construction within caves

Source: Modified from van Beynen and van Beynen (2011).

14.9.2 Karst Groundwater Vulnerability

As stated in Chapter 3 on Hydrology, approximately 25% of the world's population gets its water from karst aquifers, particularly in Asia, the Mediterranean, and the United States of America (Ford and Williams 2007, p. 441). Increasing urban and industrial expansion increases the risk of groundwater contamination from chemical spills, dumping, and landscaping. Karst aquifers are particularly vulnerable to contamination because of the rapid connections between the surface and aquifer.

There are two ways of defining the vulnerability of a karst aquifer. Intrinsic vulnerability is determined by properties of the karst environment that influence the degree of vulnerability (Daly et al. 2002). These relate to the 'plumbing' of the karst in terms of soil thickness and infiltration rates, the fracture density of the epikarst zone, the distribution of dolines, and the overall hydraulic conductivity. In combination, these determine the potential vulnerability, while adding land use and infrastructure (roads, reticulation, landfills, point pollution sources) creates the specific vulnerability. These approaches to assessing vulnerability are spatially integrated in groundwater vulnerability models (GVMs) which aim to quantify aquifer vulnerability to human-induced contamination. There are a number of these which have been reviewed by van Beynen and van Beynen (2011). EPIK (Doerfliger et al. 1999) is one of the more well-accepted GVMs specifically designed for karst aquifers and is described here.

Four GIS-based layers of epikarst (E), protective cover (P), infiltration conditions (I), and karst network development (K) are used to measure vulnerability, with each layer being scored according to characteristics seen in Table 14.3. EPIK is a point count system model, and its four attributes (layers) are weighted: $Fp = \alpha \times Ei + \beta \times Pj + \gamma \times Ik + \delta \times Kl$ with $\alpha = 3$, $\beta = 1$, $\gamma = 3$, and $\delta = 2$. The total score is termed aquifer protection (Fp), with higher values equating to lower vulnerability, as shown in Table 14.4.

Ravbar and Goldscheider (2009) compared four different intrinsic vulnerability models for a karst catchment in Slovenia as a means of delineating protection zones. These models included EPIK, PI, the Simplified Method, and the Slovene Approach. Goldscheider et al. (2000) developed PI, which incorporates protective layers above the saturated zone (P) and 'I' stands for infiltration conditions. The Simplified Method (Nguyet and Goldscheider 2006) simplifies the PI model by only measuring overlying layers and flow concentration. The Slovene Approach is the most sophisticated method and incorporates nine parameters for measuring the karst unsaturated zone, six for recharge conditions, and three for the karst saturated zone.

As might be expected, the application of these four models shows some similarities and some clear differences (Figure 14.15). Both EPIK and the Simplified Model show greater aquifer vulnerability compared to the other two methods. Nguyet and Goldscheider (2006) suggest this is because of differences in how the various models treat temporal hydrologic variations. They validated each of the models using a multi-tracer dye test. The Slovene Approach provided the most reliable results; however, its greater number of parameters may also limit its adoption due to the availability of spatial data and the sometimes-limited resources of water resource agencies.

Table 14.3 Scoring of the layers for EPIK.

Layer	Characteristic	Score
Epikarst (E)	Highly developed	1
	Moderately developed	2
	Small or absent	3
Protective cover (P)	0–20 cm soil	1
	20–100 cm	2
	100–200 cm	3
	>200 cm	4
Infiltration conditions (I)	Perennial or temporarily losing streams	1
	I1 with slope > 10% cultivated and 25% meadows	2
	I1 with slope > 10% cultivated and 25% meadows	3
	Rest of catchment	4
Karst network development (K)	Well-developed karst network	1
	Poorly developed karst	2
	Springs emerging from porous terrain	3

Source: Doerfliger et al. 1999.

Table 14.4 The total score in the EPIK model is termed aquifer protection (Fp), with higher values equating to lower vulnerability.

Vulnerability areas	Protection factor F_p
Very high	F less than or equal to 19
High	F between 20 and 25
Moderate	F higher than 25
Low	*F presence of P4*

Source: Doerfliger et al. 1999.

14.9.3 Data Availability

The detailed information on karst geomorphology and hydrology needed to populate groundwater vulnerability models may not be available for a given region. Hydrogeology maps may be compiled at a very generalised level and may treat karst as a simple, porous medium. In general, the locations and extent of cave systems may not be freely available information and may not be available to land managers. Inter-agency sharing of scientific information may also be restricted due to concerns about conservation or legal ramifications. In karst catchments, definition is always problematic and the extent of a catchment generally grossly underestimated by engineers and water resources staff unfamiliar with karst systems. There will also be dramatic changes in flow paths during floods, with older conduits becoming activated (Chapter 3). This can dramatically change the transmissivity of a karst system allowing pollutants to reach springs much sooner than conventional

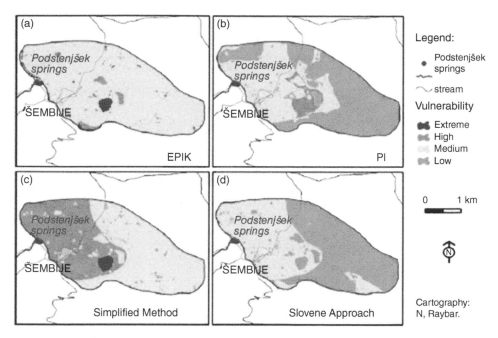

Figure 14.15 Comparison of various vulnerability models as applied to a Slovene karst catchment. Source: From Ravbar and Goldscheider 2009 © 2009 Springer Nature.

dye tracing would predict. Despite the widespread recognition of the fragility of karst ecosystems, quantitative studies of human impacts on cave biota and surface biology are generally lacking, and may not be an integral part of environmental impact assessments for developments in karst terrains (Whitten 2009).

Data voids do exist in that some government agencies and local councils do not keep accurate (or indeed any) records of such activities as logging, mining, infilling of dolines, disposal of wastewater, or of the effectiveness of rehabilitation. There may be jurisdictional anomalies in the recording of polluting or disturbance activities, with agencies saying that it's not their core business. This is one area in which citizen science can make a real contribution.

It is important to consider how environmental quality has changed over time (Van Beynen and Van Beynen 2011). Questions to be considered are:

1. What was the pristine state of the karst environment?
2. When did the disturbance start?
3. Can you clearly delineate between natural vs. anthropogenic change?
4. Has there been a steady degradation over time or has the situation improved?
5. Does the natural system recover over time, or is it permanently altered?

Determining the pre-disturbance state of a karst ecosystem may be impossible in an agricultural landscape, and it may be more effective to consider the maintenance of ecological processes rather than some ideal state. The resilience of individual species is highly variable. In a British cave Wood et al. (2008) discovered that certain species recovered within 12 months after an organic spill above the cavern, whereas other species did not. In fact, some species were not affected at all, while other species appeared that were absent prior to the event. Decadal scale investigations are needed given the slow reproductive rates of cave biota (Chapter 10; Wood et al. 2002, 2008). This does not sit well with the normal rapid assessment carried out prior to development projects.

Scale issues become important when a range of impacts are being considered. Groundwater pollution, such as from agricultural chemicals, may have a catchment-wide impact, but the biological impact of such a pollutant may be confined to an individual cave (Table 14.5).

Table 14.5 Recorded agricultural and non-agricultural impacts on British caves.

Area	Agriculture		Non-agriculture	
	Number of caves	Impact type	Number of caves	Impact type
Northern Pennines	3	Entrance infilled	4	Fly ash tipping
	4	Closed (water supply)		
	7	Farm tipping		
	3	Carcass disposal		
	4	Scrap metal/car bodies		
	1	Oil waste		
	7	Land drainage		
	1	Farm sewage		
Total	*30*		*4*	
Peak District	4	Farm tipping	2	Infilled
	1	Farm sewage	7	Fly ash tipping
			1	Oil (quarrying)
Total	*5*		*10*	
Wales	1	Entrance infilled	6	Fly ash tipping
	1	Farm tipping	1	Sewage
	1	Used as byre	2	Oil (quarrying)
Total	*3*		*9*	
Scotland	1	Farm tipping	1	Fly ash tipping
Total	*1*		*1*	
Mendips	8	Entrance infilled	2	Fly ash tipping
	11	Farm tipping	1	Oil
	6		1	Industrial effluent
Total	*25*		*4*	
Devon	0		2	Fly ash tipping
			2	Official tipping
			1	Chemical dumping
Total	*0*		*5*	

Source: Based on Hardwick and Gunn (1993).

Scale is discussed in detail in van Beynen and Townsend (2005) and North et al. (2009), with the former incorporating it into the Karst Disturbance Index (KDI) and the latter suggesting scale may be used as a weighting mechanism for scoring various disturbance indicators. Again, this emphasises the need for very good baseline data from which the significance of individual sites and their contents (speleothems, biota, archaeology, palaeontology) can be assessed.

14.10 Understanding Disputes Over Cave and Karst Resources

There is a long history of disputes over the allocation of cave and karst resources. These disputes can range from water allocation for agriculture, mining of scarce limestone resources for cement and agricultural lime, conflicts between conservation and tourism in protected areas, to the impacts of recreational caving on sensitive sites. The disputes can be viewed in terms of competition for a limited resource, the security of a given resource, and the conflicting goals and values of the actors in any resource allocation.

As has been said before, caves and karst landscapes are, in human terms, limited and non-renewable resources. Quite often the purest limestone not only forms the most economically viable location for the extractive industry but also forms the most extensive and best-decorated caves in spectacular landscapes of high relief. Limestone areas close to major cities and transport corridors are the most economical to exploit for mining, tourism, and recreation. So the potential users of the karst landscape are competing for the same limited, non-renewable resource (Osborne 1994).

It would not be so difficult to resolve this competition for a limited resource if there was a pre-existing agreement between potential users about the goals and values on which management of karst resources should be based. This is very rarely the case. Disputes over karst resources involve competition over *whose* values should form the basis of the dominant management regime. For example, government agencies, such as those covering environment & heritage, may be in competition with the mines & energy departments; recreational cavers wishing to gain access to sites may be in competition with protected area managers wishing to conserve the biology and heritage of cave sites. Developers of show caves may wish to exclude recreational cavers from sites on their leased lands for both fiscal and risk management reasons.

Miners of limestone and show cave operators both have financial interests in the limestone resource, while recreational cavers could assert that they have a right to enjoy public assets and that this right should not be restricted to certain groups on the basis of perceived merit or their ability to pay large amounts of money to a commercial operator. There will, therefore, be competition between actors who see them as a financial resource to be exploited, as a recreational resource to enjoy, those who have a scientific interest in them, or those who wish to conserve the resource and limit human access to fragile sites.

A further dimension to all of this is the need for resource or tenure security. This is usually couched in terms of the needs of primary industries such as agriculture, forestry, or mining. It does, however, emerge as an important issue in many areas of karst management and has been the basis of many disputes. A company mining limestone for cement or agricultural lime could reasonably expect security of tenure if they have complied with existing environmental legislation and are operating in line with mandated industry standards. This is often not the case when inadvertent discoveries of caves in the quarry arouse the attention of conservation groups or other actors. Conversely, conservation actors could reasonably expect that once a limestone area attains protected area status activities which degrade or destroy the resource would not occur. However, in many areas of Australia mining legislation overrides other titles and so leases can be granted over reserves. This was the case at the Wellington, Wombeyan, and Yessabah karst areas in New South Wales, Mount Etna in Queensland, and Ida Bay in Tasmania (Osborne 1994).

Tensions between recreational cavers and management authorities often arise. Cavers wish to maintain traditional access to caves and feel they have a degree of 'ownership' over caves they have explored, mapped, and maybe carried out gating and cave cleaning. Often caving groups have provided data and maps to management authorities, such as parks services, on the basis of continued access, though these arrangements may not be formalised. Conversely, the management authorities feel they have a responsibility to protect the caves from damage and that they have a legal responsibility to do so. These considerations outweigh any traditional use rights that may have arisen from initial exploration and mapping. Given these tensions and conflicts, there is a need to develop policies and procedures that produce a reasonable balance between the needs of different actors and the conservation of karst environments. Two approaches that may help in this regard are ensuring that reliable and relevant data are available to inform decisions and that lines of communication remain open and cordial between the various interested parties.

The provision of reliable and relevant data on karst areas has traditionally been the preserve of geological surveys and mines departments. In general, this information is rarely up to date and comprehensive, and may only present very generalised views of the karst resource. Information about hydrology and caves is frequently lacking. The siloisation of scientific literature also ensures that recent geomorphological literature, especially the journal literature, is often not referenced. Biological and ecological information can be of high quality where surveys have been conducted, and information on endangered species may be available. Most State Parks Services have compiled data on the caves and karsts in their areas of responsibility, and some do employ karst geomorphologists. Even if good quality information is available, most limestone miners and mines departments are not aware of the vulnerability of karst environments. There is, therefore, a need for a comprehensive database which covers all karst areas in a jurisdiction and provides access to bibliographies and data sources.

All too often, mining interests and conservationists only see each other in the media or on the steps of the local courthouse. Neither location is very conducive to meaningful dialogue and conflict resolution. It would be useful if an inclusive forum were to be created in each region that allowed representatives of government departments, nature conservation agencies, industry groups, recreational users, and traditional owners to meet regularly or in response to new proposals affecting karst areas. This might occur under the aegis of catchment management or natural resource management groups.

14.11 The IUCN Guidelines for Cave and Karst Protection

The WCPA (World Commission on Protected Areas) is one of six Commissions of the International Union for the Conservation of Nature (IUCN). It is the World's leading global network of protected area experts with over 1000 members in 160 countries working in a voluntary capacity. In 1995, the WCPA Working Group on Cave and Karst Protection commenced writing a set of guidelines (Figure 14.16), with input from a large number of cave experts globally.

The published guidelines (Watson et al. 1997) have been translated into 12 languages and have been adopted as providing a sound basis for the management of cave and karst protected areas around the world. A second edition of the guidelines is now being prepared under the auspices of the Cave and Karst Working Group of the Geoheritage Specialist Group of IUCN/WCPA. The original guidelines are presented here as they remain the best encapsulation of principles for the management of karst terrains:

Figure 14.16 Cover of Guidelines for Cave and Karst Protection. Photo of karst window south of Guilin along the Li River, China.

1. Effective planning for karst regions demands a full appreciation of all their economic, scientific and human values, within the local cultural and political context.
2. The integrity of any karst system depends upon an interactive relationship between land, water and air. Any interference with this relationship is likely to have undesirable impacts and should be subjected to thorough environmental assessment.
3. Land managers should identify the total catchment area of any karst lands, and be sensitive to the potential impact of any activities within the catchment, even if not located on the karst itself.
4. Destructive actions in karst, such as quarrying or dam construction, should be located so as to minimise conflict with other resource or intrinsic values.
5. Pollution of groundwater poses special problems in karst and should always be minimised and monitored. This monitoring should be event-based rather than at merely regular intervals, as it is during storms and floods that most pollutants are transported through the karst system.
6. All other human uses of karst areas should be planned to minimise undesirable impacts and monitored in order to provide information for future decision-making.
7. While recognising the non-renewable nature of many karst features, particularly within caves, good management demands that damaged features be restored as far as is practicable.
8. The development of caves for tourism purposes demands careful planning, including consideration of sustainability. Where appropriate, restoration of damaged caves should be undertaken, rather than opening new caves for tourism.
9. Governments should ensure that a representative selection of karst sites be declared as protected areas under legislation which provides secure tenure and active management.
10. Priority in protection should be given to areas or sites having high natural, social, or cultural value; possessing a wide range of values within the one site; which have suffered minimal environmental degradation; and/or of a type not already represented in the protected areas system of their country.
11. Where possible, a protected area should include the total catchment area of the karst.
12. Where such coverage is not possible, environmental controls, or total catchment management agreements under planning, water management, or other legislation should be used to safeguard the quantity and quality of water inputs to the karst system.
13. Public authorities should identify karst areas not included within protected areas and give consideration to safeguarding the values of these areas by such means as planning controls, programs of public education, heritage agreements, or covenants.
14. Management agencies should seek to develop their expertise and capacity for karst management.
15. Managers of karst areas and specific cave sites should recognise that these landscapes are complex three-dimensional integrated natural systems comprised of rock, water, soil, vegetation, and atmosphere elements.
16. Management in karst and caves should aim to maintain natural flows and cycles of air and water through the landscape in balance with prevailing climatic and biotic regimes.
17. Managers should recognise that in karst, surface actions may be sooner or later translated into impacts directly underground or further downstream.

18. Pre-eminent amongst karst processes is the cascade of carbon dioxide from low levels in the external atmosphere through greatly enhanced levels in the soil atmosphere to reduced levels in cave passages. Elevated soil carbon dioxide levels depend on plant root respiration, microbial activity, and a healthy soil invertebrate fauna. This cascade must be maintained for the effective operation of karst solution processes.

19. The mechanism by which this is achieved is the interchange of air and water between surface and underground environments. Hence, the management of quality and quantity of both air and water is the keystone of effective management at regional, local and site-specific scales. Development on the surface must take into account the infiltration pathways of water.

20. Catchment boundaries commonly extend beyond the limits of the rock units in which the karst has formed. The whole karst drainage network should be defined using planned water tracing experiments and cave mapping. It should be recognised that the boundary of these extended catchments can fluctuate dramatically according to weather conditions, and that relict cave passages can be reactivated following heavy rain.

21. More than in any other landscape, a total catchment management regime must be adopted in karst areas. Activities undertaken at specific sites may have wider ramifications in the catchment due to the ease of transfer of materials in karst.

22. Soil management must aim to minimise erosive loss and alteration of soil properties such as aeration, aggregate stability, organic matter content, and healthy soil biota.

23. A stable natural vegetation cover should be maintained as this is pivotal to the prevention of erosion and maintenance of critical soil properties.

24. Establishment and maintenance of karst protected areas can contribute to the protection of both the quality and quantity of groundwater resources for human use. Catchment protection is necessary both on the karst and on contributing non-karst areas. Activities within caves may have detrimental effects on regional groundwater quality.

25. Management should aim to maintain the natural transfer rates and quality of fluids, including gases, through the integrated network of cracks, fissures, and caves in the karst. The nature of materials introduced must be carefully considered to avoid adverse impacts on air and water quality.

26. The extraction of rocks, soil, vegetation, and water will clearly interrupt the processes that produce and maintain karst, and therefore such uses must be carefully planned and executed to minimise environmental impact. Even the apparently minor activity of removing limestone pavement or other karren for ornamental decoration of gardens or buildings has a drastic impact and should be subject to the same controls as any major extractive industry.

27. Imposed fire regimes on karst should, as far as is practicable, mimic those occurring naturally.

28. While it is desirable that people should be able to visit and appreciate karst features, such as caves, the significance and vulnerability of many such features means that great care must be taken to minimise damage, particularly when cumulative over time. Management planning should recognise this fact, and management controls should seek to match the visitor population to the nature of the resource.

29. International, regional, and national organisations concerned with aspects of karst protection and management should recognise the importance of international co-operation and do what they can to disseminate and share expertise.
30. The documentation of cave and karst protection/management policies should be encouraged, and such policies made widely available to other management authorities.
31. Databases should be prepared; listing cave and karst areas included within protected areas, but also identifying major unprotected areas which deserve recognition. Karst values of existing and potential World Heritage sites should be similarly recorded.

References

Aley, T. (1994). Some thoughts on environmental management as related to cave use. *Journal of the Australasian Cave and Karst Management Association* 17: 4–10.

Balogh, J.C. and Walker, W.J. (1992). *Golf Course Management and Construction: Environmental Issues*. Ann Arbor, MI: Lewis Publishing.

Baran, N., Lepiller, M., and Mouvet, C. (2008). Agricultural diffuse pollution in a chalk aquifer (Trois Fontaines, France): influence of pesticide properties and hydrodynamic constraints. *Journal of Hydrology* 358: 56–69.

Berryhill, W.S.J. (1989). The impact of agricultural practices on water quality in karst regions. In: *Engineering and Environmental Impacts of Sinkholes and Karst; Proceedings of the Third Multidisciplinary Conference* (ed. B.F. Beck), 159–164. Rotterdam: A. A. Balkema.

Bou Kheir, R., Abdallah, C., and Khawlie, M. (2008). Assessing soil erosion in Mediterranean karst landscapes of Lebanon using remote sensing and GIS. *Engineering Geology* 99: 239–254.

Boulton, A.J., Humphreys, W.F., and Eberhard, S.M. (2003). Imperilled subsurface waters in Australia: biodiversity, threatening processes and conservation. *Aquatic Ecosystem Health and Management* 6: 41–54.

Boyer, D.G. and Pasquarell, G.C. (1999). Agricultural land use impacts on bacterial water quality in a karst groundwater aquifer. *Journal of the American Water Resources Association* 35: 291–300.

Calò, F. and Parise, M. (2006). Evaluating the human disturbance to karst environments in southern Italy. *Acta Carsologica* 35 (2-3).

Campbell, B.L., Elliott, G.L., and Loughran, R.J. (1986). Measurement of soil erosion from fallout of Caesium-137. *Search* 17: 148–149.

Celico, F., Musilli, I., and Naclerio, G. (2004). The impacts of pasture and manure-spreading on microbial groundwater quality in carbonate aquifers. *Environmental Geology* 46: 233–236.

Chapman, P. (1993). *Caves and Cave Life*. London: HarperCollins.

Costin, A.B., Greenaway, M.A., and Wright, L.G. (1984). *Harvesting Water from Land: Land Use Hydrology of the Upper Shoalhaven Valley of New South Wales*. Centre for Resource and Environmental Studies.

Coxon, C. (1999). Agriculturally induced impacts. In: *Karst Hydrogeology and Human Activities: Impacts, Consequences and Implications* (eds. D.P. Drew and H. Hotzl), 37–80. Rotterdam: Balkema.

Coxon, C. (2011). Agriculture and karst. In: *Karst Management* (ed. P.E. Van Beynen), 103–138. Dordrecht: Springer.

Crain, A.S. (2002). Pesticides and nutrients in karst springs in the Green River Basin, Kentucky. *USGS Fact Sheet*. U.S. Geological Survey.

Crawford, N.C. (1984). Sinkhole flooding associated with urban development upon karst terrain Bowling Green, Kentucky. In: *Sinkholes: Their Geology, Engineering and Environmental Impact* (ed. B.F. Beck), 283–292. Orlando: Multidisciplinary Conference on Sinkholes.

Dalton, M.S. and Frick, E.A. (2008). Fate and transport of pesticides in the ground water systems of Southwest Georgia, 1993–2005. *Journal of Environmental Quality* 37: S-264–S-272.

Daly, D., Dassargues, A., Drew, D. et al. (2002). Main concepts of the European approach for karst groundwater vulnerability assessment and mapping. *Hydrogeology Journal* 10: 340–345.

D'Antonio, R.G., Winn, R.E., Taylor, J.P. et al. (1985). A waterborne outbreak of cryptosporidiosis in normal hosts. *Annals of Internal Medicine* 103: 886–888.

Davey, A.G. and White, S. (1986). Management of Victorian caves and karst: a report to the caves classification committee. Department of Conservation, Forests and Lands.

Doerfliger, N., Jeannin, P.Y., and Zwahlen, F. (1999). Water vulnerability assessment in karst environments: a new method of defining protection areas using a multi-attribute approach and GIS tools (EPIK method). *Environmental Geology* 39: 165–176.

Dolliver, H. and Gupta, S. (2008). Antibiotic losses in leaching and surface runoff from manure-amended agricultural land. *Journal of Environmental Quality* 37: 1227–1237.

Dougherty, P.H. (1981). The impact of the agricultural land-use cycle on flood surges and runoff in a Kentucky karst region. In: *Proceedings of the 8th International Congress of Speleology, 18–24 July 1981 Bowling Green, Kentucky* (ed. B.F. Beck), 267–269. Department of Geography and Geology, Western Kentucky University.

Drew, D. (1996). Agriculturally induced environmental changes in the Burren karst, Western Ireland. *Environmental Geology* 28: 137–144.

Drew, D. and Coxon, C.E. (1988). The effects of land drainage on groundwater resources in karst areas of Ireland. In: *Proceedings of the 21st International Association of Hydrogeology Congress, 1988 Guilin, China* (ed. Y. Daoxian), 204–209. Beijing: Geological Publishing House.

Ford, D.C. and Williams, P.W. (2007). *Karst Hydrogeology and Geomorphology*, 441. Chichester, England: Wiley.

Gams, I., Nicod, J., Sauro, U. et al. (1993). Environmental change and human impacts on the Mediterranean karsts of France, Italy and the Dinaric region. In: *Karst Terrains: Environmental Changes and Human Impact* (ed. P.W. Williams), 59–98. Catena.

Gill, D. (1993). Guidelines for caving and research expeditions to the Mulu National Park, Sarawak. *International Caver* 8: 41.

Gillieson, D.S. (1980). Pit structures from Selminum Tem Cave, Western Province, Papua New Guinea. *Australian Archaeology*: 10, 26–10, 32.

Gillieson, D.S. (1989). Limestone soils in the New Guinea Highlands: a review. In: *Resource Management in Limestone Landscapes: International Perspectives* (eds. D. Gillieson and D. Ingle Smith), 191–200. Department of Geography and Oceanography, University College, Australian Defence Force Academy.

Gillieson, D., Cochrane, A., and Murray, A. (1994). Surface hydrology and soil erosion in an arid karst: the Nullarbor Plain, Australia. *Environmental Geology* 23(2): 125–133.

Gillieson, D. and Thurgate, M. (1999). Karst and agriculture in Australia. *International Journal of Speleology* 28B: 149–168.

Gillieson, D.S., Gorecki, P., Head, J., and Hope, G. (1986). Soil erosion and agricultural history in the Central Highlands of New Guinea. In: *International Geomorphology* (ed. V. Gardiner), 507–522. London: Wiley.

Gillieson, D., Wallbrink, P., and Cochrane, A. (1996). Vegetation change, erosion risk and land management on the Nullarbor Plain, Australia. *Environmental Geology* 28 (3): 145–153.

Goldie, H.S. (1993). The legal protection of limestone pavements in Great Britain. *Environmental Geology* 28: 160–166.

Goldscheider, N., Klute, M., Sturm, S., and Hötzl, H. (2000). The PI method – a GIS based approach to mapping groundwater vulnerability with special consideration on karst aquifers. *Zeitschrift für angewandte Geologie* 46: 157–166.

Harding, K.A. and Ford, D.C. (1993). Impact of primary deforestation upon limestone slopes in northern Vancouver Island, British Columbia. *Environmental Geology* (3): 137–143.

Hardwick, P. and Gunn, J. (1993). The impact of agriculture on limestone caves. In: *Karst Terrains: Environmental Changes and Human Impact*, vol. 25 (ed. P.W. Williams), 235–250. Catena Supplement.

Head, L. (1989). Prehistoric aboriginal impacts on Australian vegetation: an assessment of the evidence. *Australian Geographer* 20: 37–46.

Holland, E.A. (1991). *Borenore Caves Reserve: Resource Guidelines for the Mt. Canobolas Trust*. Jenolan Caves Reserve Trust.

Holland, E. (1994). The effects of fire on soluble rock landscapes. *Helictite* 32: 3–9.

Huang, Q., Cai, Y., and Xing, X. (2008). Rocky desertification, antidesertification and sustainable development in the karst mountain region of Southwest China. *Ambio* 37: 390–392.

Jennings, J.N. (1983). Karst landforms. *American Scientist* 71: 578–586.

Jiang, Y. (2006). The impact of land use on soil properties in a karst agricultural region of Southwest China: a case study of Xiaojiang watershed, Yunnan. *Journal of Geographical Sciences* 16: 69–77.

Kiernan, K. (1988). *The Management of Soluble Rock Landscapes: An Australian Perspective* 61 pp. Sydney: Sydney Speleological Society.

Kiernan, K. (1989). Human impacts and management responses in the karsts of Tasmania. In: *Resource Management in Limestone Landscapes: International Perspectives* (eds. D. Gillieson and D. Ingle Smith), 69–92. Canberra: Department of Geography and Oceanography, University College, Australian Defence Force Academy.

King, K.W., Balogh, J.C., Hughes, K.L., and Harmel, R.D. (2007). Nutrient load generated by storm event runoff from a golf course watershed. *Journal of Environmental Quality* 36: 1021–1030.

Kingsbury, J.A. (2008). Relation between flow and temporal variations of nitrate and pesticides in two karst springs in northern Alabama. *Journal of the American Water Resources Association* 44: 478–488.

Kozar, M.D. and Mathes, M.V. (2001). Occurrence of coliform bacteria in a karst aquifer, Berkeley County, West Virginia, USA. In: *Geotechnical and Environmental Applications of Karst Geology and Hydrology* (eds. B.F. Beck and J.G. Herring). Lisse: Balkema.

Kranjc, A. (1979). The influence of man on cave sedimentation. In: *Actes Symp. Intern. Erosion Karstique Aix–Marseille–Nîmes*, 117–123.

Meiman, J. (1991). The effects of recharge basin land-use practices on water quality at Mammoth Cave National Park, Kentucky. In: *Proceedings of the 1991 Cave Management Symposium, 1991 Bowling Green, Kentucky*, 105–115. American Cave Conservation Association.

Metcalfe, C.D., Beddows, P.A., Bouchot, G.G. et al. (2011). Contaminants in the coastal karst aquifer system along the Caribbean coast of the Yucatan Peninsula, Mexico. *Environmental Pollution* 159: 991–997.

Mihevc, A., Gabrovšek, F., Knez, M. et al. (2016). Karst in Slovenia. *Boletín Geológico y Minero* 127: 79–97.

Milanovic, P. (2002). The environmental impacts of human activities and engineering constructions in karst regions. *Episodes* 25: 13–21.

Murray, A.S., Caitcheon, G., Olley, J., and Crockford, H. (1990). Methods for determining the sources of sediments reaching reservoirs: targeting soil conservation. *ANCOLD Bulletin* 85: 61–70.

National Parks Wildlife Service (1983). *Harvesting and Rehabilitation of Jounama Pine Plantation, Kosciusko National Park: Environmental Impact Statement.* NPWS (NSW) and Forestry Commission (NSW).

Nguyet, V. and Goldscheider, N. (2006). A simplified methodology for mapping groundwater vulnerability and contamination risk, and its first application in a tropical karst area, Vietnam. *Hydrogeology Journal* 14: 1666–1675.

North, L., Van Beynen, P., and M., P. (2009). Interregional comparison of karst disturbance: West Central Florida and Southeast Italy. *Journal of Environmental Management* 90: 1770–1781.

Northcote, K.H. (1979). A factual key for the recognition of Australian soils. 123pp. *Rellim Tech. Publ., Glenside, South Australia.*

O'Neill, A.L., Head, L.M., and Marthick, J.K. (1993). Integrating remote sensing and spatial analysis techniques to compare aboriginal and pastoral fire patterns in the East Kimberley, Australia. *Applied Geography* 13: 67–85.

Osborne, R.A.L. (1994). Caves, cement, bats and tourists: karst science and limestone resource management in Australia. *Journal and Proceedings of the Royal Society of New South Wales* 127: 01–22.

Page, D., Wall, B., and Crowe, M. (2006). *The Quality of Drinking Water in Ireland, a Report for the Year 2005.* Wexford: Environmental Protection Agency.

Panno, S.V. and Kelly, W.R. (2004). Nitrate and herbicide loading in two groundwater basins of Illinois' sinkhole plain. *Journal of Hydrology* 290: 229–242.

Pasquarell, G.C. and Boyer, D.G. (1995). Agricultural impacts on bacterial water quality in karst ground-water. *Journal of Environmental Quality* 24: 959–969.

Pasquarell, G.C. and Boyer, D.G. (1996). Herbicides in karst groundwater in Southeast West Virginia. *Journal of Environmental Quality* 25: 755–765.

Pichler, T., Brinkmann, R., and Scarzella, G.I. (2008). Arsenic abundance and variation in golf course lakes. *Science of the Total Environment* 394: 313–320.

Pronk, M., Goldscheider, N., and Zopfi, J. (2006). Dynamics and interaction of organic carbon, turbidity and bacteria in a karst aquifer system. *Hydrogeology Journal* 14: 473–484.

Pronk, M., Goldscheider, N., and Zopfi, J. (2007). Particle-size distribution as indicator for faecal bacteria contamination of drinking water from karst springs. *Environmental Science & Technology* 41: 8400–8405.

Ravbar, N. and Goldscheider, N. (2009). Comparative application of four methods of groundwater vulnerability mapping in a Slovene karst catchment. *Hydrogeology Journal* 17: 725–733.

Rosewell, C.J. and Edwards, K. (1988). Soiloss: a program to assist in the selection of management practices to reduce erosion. *Soil Conservation Service of NSW Technical Handbook,* 11: 71 pp plus disk.

Sauro, U. (1993). Human impact on the karst of the Venetian Fore-Alps, Italy. *Environmental Geology* 21: 115–121.

Stanton, R.K., Murray, A.S., and Olley, J.M. (1992). Tracing recent sediment using environmental radionuclides and mineral magnetics in the karst of Jenolan caves, Australia. In: *Erosion and Sediment Transport Monitoring Programmes in River Basins*. Proceedings of a symposium held at Oslo, August 1992. IAHS Publ. no. 210 (eds. J. Bogen, D.E. Walling and T.J. Day). Wallingford, Oxford, England: IAHS Press.

Suzuki, T., Kondo, H., Yaguchi, K. et al. (1998). Estimation of leachability and persistence of pesticides at golf courses from point-source monitoring and model to predict pesticide leaching to groundwater. *Environmental Science & Technology* 32: 920–929.

Urich, P. (1991). Stress on tropical karst resources exploited for the cultivation of wet rice. In: *Proceedings of the International Conference on Environmental Changes in Karst Areas, 15–27 Sept 1991* (eds. U. Sauro, A. Bondesan and M. Meneghel), 39–48. Department of Geography, University of Padua, Italy.

Urushibara-Yoshino, K. (1993). Human impact on karst soils: Japanese and other examples. In: *Karst Terrains: Environmental Changes and Human Impact* (ed. P.W. Williams), Catena Supplement 25: 219–233.

Van Beynen, P.E. and Townsend, K.M. (2005). A disturbance index for karst environments. *Environmental Management* 36: 101–116.

Van Beynen, P.E. and Van Beynen, K.M. (2011). Human disturbance of karst environments. In: *Karst Management* (ed. P.E. Van Beynen), 379–397. Dordrecht: Springer.

Walling, D.E. and Quine, T.A. (1990). Calibration of caesium-137 measurements to provide quantitative erosion rate data. *Land Degradation & Development* 2: 161–175.

Wang, S.J., Liu, Q.M., and Zhang, D.F. (2004). Karst rocky desertification in Southwestern China: geomorphology, landuse, impact and rehabilitation. *Land Degradation and Development* 15: 115–121.

Watson, J., Hamilton-Smith, E., Gillieson, D., and Kiernan, K. (1997). *Guidelines for Cave and Karst Protected Areas*. Gland: World Conservation Union (IUCN).

White, W.B. (1988). *Geomorphology and Hydrology of Karst Terrains*. New York: Oxford University Press.

Whitten, T. (2009). Applying ecology for cave management in China and neighbouring countries. *Journal of Applied Ecology* 46: 520–523.

Williams, P.W. (1993). Environmental change and human impact on karst terrains: an introduction. In: *Karst Terrains: Environmental Changes and Human Impact* (ed. P.W. Williams), Catena Supplement 25: 1–19.

Williams, J.E., Whelan, R.J., and Gill, A.M. (1994). Fire and environmental heterogeneity in southern temperate forest ecosystems: implications for management. *Australian Journal of Botany* 42: 125–137.

Wood, P.J., Gunn, J., and Perkins, J. (2002). The impact of pollution on aquatic invertebrates with a subterranean ecosystem – out of site out of mind. *Archiv für Hydrobiologie* 155: 223–237.

Wood, P.J., Gunn, J., and Rundle, S.D. (2008). Response of benthic cave invertebrates to organic pollution events. *Aquatic Conservation: Marine and Freshwater Ecosystems* 18: 909–922.

Worthington, S.R.H., Smart, C.C., and Ruland, W.W. (2002). Assessment of groundwater velocities to the municipal wells at Walkerton. In: *Proceedings of the 55th Conference of the Canadian Geotechnical Society, 2002 Niagara Falls, Ontario*, 1081–1086.

Xiong, Y.J., Qiu, G.Y., Mo, D.K. et al. (2009). Rocky desertification and its causes in karst areas: a case study in Yongshun County, Hunan Province, China. *Environmental Geology* 57: 1481–1488.

Young, E.A. (1992). Aboriginal land rights in Australia: expectations, achievements and implications. *Applied Geography* 12: 146–161.

Yuan, D. (1993). Environmental change and human impact on karst in southern China. In: *Karst Terrains: Environmental Changes and Human Impact* (ed. P.W. Williams), Catena Supplement 25: 99–107.

15

Documentation of Caves

Since the first edition of this book was published in 1996, there have been significant advances in the ways that we document caves. These can be broadly divided into two categories, technological and methodological. The rapid growth of improved means of surveying natural and built environments using laser rangefinders, digital photography, 3D laser scanning, and drones has meant that the basic process of surveying caves has become quicker, but also that areas of cave – such as in the roof and flooded passages – can now be documented in detail. The advent of high-resolution digital photography and very sensitive photoreceptors (CMOS) means that stunning cave photos can be captured using individual cavers' lights, flashes, and slave units, even in the very large cave chambers being discovered and documented in Southeast Asia. All of these technological advances are now being used in cave research for the first time, providing much better morphological data and generating new hypotheses about the processes of cave development.

For a long time, caves and karsts were viewed by managers as being outside the normal range of natural resources management, but the growth in visitor numbers in many protected karst areas and the establishment of specialist cave manager positions have brought caves more securely into the normal range of management activities. Thus, in many countries, there are now cave classification schemes for use in management and schemes for geoheritage assessment are now widespread. This has been facilitated in part by the growth in the number of cave and karst World Heritage sites and the development of documentation, evaluation, and monitoring schemes for them.

15.1 Geoheritage Assessment

Geodiversity refers to the variety of the geological and physical elements of nature, such as minerals, rocks, soils, fossils, and landforms, and active geological and geomorphological processes. Together with biodiversity, geodiversity constitutes the natural diversity of planet Earth (Gray 2013). Geoheritage comprises those elements of natural geodiversity, which are of significant value to humans, but do not involve the depletion or degradation of those elements (Houshold et al. 1997; Sharples 2002). The importance of this definition is that it implies a distinction between the utilitarian resource values derived from the removal, processing or manipulation of rocks, landforms, and soils by means such as mining, engineering, or agriculture, and the conservation values of rocks, landforms, and soils as heritage in their natural state.

Gordon et al. (2018) have emphasised the importance of including geodiversity in protected area management planning. Gray (2019) further emphasises the value of geodiversity in World Heritage and Global Geoparks, and that these values are highly valued by society and contribute to the wellbeing of society as a whole. This goes some way to addressing the fundamental schism between the widespread economic use of geological resources and the use of those resources for passive recreation and conservation.

Geoheritage forms part of the natural capital of planet Earth. It may be of value to humans as:

- Providing scientific evidence of the past development of the Earth, and of the evolution of life on Earth
- Sites of importance for research and education
- Features which inspire us because of their aesthetic qualities
- Features of recreational or tourism significance (e.g. mountains, cliffs, caves, beaches, etc.)
- Features which form the basis of landscapes that have contributed to the 'sense of place' of particular human communities, and
- Features which play a role in the cultural or spiritual values of human communities (e.g. sacred caves and mountains).

Each of these points can form a theme or themes which can be developed and used to compare sites as part of a process of inscription on a national heritage list or the World Heritage list. In 2006, the Australian Department of Environment convened a workshop to identify the most significant karst sites in Australia, and assess their values and significance. This was a precursor to listing some sites on the National Heritage List. The sites were discussed in broad groups to reflect karst 'types', which developed in different climatic and physiographic regimes. Within each type, karst sites have similar characteristics or developmental history; however each karst area or site also has unique aspects. The delineation of broad types gives a basis for comparison for many, but not all, of the features and values found in karst landscapes. Types included:

- Temperate Eastern Highland impounded karsts
- Monsoonal tropical karsts
- Southern Tertiary basin karsts
- Coastal zone karsts
- Island karsts, and
- Karsts of the arid zone.

Some karsts fall into more than one category. For example, the extensive Chillagoe and Mitchell-Palmer karst of north Queensland (Australia) is part of the Eastern Highland impounded karsts, as well as representing a monsoonal tropical karst (Gillieson 2016). Nevertheless, these groupings provided a useful starting point in which to highlight and compare values. Further to the identification of characteristics inherent within each broad type, a matrix was developed during the workshop which set out a series of heritage themes against which cave and karst values could be grouped, highlighted, and compared, and within each major type. This provides a sieving methodology for the assessment of individual geosites. The themes fell into two main categories, each with a number of subcategories. These are:

Geodiversity, as described by:

- The evolution of Australia's geodiversity
- Presence of palaeokarst and multi-phase caves
- Complexity of the hydrology
- Bedrock type and complexity
- Karst geomorphology, and
- Importance for research.

Biodiversity, as described by:

- Importance in illustrating evolutionary processes
- Importance as refugia
- Consideration of biogeography and isolating factors
- Species diversity, and
- Importance for research.

The practical implementation of geoconservation requires that significant elements of geodiversity – those requiring special management prescriptions – be identified on the ground through a process of inventory. The most detailed approach to developing systematic thematic inventories involves:

- Developing or adopting a classification scheme for a theme under consideration.
- Using available data and further fieldwork as necessary to identify all known examples of each classified group within a defined study area.
- Comparing the known examples in each classified group to identify which are the best expressed or developed examples of their type.

There are a number of classification schemes in use for cave and karst geoheritage. Grimes (1995) has developed a broad scheme of geological and geomorphological types that can be used at a regional or national level. At the level of the individual cave, it is important to recognise that the values can fall into three main categories: geological, biological, or cultural. An essential first step in the assessment of geological heritage is the compilation of a cave inventory. Any cave inventory should first consider the context in which the cave occurs. Thus, the cave needs to be placed in its relationship with local geology and geomorphology and the extent to which it presents any typical or unusual features, such as a cave developed in fault gangue, or a mineralised void related to skarn rocks. The second task is to compile an inventory of the cave contents. This should include the type and extent of calcite formations or speleothems, sediment deposits and bone deposits, water bodies, cave solutional features such as pendants and anastomosing tubes, and rare or unusual minerals. Any natural hazards should also be noted at the time. Cave mapping is a very useful adjunct to record special features, determine potential impacts of either future visitors or local economic developments. A set of accepted symbols endorsed by the International Speleological Union is used for mapping, coupled with accepted survey accuracy grades. In Australia, the Australian Speleological Federation has published a

Karst Index (Matthews 1985) which is maintained online and contains data on over 10 000 caves in Australia. The Index also has details of many cave maps that have been produced and curated by individual speleological clubs. The Index does provide a basis by which individual caves and karst sites can be classified and then compared for geoheritage evaluation. Similarly, there is a State karst inventory for Tasmania and New South Wales. In all cases, a proforma based on a classification scheme is used for recording basic data.

Karst areas may have sufficient significance to be inscribed on a Heritage List. National Heritage listing does not necessarily change land tenure or ownership, and is not the same as an area becoming a National Park or equivalent Protected Area. In Australia, if a place is listed on the National Heritage, its values will be protected under the *Environment Protection and Biodiversity Conservation Act 1999* (EPBC Act). Approval under the EPBC Act is required for any action that could have a significant impact on the National Heritage values of a listed place.

Key tools used to decide a place's heritage significance are criteria and thresholds. *Criteria* are a collection of principles or characteristics used to help decide if a place has heritage values. There will usually be several criteria that might be applied to a place being considered:

a) The place has outstanding heritage value to the nation because of the place's importance in the course, or pattern, of Australia's natural or cultural history.
b) The place has outstanding heritage value to the nation because of the place's possession of uncommon, rare or endangered aspects of Australia's natural or cultural history.
c) The place has outstanding heritage value to the nation because of the place's potential to yield information that will contribute to an understanding of Australia's natural or cultural history.
d) The place has outstanding heritage value to the nation because of the place's importance in demonstrating the principal characteristics of: (i) a class of Australia's natural or cultural places, or (ii) a class of Australia's natural or cultural environments;
e) The place has outstanding heritage value to the nation because of the place's importance in exhibiting particular aesthetic characteristics valued by a community or cultural group.
f) The place has outstanding heritage value to the nation because of the place's importance in demonstrating a high degree of creative or technical achievement at a particular period.
g) The place has outstanding heritage value to the nation because of the place's strong or special association with a particular community or cultural group for social, cultural, or spiritual reasons.
h) The place has outstanding heritage value to the nation because of the place's special association with the life or works of a person, or group of persons, of importance in Australia's natural or cultural history.
i) The place has outstanding heritage value to the nation because of the place's importance as part of Indigenous tradition.

Thresholds relate to the level or ranking of the heritage values that a place must possess in order to be placed on a heritage list. Usually a comparative analysis of similar places in Australia needs to be carried out. Criteria of integrity and authenticity of the place may also be important. Threshold determination may also need to rely heavily on relevant experts with access to a range of unpublished literature or relevant data. Although there are tools for assessing the biological values of a place, until recently there has been no comparable tool for comparative analysis of geological or geomorphological values. White and Wakelin-King (2014) have developed a semi-quantitative methodology for this purpose. The Earth Sciences Comparative Matrix (ESCoM) groups sites in process themes. Each site is assessed against National Heritage criteria and compared with other similar places according to their degree of unusualness, integrity, and authenticity. A site scoring well across multiple themes has increased heritage significance. The overall values of a site are quantified, leading to a ranking which enables a qualitative judgement on whether it achieves the threshold of outstanding heritage value. Gap analyses are a well-tested method for comparative analysis of a suite of possible sites (Sharples 2014). They have been applied to the assessment of potential World Heritage sites in karst (Gillieson 2019; Williams 2008) and provide guidance for State Party nominators. In Spain, Martín-Duque et al. (2012) have developed a comprehensive methodology for the categorisation and assessment of karstic landscapes, as precursor to regional and recreational planning. Also in the Picos de Europa region of Spain, Ballesteros et al. (2019) have developed a thorough methodology for the assessment and management of deep caves (>1 km deep) based on groundwater flow regime, cave climate, and the needs of scientific research. A range of digital resources, including online GIS mapping, 3D imaging, and interactive route mapping are now available to interpret and assist visitors interested in geoheritage (Cayla 2014).

15.2 Cave Mapping

Humans have used caves for a variety of purposes for tens of millennia (Chapter 12), both for religious and for more utilitarian purposes. The earliest known cave image comes from eastern Turkey and depicts a cave thought to be a source of the River Tigris (Figure 15.1; Waltham 1976).

Early attempts to depict and map caves verged on the fanciful with the depiction of cave decorations as anthropomorphs and lithographed maps being both inaccurate and

Figure 15.1 Earliest known cave image – 2800 years ago, tablet made for Assyrian King Shalmaneser III.

Figure 15.2 View of the Antiparos Cave, Naxos, Greece. Visit of the Marquis de Nointel, in 1673. He postulated the vegetative growth of stalactites.

limited in their coverage. For example, the Antiparos Cave on Paros, Greece was held to be an example of the vegetative growth of speleothems, with forms analogous to carrots, cauliflowers, and other domestic vegetables (Figure 15.2). In December 1673, the Marquis de Nointel held Christmas mass in the cave. He served as French ambassador to the Ottoman court in Istanbul and was an admirer of ancient Greek civilisation. On Paros, he was the guest of the famous pirate Daniel (Ollier, 1893).

The first known map to attempt to accurately depict a cave in both plan and section was Pen Park Hole near Bristol in England. The cave is developed on several levels and attains a depth of 37 m. It is likely to be hydrothermal in origin (Ford and Williams 1989). The earliest survey was carried by Captain Greenville Collins of the survey yacht 'Merlin'. In September 1682, Captain Collins was visiting Sir Robert Southwell, who persuaded him to undertake exploration and survey of the cave. Sir Robert was responsible for the publication of a plan of the chamber and a section of the whole cave in the *Philosophical Transactions of the Royal Society* in January 1683 (Southwell 1683). Further details of the cave are found at www.penparkhole.org.uk/index.php.

A more comprehensive exploration and mapping of caves in Austria, Czechoslovakia, and Slovenia was carried by J. A. Nagel in 1748. Commissioned by Emperor Franz I, the Emperor

Figure 15.3 Nagel's exploration of caves in Slovenia included the novel (and unrepeated) use of ducks with lit candles attached to explore cave lakes.

of the Holy Roman Empire of German Nation (reigned 1745–1765), the surveys produced a 100-page manuscript with maps and watercolours of caves (Figure 15.3). In 17 chapters, Nagel describes such important caves as Postojnska jama, the caves near Planina, Vilenica cave, and Predjama cave in Slovenia, and the cave at Sloup and the Machocha abyss in Moravia.

In Australia, Major Thomas Mitchell carried out extensive exploration and surveys of the caves, rivers, and lakes of the new colony of New South Wales, starting in 1829 when he was appointed Surveyor-General. On his remarkable 65 day private excursion in 1830, Mitchell explored, diarised, and sketched caves at Wellington, Molong, and Boree (Borenore), and visited other sites nearby. Nine days were based at Wellington, of which six were spent extracting earth-coated bones from the Large or Big (i.e. Cathedral), Breccia (Mitchell), and Bone Caves – exploring, digging, surveying, and sketching (Figures 15.4 and 15.5).

> *Monday 29th June: We set out for the caves early determined to have a good day's work, I surveyed first the large cave with the compass and a line of 20 feet – then I commenced a view of the large gallery with the great altar & then I measured to the bone cave (80 feet), and surveyed it, commencing also a view of the little chamber already mentioned. I this day set men to dig where the brecchia seemed to come to the surface at some distance from the bone cave, and there also they soon found bones – the brecchia being very hard, seemed only a species of limestone rock.* (Mitchell 1830).

The outcome of the investigations at Wellington Caves appeared as one chapter in his well-known account in *Three Expeditions into Eastern Australia* (Dunkley 2016; Mitchell

Figure 15.4   The Altar in Cathedral Cave at Wellington, New South Wales, Australia, from a sketch made by Thomas Mitchell on 29th June 1830 (Mitchell 1838: 360).

1838). His work provided the first descriptions of fossil bone deposits at Wellington Caves, NSW, Australia, as well as other sites. Substantial amounts of material from his excavations were sent to England and had a huge influence on the thinking of noted palaeontologists and geologists such as Lyell, Owen, and Cuvier. More details of his surveys and discoveries can be found at http://gutenberg.net.au/ebooks/e00036.html#CHAPTER%203.15.

Major Mitchell used a compass and chain to survey the caves in Australia. By the twentieth century, cavers were using Army surplus compasses, Abney levels, and metal tapes to survey caves. The process was slow and involved placing survey markers, either cairns or carbide map marks on the walls. By the 1960s, more compact equipment was available. Waterproof and shockproof Suunto compasses and clinometers, fibreglass tapes and – for large chambers – Japanese forestry compasses on a tripod were all in use and made the process quicker. The French cavers introduced the Topofil, a spool of string in a plastic box which fed out through a metric counter so distances could be read from the counter. In some cases, the string was left in the caves, but this was a rare and undesirable practice. The recording of wall details and cave contents was done on waterproof slates or notebooks, and then the survey results integrated with the details on paper or acetate sheets. I recall a caving expedition in Papua New Guinea in the late 1970s where teams explored further into the cave and mapped their way out to join the developing survey.

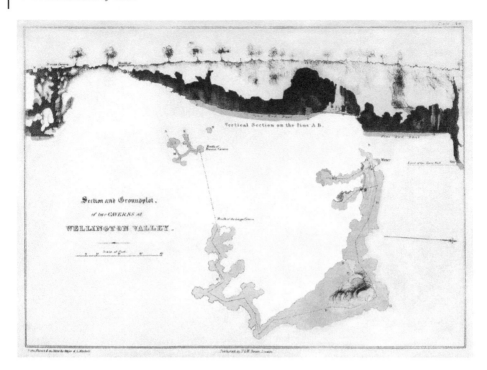

Figure 15.5 Mitchells' 1830 survey of Cathedral Cave, Wellington, New South Wales, Australia. Source: From Mitchell (1838: 361).

The survey party was not allowed to retire to a well-earned rest before the day's survey was drawn up, often in the early morning.

The need to better map caves and their resources led to the establishment of specialised cave research institutes in America, China, Cuba, France, Hungary, and Slovenia, initiating the profession of cave surveyor. Recreational use of caves from the early decades of the twentieth century has led to widespread adoption of more conventional survey instruments and standardisation of cave mapping standards and symbols through the Union Internationale de Speleologie (UIS). The UIS has produced standard cave mapping symbols (Figure 15.6) and survey standards (Fabre and Audetat 1978; Grossenbacher 1992).

From Table 15.1, the majority of cave surveys fall into categories 4 and 5 with show caves often having surveys to category 6 or 7 to facilitate the location of infrastructure.

There are now a number of very useful software programs available, which allow for direct data entry on tablets or smartphones. Details of wall topography and cave contents can also be entered directly. The perennial problem of multilevel caves overlapping in plan and section (Figure 15.7) can be resolved using these tools, through three-dimensional models and rotation in AutoCAD derived software.

Using GIS, it is now possible to map both the surface and underground features of karst terrains and this facility can provide new insights into their development and precise definition of their extent for the first time. In many cases, the integration of cave surveys with surface topography and geology (Figures 15.8 and 15.9) has identified potential areas for exploration.

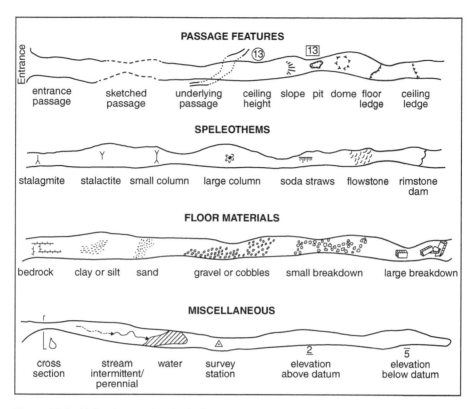

Figure 15.6 Union Internationale de Speleologie (UIS)standard cave mapping symbols.

Table 15.1 UIS survey standards.

1	**Sketch or diagram from memory - not to scale.**
2	Map compiled from notes, sketches and estimates of directions and distances made in the cave.
3	Significant directions measured by compass. Distances measured by a cord of known length - precision 0. 5 m
4	Compass and tape traverse, using deliberately chosen stations. Distances by tape or topofil or rangefinder - precision 0.1 m
5	Compass and tape traverse. Directions by calibrated compass, vertical angles by calibrated clinometer - precision 5 cm
6	Traverse and/or triangulation using calibrated, tripod-mounted instruments for directions and vertical angles - precision 2 cm
7	Controlled traverse and/or triangulation using theodolite - precision 1 cm
8	Conventional theodolite traverse and/or triangulation conforming with requirements for an acceptable cadastral survey - precision 5 mm

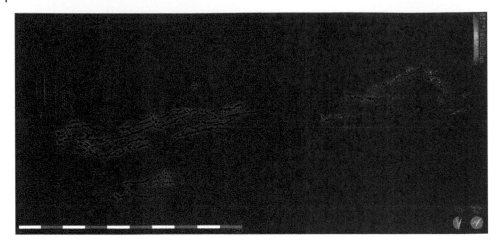

Figure 15.7 Survex 3D plot of caves in Gunung Mulu National Park, Malaysia. Source: Data from the Mulu Caves Project.

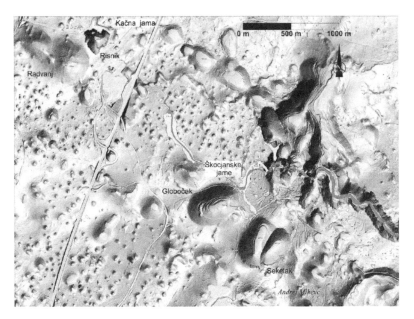

Figure 15.8 Integration of cave surveys with LIDAR terrain mapping at Skocjanske Jame, Slovenia. Source: Image by Andrej Mihevc.

15.3 Cave Photography

Cave photography evolved in parallel with surface photography with the main limitations being the size and weight of cameras, and the need for adequate lighting given slow exposures (Howes 1989). The first cave photographs were taken in the Blue John Mine of Derbyshire, UK, in 1865 (Figure 15.10), while some of the earliest flashlit photographs (using magnesium powder) were taken in Australian caves from the 1880s onwards. Stereopairs were also taken at an early date and were commercially available.

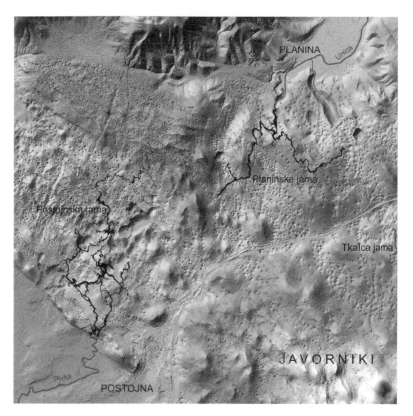

Figure 15.9 Integration of cave mapping (Postojnska jama and Planinska jama) with geomorphological mapping based on LIDAR imagery, Slovenia. Image by Andrej Mihevc.

Figure 15.10 The first cave photos of a chandelier in the Blue John mine, Derbyshire, UK, in 1865. Photo by Alfred Brothers, reprinted courtesy of Chris Howes.

Figure 15.11 Photographic equipment used in Mammoth Cave, Kentucky, USA, by Charles Waldack 1866. Note the stereoscopic camera with twin lenses and the large reflector used with magnesium powder. Waldack is the middle figure. Source: Library of Congress Collection.

The equipment was cumbersome with heavy plate cameras, magnesium powder or ribbon burners and reflectors (Figure 15.11). Stereo photography was introduced early and stereo view postcards were popular with the public (Figure 15.12). The photographic plates had to be developed soon after exposure, and makeshift darkrooms were often set up in drier parts of a cave.

Charles Waldack was born in Belgium, then emigrated to America and settled in Cincinnati. He was a trained chemist, and so opened a photography business. Practical photography had been around since 1839 when French physicist and painter Louis Daguerre developed the daguerreotype, the first successful form of photography. Waldack's photos in Mammoth Cave were amongst the first in the world and were certainly the first to be commercially available.

The first recorded cave photographs in Australia were made at Naracoorte in South Australia by Thomas Hannay of Maldon, Victoria, Australia (Reed and Bourne 2018). Mr. Hannay was an itinerant photographer who captured images across western Victoria and was particularly active in Portland. The 1860 photographs of Blanche Cave (Figure 15.13) are some of the first photographs in the world to show a cave entrance. They were also used as the basis for a popular engraving.

In Australia, photographs were also taken in many show caves and similarly were made available as postcards. Many thousands were taken by Charles Kerry in Sydney (NSW; Figure 15.14) and also for the company of Murray Views in Tumut, NSW, a firm still in existence. Kerry was probably the most prolific cave photographer and his photos were used as the basis for engravings in the *Picturesque Atlas of Australia* (Garran 1884: 149–156). His partnership with Oliver Trickett at Jenolan Caves meant that new discoveries were often photographed. These provide an invaluable record of the caves in pristine condition (Figures 15.14 and 15.15).

Modern cave photography uses digital cameras almost exclusively and coupled with flash and slave units can produce outstanding results (Howes 1997). There are essentially

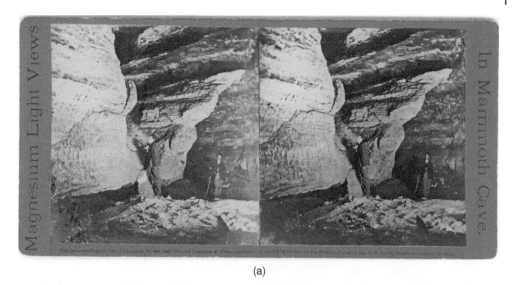

(a)

(b)

Figure 15.12 Two of Waldack's stereo views taken in Mammoth Cave, Kentucky, USA (Waldack 1866). Source: Library of Congress Collection.

three types of digital cameras in use: Compact, Prosumer, and Digital Single Lens Reflex (DSLR). Most compact digital cameras have a resolution of at least 16 megapixels, and many are both waterproof and shockproof. They have both manual controls (aperture, shutter speed, and sensitivity and preset or automatic settings. There are two main types of photosensor element; Charge-Coupled Device (CCD) and Complementary Metal–Oxide–Semiconductor (CMOS) and storage is on removable SD cards. Flash is built-in, and there is a viewing screen.

Prosumer or Bridge Digital cameras have a nonremovable zoom lens which is reasonably sealed against dust. They are comparable in size and weight to the smallest DSLR cameras

Figure 15.13 Blanche Cave (South Australia, Australia), 1860, second roof window entrance. Geologist Julian Tenison-Woods is seated above the cave. Photographer: Thomas Hannay. Source: State Library of South Australia, B36858.

but lack the removable lenses, larger sensors, mirror, and reflex system which characterise DSLRs. The viewing screen may be moveable, and the optics will generally be of better quality.

DSLR cameras are generally used by professionals and serious photo enthusiasts. The cameras basically use a mirror and pentaprism or pentamirror system to direct light from the lens to the viewfinder eyepiece. This through-the-lens viewfinder system ensures that what is seen through the eyepiece is captured by the light sensor array. They are heavier than other types, but the combination of high-quality optics and large sensor arrays (typically up to 36 million pixels) provide excellent image definition when the subject matter is correctly exposed and in focus. Their manual features allow for great creativity in photography. The liquid crystal display (LCD) screen at the back of a DSLR camera is used to display the image directly after taking the shot or viewing saved images previously saved on the memory card. Thus, it may be difficult to compose a photo with a DSLR in very dark caves while looking through the viewfinder and using a torch to locate the boundaries of the scene.

Intermediate between the prosumer and the DSLR is the compact mirrorless camera. One model of this type has a moveable viewing screen, wireless connectivity, and controls. This type can also use a range of lenses with an adapter (J. Garnett pers. comm.).

Flash units vary in size and power and those on compact cameras are generally of low power. Using a flashgun off-camera is possible with Prosumer and DSLR cameras, and will generally result in better composition and avoid the flat lighting produced by flash on-camera. A flash slave unit, triggered by the flash on or off the camera, can add depth

Figure 15.14 Massive speleothems in Nettle Cave, Jenolan, Australia, May 1884. These steps and guardrail are still in place, with Nettle Cave being used for adventure tours. Source: National Library of Australia.

and backlighting to a subject. For large chambers an array of slave units can be used to provide both illumination and depth.

Incorporating as many of the following elements as possible will increase the chance of making a photo more interesting to the viewer (Smith 2009):

- Include both near and far objects.
- Include some aspect which will give a comparison of scale.
- A dominant feature in a photo should catch the observer's eye and lead it towards the rest of the photo. An example is a person looking towards the rest of the scene or at a dominant feature.
- The 'thirds principle' is a good rule of thumb to use. Try to position a person one-third of the way across the photo.
- Don't cut a person off at the knees, ankles, or neck. Either include a person's head and shoulders to mid-chest level or the whole person.
- Make sure the model has an acceptable facial expression and is positioned in a good pose.
- Get the exposure correct by adjusting the flash units or camera aperture.

Cave photography is best carried out on a dedicated caving trip. Some photographs can be taken using compact cameras on exploratory trips, but setting up scenes and lighting

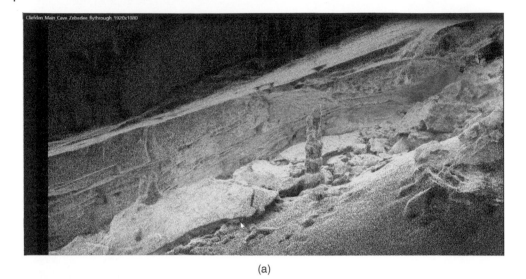

(a)

(b)

Figure 15.15 (a) Internal view of Cathedral Cave, Wellington, New South Wales, Australia. Speleothem is same as that in Figure 15.4. Source: Data provided by Robert Zlot (CSIRO) 2014, under Creative Commons license. (b) Internal view of Cathedral Cave, Wellington, New South Wales, Australia. Large chamber with roof collapse blocks. Source: Data provided by Robert Zlot (CSIRO) 2014, under Creative Commons license.

is time-consuming and can try the patience of cave models. There may be delays while the positions of flashes or slave units are adjusted, and the models may have to change positions to improve the composition. The cave model must be relaxed and comfortable with the reality that they may be in the same place for some time, and bored or contorted facial expressions should be avoided.

15.4   3D Scanning of Caves

The survey of very large cave systems is time-consuming. In Australia, the Jenolan Caves Survey Project (James et al. 2009) produced a comprehensive 3D model of more than 20 km of cave. The survey information acquired included total station and laser distance measurements. Large chambers were recorded using distance measurements taken in twelve-point cross-sections spaced at 10 m intervals. The final model is highly detailed and accurate, but the whole project involved a very large amount of surveying and data processing effort over nearly 20 years (1987–2005).

Over the last decade, medium and long-range terrestrial laser scanning (TLS) technology has developed that can collect high point densities with unprecedented accuracy and speed. Accurate surveying of caves has always been fundamental to understanding their origin and the processes that lead to their current state, as well as providing tools for management. The availability of very accurate 3D models of caves has been taken up by scientists interested in speleogenesis, as areas inaccessible to cavers in the roof of a cave can now be rendered in great detail.

Initially, the TLS instruments were mounted on tripods and were best suited to larger chambers such as show caves. Recent devices have reduced in size and weight, making it possible to have hand-held scanners suitable for use in any cave. This has been accompanied by increases in point cloud resolution of great accuracy. Software packages have become more efficient in terms of handling large data volumes and have improved functions for visualisation and product generation.

Examples of 3D caves can be seen in Figures 15.14–15.17.

Terrestrial scanners are typically mounted on a stationary surveying tripod and acquire millions of precise range measurements of the surfaces surrounding the station over a period of a few minutes. The resulting point cloud can then be imported into Open Source

Figure 15.16   Point cloud from 3D laser Plan scanning, Cathedral Cave, Wellington, New South Wales, Australia. Source: Data provided by Robert Zlot (CSIRO) 2014, under Creative Commons license.

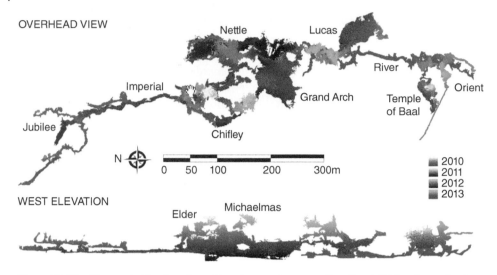

OVERHEAD VIEW

Nettle
Lucas
River
Imperial
Orient
Grand Arch
Temple
of Baal
Jubilee
Chifley

N

0   50   100   200   300m

■ 2010
■ 2011
■ 2012
■ 2013

WEST ELEVATION
Elder   Michaelmas

Figure 15.17   Plan and side elevation of Jenolan tourist caves, New South Wales, Australia. Source: Derived mobile 3D laser scanning, from Zlot and Bosse (2014). Source: Used with permission of the National Speleological Society (www.caves.org).

software, such as Meshlab (http://www.meshlab.net) or CloudCompare (https://www.danielgm.net/cc), within which rendered 3D objects can be created. In the case of caves, rendered objects can be external or internal views. Contouring and overlying photography can also be carried out. Combining data from multiple stations can be done if there is sufficient overlap between the scanned surfaces. In practice, it is more common to relate consecutive stations using standard surveying techniques or to include known survey targets in the scans (Rüther et al. 2009). The technique produces high-quality data, but limitations include the size and weight of the equipment, its cost and relative fragility, and the difficulty of transporting it through tight squeezes.

> *The most time-consuming part of the scan was moving from station to station and shooting in the targets. Every setup was a challenge to determine where the previous and upcoming scans' shadows would occur and locating the best combination of scan coverage and setup efficiency.* McIntire (2010)

In contrast, mobile mapping is a technique where measurements of the environment are acquired while moving continuously through it (Zlot and Bosse 2014). There are systems using LIDAR and GNSS which are aircraft-borne, and some wheeled systems used in mines and inside buildings. More recently, a handheld device called Zebedee has been developed and used successfully by CSIRO scientists in Australia (Figure 15.18: Zlot and Bosse 2014). The components of this system are a lightweight handheld laser-scanning device coupled with data processing software capable of accurately estimating the position and orientation of the scanner over time as it is moved through any cave.

The scanner measures tens of thousands of distance ranges per second from the sensor origin to points on cave surfaces using narrow infrared laser pulses. Given an accurate estimate of the scanner's motion, the set of range measurements can be projected into

Figure 15.18 (a) Mapping Koonalda Cave, South Australia, with the Zebedee 3D mapping system. The scanning device is held in the operator's right hand, with a battery pack and a small laptop for recording data carried in a backpack. (b) The components of the Zebedee system. Source: From Zlot and Bosse (2014). Source: Used with permission of the National Speleological Society (www.caves.org).

(x,y,z) points in a common coordinate frame, thereby generating a consistent point cloud model of the cave and its surrounds. The maximum range of the scanner is about 35 m, and the precision is typically 1–3 cm, depending on the distance and incidence angle to the surface, as well as the reflectivity of the surface. A unique design feature of Zebedee is the spring on which the laser scanner is mounted. This converts the natural motion of the operator carrying the device into a 3D rotational motion of the scanner.

Once millions of points have been collected into a point cloud, it remains to process these into a coherent image. Data processing stage goes through these three steps in their order of sequence: filtering and registration, 3D meshing, and post-processing (Idrees and Pradhan 2016). Combining multiple scans by registration into an aligned single point cloud requires registration points which are visible in each scan. From these, the software computes accurate transformation parameters (translation, rotation, and scale factors: González-Aguilera et al. 2009, 2011). All point clouds processing software provides the flexibility of using the automatic and manual point correspondence to create high quality registered point clouds. Once scans are aligned, subsequent steps eliminate invalid data caused by instruments, lessen data redundancy, and maximise data processing speed and efficiency. The depiction of 3D surfaces is usually by the creation of a triangulated irregular network (TIN), created from a set of x, y, and z coordinates values, is the most widely used topological data structure to depict 3D surfaces. According to Pucci and Marambio (2009), triangulation performs three key functions: transform point clouds to a more visually perceptive facsimile, reduce data size, and permit interactivity within and across platforms.

Point clouds are basically the primary output of laser scanning, which are by themselves not useful without software packages to process them. Though not at an equal pace with hardware development, point cloud processing packages (Table 15.2) have improved dramatically, and many are now available as Open-Source.

The 3D mapping of caves allows for greater definition of wall forms and features relevant to an understanding of speleogenesis. Roncat et al. (2011) used TLS in the Marchenhohle, Austria to define features characteristic of hypogene caves. They concluded that the highly accurate scans made it possible to define areas of cupolas and areas where flat roof sections and fractures dominated. In the Gomantong Cave, Sabah, Malaysia, Lundberg et al. (2017) used laser scanning to map wall scallops formed by water previously filling phreatic passages. They concluded that the original flows in a series of sub-parallel passages were artesian, fed by water sinking in distant highland terrain and moving towards the coast. In the same cave, McFarlane et al. (2015) used the same technique to map the distribution of cave swiftlet nests and roosting bats. High-resolution TLS data were imported into ArcGIS software, allowing for semi-automated counting of nests based on resolved geometry and laser return intensity. Nest resolution and counting accuracy were better than 2%. Spatial analysis of nest locations has established a maximum packing density of 268 nests/m^2 in optimum locations, which correspond to roof slopes of >20°. Co-occurring *Rhinolophid* bats roost adjacent to, but not within nest locations, preferring roof surfaces close to horizontal. This 'proof of concept' study clearly indicates the utility of TLS for ecological studies of wildlife roosting in very inaccessible locations.

Table 15.2   Point cloud processing software packages.

Package	Developer	Availability	Basic Point Cloud Management Capability	3D Modelling
FARO Scene	FARO Technologies	Commercial	Automated (target & target-less) registration. Powerful processing & editing capability; colour coded 3D point cloud view. Map external photo to point cloud, data sharing via the internet with SCENE WebShare	3D meshing & editing
Leica Cyclone	Leica Geosystems	Commercial	Automated (target & target-less) registration. Efficient processing & editing. Full 3D point clouds visualisation, map external photo to point cloud. Web-based data sharing with Cyclone-PUBLISHER. Direct point cloud importing from other products. Supports distributed parallel computing with Cyclone-SERVER.	3D meshing and 3D CAD geometry/meshes from point clouds as for engineering applications
RiSCAN PRO & Ri Profile	Riegl LMS	Commercial	Automated (target & target-less) registration, support hue, saturation, and brightness filtering, colour mapping 3D view.	3D meshing & editing, 3D CAD geometry model of pipe from point cloud
RealWorks	Trimble	Commercial	Automated (target & target-less) registration, 3D point cloud visualisation, Advanced-Tank edition. Data sharing with the publisher.	3D meshing & editing. 3D CAD geometry model of pipe from point cloud
MeshLab	ISTI-CNR, Italy	Open Source	Scan registration and editing	3D meshing & polygon editing
CloudCompare	Cloud Compare Project	Open Source	Scan registration and edit point cloud. 3D colour view	Meshing & polygon editing

Source: Modified from IDREES and PRADHAN 2016.

15.5   Drones

The use of drones for aerial photography and the generation of digital elevation models (DEM) has now become widespread, both recreationally and commercially. The availability of cheap quadracopters with high-quality cameras has led to their use in cave photography. This provides viewpoints otherwise inaccessible to cavers, and the integration of photos through structure through motion software (Alessandri et al. 2019) creates accurate photo mosaics of cave floors. These products may be combined with 3D models created using TLS data.

15.6 Mapping World Heritage Caves in Gunung Mulu National Park, Malaysia

The Gunung Mulu National Park is located about 100 km east of Miri, a coastal city in northern Sarawak, Malaysia. The small town of Mulu is reached by daily air services from Miri and Kuching and can also be reached by boats up the Baram and Tutoh rivers. Gunung Mulu National Park has an area of 90 000 ha and most visitors concentrate on the southernmost karst, accessible from the park headquarters adjacent to the town. Over 90% of the park remains unvisited and is in pristine condition (Gill 1999).

The karst mountains exhibit classical tropical karst features, pockmarked with dolines, closed depressions, valleys, and caves (Gillieson and Clark 2010). In the southern hills, 'The Garden of Eden' is a fine example of a tiankeng, being over 1000 m in diameter and several hundred metres deep. The Melinau, Melinau Paku, and Medalam Rivers have dissected the limestones, forming deeply incised gorges with towering 300 m high cliffs and remnants of high-level caves. Some of the world's finest examples of pinnacle karst can be found on the karst mountains of Gunung Benarat and Api.

The caves of the park are now world-famous. Over 100 caves are now known with a total mapped length of over 330 km (Figure 15.19). The caves are some of the largest to be found in the world, with 30–40 m diameter passages being common. They are amongst the finest examples of tropical river caves known, with well-developed flood incuts, extensive clastic sediment deposits, and circular or elliptical tubes linking different cave levels. Deer Cave is one of the world's largest natural cave passages, measuring 120–150 m in diameter and 1000 m in length. The Clearwater Cave System is presently 227 km in length, the longest cave so far discovered in Asia. It displays outstanding tubular passages and canyons developed at many levels. Gua Nasib Bagus (Good Luck Cave) contains 'Sarawak Chamber', the world's largest natural single chamber within a cave. The chamber measures 600 m long by 415 m wide and 80 m high, with a floor area of 162 700 m^2 and a volume of 12 million cubic meters. Reversal of the earth's magnetic field, recorded in the sediments at 1.8 million years before present, indicates that the caves are at least 2 million years old, possibly as much as 3 million years. Notches in the walls of the caves at various levels can be correlated with interglacial periods. The caves and associated karsts are therefore of outstanding significance in recording the major changes in the earth's history and have thus joined the rich array of karst sites inscribed on the World Heritage List.

A World Heritage nomination for the Gunung Mulu National Park was submitted in 1999 (Gill 1999) and the park inscribed on the World Heritage List in 2000. Subsequently, a new management plan was drawn up for the park (Manidis Roberts Consultants 2000). Integral to the management plan was the need to map the caves and support scientific research to improve knowledge of the park's natural and cultural values.

The approaches to sustainable management through the Visitor Impact Management Process (VIM) first establish precise objectives for managing both the environment and the

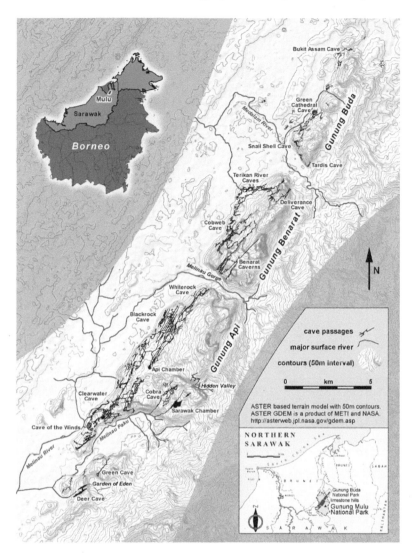

Figure 15.19 Plan of the caves of Gunung Mulu National Park, Malaysia, including those of Gunung Buda. Source: http://www.mulucaves.org/wordpress/surveying/new-mulu-map.

visitor experience. A range of environmental and social indicators measure the extent to which objectives are being achieved. The periodic review of these indicators allows for the development of appropriate management responses. Any new developments are subject to a review process involving these indicators, and the plan itself is reviewed every five years.

The ongoing vision for the park has been that it should be a role model for sustainable management of natural areas and a pre-eminent international tourism destination in Malaysia. This has been achieved, and other parks are following the lead provided by Mulu. The primary challenge for management is to establish an effective management team and develop capabilities of staff in both conservation and ecotourism. It is critical that visitor numbers do not exceed human capacity building and the infrastructure development in the park. Opportunities for employment of local people have been provided, and today 97% of the staff are of local origin, with 80 local people employed directly at the park headquarters.

The Mulu Caves Project (http://www.mulucaves.org/wordpress) is a collaboration between UK speleologists and the Sarawak Authorities, particularly the Sarawak Forestry Corporation, working in association with Gunung Mulu National Park management and staff. The mapping of the Mulu caves has taken place over the last thirty years thanks to a sustained effort by numerous British cavers, led by Andy Eavis. The surveys

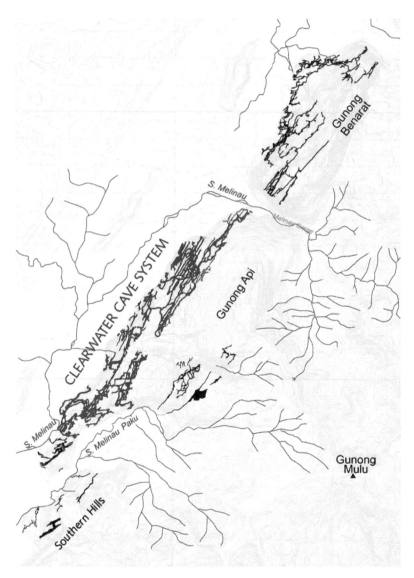

Figure 15.20 Plan of Clearwater Cave system and caves of Gunong Benarat, Gunung Mulu, Malaysia National Park, Malaysia. Source:http://www.mulucaves.org/wordpress/the-caves-2/the-major-cave-systems/the-clearwater-cave-system.

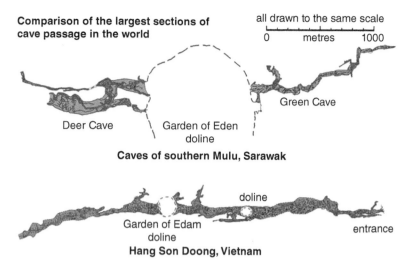

Figure 15.21 Comparison between the Deer and Green Caves system at Gunung Mulu National Park, Malaysia, and Hang Son Doong Cave in Vietnam. Source: http://www.mulucaves.org/wordpress/articles/the-largest-passage-on-earth.

were needed to document the ongoing discoveries, and also to facilitate the scientific development research in the park. Initially, the cave passages were surveyed as they were being discovered by small teams of two or three surveyors. Typically, the instruments used were Suunto compass and clinometers with 30 m fibreglass tapes. Topofil instruments were tried on one expedition, but the bio-degradable cotton degraded before it reached the caves. Initial attempt to use laptops in the field had limited success as the high humidity caused electronic failure. Laser distance measuring instruments have now replaced tapes, and field data are entered into survey packages, such as Survex (Figure 15.20). The centre line survey can be printed out at base camp and the passage detail from the cave notes hand-drawn onto this. These drawings can then form the basis of a hand-drawn survey back in the UK or can be scanned into cave drawing packages, such as Therion, which produce finished composite surveys. These surveys can then be compared with other caves (Figure 15.21) to determine global ranking of cave passages.

References

Alessandri, L., Baiocchi, V., Del Pizzo, S., Rolfo, M. F. & Troisi, S. 2019. Photogrammetric survey with fisheye lens for the characterization of the La Sassa cave. *Int. Arch. Photogramm. Remote Sens. Spatial Inf. Sci.,* XLII-2-W9-25-2019.

Ballesteros, D., Fernández-Martínez, E., Carcavilla, L., and Jiménez-Sánchez, M. (2019). Karst cave geoheritage in protected areas: characterisation and proposals of management of deep caves in the Picos de Europa National Park (Spain). *Geoheritage* 11: 1919–1939.

Cayla, N. (2014). An overview of new technologies applied to the management of geoheritage. *Geoheritage* 6: 91–102.

CSIRO (2014). *Wellington Caves: Zebedee 3D Data Collection*, Commonwealth Scientific and Industrial Research Organisation, Sydney. Creative Commons Attribution 3.0 License, doi 10.4225/08/553EEE63CC584.

Dunkley, J.R. (2016). The 1830 cave diaries of Thomas Livingstone Mitchell. *Helictite* 42: 21–37.

Fabre, G. & Audetat, M. 1978. Signes speleologiques conventionnels. *UIS/AFK/FFS CERGH Memoires,* 14.

Ford, D.C. and Williams, P.W. (1989). *Karst Geomorphology and Hydrology*. London: Unwin Hyman.

Garran, A. (1884). *Picturesque Atlas of Australasia*. Sydney: The Picturesque Atlas Publishing Company Ltd.

Gill, D. 1999. Nomination of the Gunung Mulu National Park, Sarawak, Malaysia for World Heritage Listing. *Report to UNESCO World Heritage Committee*. Kuching: Sarawak Forestry Department.

Gillieson, D. (2016). Natural heritage values of the Chillagoe and Mitchell-palmer karst and caves. *North Queensland Naturalist* 46: 71–85.

Gillieson, D. (2019). Global distribution of cave and karst world heritage properties - a review. *Journal of the Australasian Cave & Karst Management Association* 116: 5–13.

Gillieson, D.S. and Clark, B. (2010). Mulu: the World's Most spectacular tropical karst. In: *Geomorphological Landscapes of the World* (ed. P. Migon). Springer.

González-Aguilera, D., Muñoz-Nieto, A., Gómez-Lahoz, J. et al. (2009). 3D digital surveying and modelling of cave geometry: application to paleolithic rock art. *Sensors* 9: 1108–1127.

González-Aguilera, D., Rodriguez-Gonzalvez, P., Mancera-Taboada, J. et al. (2011). Application of non-destructive techniques to the recording and modelling of palaeolithic rock art. In: *Laser Scanning, Theory and Applications* (ed. C.-C. Wang). InTech.

Gordon, J.E., Crofts, R., Díaz-Martínez, E., and Woo, K.S. (2018). Enhancing the role of geoconservation in protected area management and nature conservation. *Geoheritage* 10: 191–203.

Gray, M. (2013). *Geodiversity: Valuing and Conserving Abiotic Nature*. Chichester, UK: Wiley Blackwell.

Gray, M. (2019). Geodiversity, geoheritage and geoconservation for society. *International Journal of Geoheritage and Parks*.

Grimes, K. G. 1995. A Classification of Geological and Geomorphological Features: Accompanying volume of a report prepared for the Australian Heritage Commission by the Standing Committee for Geological Heritage of the Geological Society of Australia Inc. Canberra, ACT: Australian Heritage Commission.

Grossenbacher, Y. (1992). Höhlenvermessung. *SGH-Kurs* 4.

Houshold, I., Sharples, C., Dixon, G., and Duhig, N. (1997). Georegionalisation – a more systematic approach for the identification of places of geoconservation significance. In: *Pattern and Process: Towards a Regional Approach to National Estate Assessment of Geodiversity* (ed. R. Eberhard). Canberra: Environment Australia and the Australian Heritage Commission Environment Australia.

Howes, C. (1989). *To Photograph Darkness: The History of Underground and Flash Photography*. Gloucester, UK: Alan Sutton Publishing.

Howes, C. (1997). *Images Below*. Cardiff, UK: Wild Places Publishing.

Idrees, M. and Pradhan, B. (2016). A decade of modern cave surveying with terrestrial laser scanning: a review of sensors, method and application development. *International Journal of Speleology* 45: 71–88.

James, J.M., Martin, D.J., Tunnock, G.M., and Warild, A.T. (2009). A cave survey for research and tourist cave management. In: *Proceedings of the 15th International Congress of Speleology* (ed. W.B. White), 1381–1387. Huntsville: National Speleological Society.

Lundberg, J., Carroll, W., Roberts, W. et al. (2017). Analysis of scallops in Gomantong caves, by GIS processing of 3D terrestrial laser scanner data. In: *Proceedings of the 17th International Congress of Speleology*, 285–288. Sydney.

Manidis Roberts Consultants 2000. Gunung Mulu National Park Integrated Development and Management Plan. Sydney.

Martín-Duque, J.F., García, J.C., and Urquí, L.C. (2012). Geoheritage information for geoconservation and geotourism through the categorization of landforms in a karstic landscape: a case study from Covalagua and Las Tuerces (Palencia, Spain). *Geoheritage* 4: 93–108.

Matthews, P.B. (1985). *Australian Karst Index*. Australian Speleological Federation: Melbourne.

Mcfarlane, D., Roberts, W., Buchroithner, M. et al. (2015). Terrestrial LiDAR-based automated counting of swiftlet nests in the caves of Gomantong, Sabah, Borneo. *International Journal of Speleology* 44: 191–195.

McIntire, D. (2010). Laser scanning Mushpot Cave. *The American Surveyor* 7: 18–27.

Mitchell, T.L. (1830). Bathurst road (diary entries 29 May to 3 August 1830). In: *C42: Field Note and Sketchbook, 21 May 1828–3 August 1830* (ed. S.T. Mitchell). New South Wales: State Library of NSW.

Mitchell, T. L. 1838. *Three Expeditions into the Interior of Australia, with Descriptions of the Recently Explored Region of Australia Felix, and the Present Colony of New South Wales,* London, T. and W. Boone.

Ollier C-F. 1893. Relation de la Visite Du Marquis de Nointel À La Grotte d'Antiparos en 1673, Paris, Paris, Hachette Livres (reprinted)

Pucci, B. & Marambio, A. Olerdola's Cave, Catalonia, past and present: A virtual reality reconstruction from terrestrial laser scanner and GIS data, Proceedings of 3D ARCH'2009: 3D Virtual Reconstruction and Visualization of Complex Architectures, 25–28 February 2009 Trento, Italy. 1–6.

Reed, E. and Bourne, S. (2018). New evidence confirms Thomas Hannay as the first photographer of Naracoorte caves and emphasises the importance of historical writing in caves. *Helictite* 44: 45–58.

Roncat, A., Dublyansky, Y., Spötl, C., and Dorninger, P. (2011). Full-3D surveying of caves: a case study of Märchenhöhle (Austria). In: *Proceedings of the International Association for Mathematical Geosciences 2011* (eds. R. Marschallinger and G. Zobl), 1393–1403. Salzburg, Austria.

Rüther, H., Chazan, M., Schroeder, R. et al. (2009). Laser scanning for conservation and research of African cultural heritage sites: the case study of Wonderwerk cave, South Africa. *Journal of Archaeological Science* 36: 1847–1856.

Sharples, C. (2002). *Concepts and Principles of Geoconservation*. Hobart: Tasmanian Parks & Wildlife Service.

Sharples, C. (2014). A Thematic Gap Analysis of the Tasmanian Geoconservation Database: Glacial and Periglacial Landform Listings in the Tasmanian Wilderness World Heritage Area; Resource Management and Conservation Division, Department of Primary Industries Parks Water and Environment, Hobart, Nature Conservation Series 14/4.

Smith, G. A. Cave Photography with Digital Cameras. Proceedings of the 27th ASF Biennial Conference, 2009 Sale, Victoria.

Southwell, R. (1683). A description of Pen-Park-Hole in Gloucestershire. *Philosophical Transactions of the Royal Society* 13: 2–6.

Waldack, C.L. (1866). *Magnesium Light Views in Mammoth Cave*. New York: Edward and Henry T. Anthony.

Waltham, A.C. (1976). The Tigris tunnel and Birkleyn caves, Turkey. *British Cave Research Association Bulletin* 14: 31–34.

White, S. and Wakelin-King, G. (2014). Earth sciences comparative matrix: a comparative method for geoheritage assessment. *Geographical Research* 52: 168–181.

Williams, P.W. (2008). *World Heritage Caves and Karst: A Thematic Study*. Gland: IUCN.

Zlot, R. and Bosse, M. (2014). Three-dimensional mobile mapping of caves. *Journal of Cave and Karst Studies* 76: 191–206.

Glossary of Cave and Karst Terminology

Modified and extended from that compiled by J. N. Jennings for the Australian Speleological Federation.

Abbreviations and Conventions

Abb. = abbreviation
Syn. = synonym (word with same meaning)
Cf. = confer (compare) with the following term which is not identical but related to it
n. = noun
v. = verb

A

Accidental (n.) An animal accidentally living in a cave.

Active Cave A cave which has a stream flowing in it.

Adaptation An inherited characteristic of an organism in structure, function, or behaviour which makes it better able to survive and reproduce in a particular environment. Lengthening of appendages, loss of pigment, and modification of eyes are considered adaptations to the dark zone of caves.

Aeolian Calcarenite A limestone formed on land by solution and redeposition of calcium carbonate in coastal dune sands containing a large proportion of calcareous sand from mollusc shells and other organic remains.

Aggressive Referring to water which is still capable of dissolving more limestone, other karst rock, or speleothems.

Allogenic Referring to water or sediment which has a source on nonkarstic rocks.

Anastomosis A mesh of tubes or half-tubes.

Aragonite A less common crystalline form of calcium carbonate than calcite, denser, and orthorhombic.

Arthropods The most common group of animals inhabiting caves, including insects, crustaceans, spiders, millipedes, etc. They have jointed limbs and external skeletons.

Autogenic Referring to water or sediment which is derived from karstic rocks.

Caves: Processes, Development, and Management, Second Edition. David Shaw Gillieson.
© 2021 John Wiley & Sons Ltd. Published 2021 by John Wiley & Sons Ltd.

Autotroph A green plant, bacterium, or protist which manufactures complex organic compounds (food) from simple inorganic raw materials, using a source of energy from light or chemical compounds.

B

Bare Karst Karst with much exposed bedrock.

Bat A member of the order Chiroptera, the only mammals capable of true flight as they have membranes between the toes of their forefeet.

Bathyphreatic Referring to water moving with some speed through downward looping passages in the phreatic zone.

Bed A depositional layer of sedimentary bedrock or unconsolidated sediment.

Bedding-Grike A narrow, rectilinear slot in a karst rock outcrop as a result of solution along a bedding-plane.

Bedding-Plane A surface separating two beds, usually planar.

Bedding-Plane Cave A cavity developed along a bedding-plane and elongate in cross-section as a result.

Benthic Bottom dwelling, usually on the bed of a stream, pond, lake, or the sea.

Biomass The total weight of living matter in a given area, or in a community, at a particular trophic level, or of a particular type of organism at a site.

Biospeleology The scientific study of organisms living in caves.

Blind Shaft A vertical extension upwards from part of a cave but not reaching the surface; small in area in relation to its height.

Blind Valley A valley that is closed abruptly at its lower end by a cliff or slope facing up the valley. It may have a perennial or intermittent stream which sinks at its lower end or it may be a dry valley.

Blowhole (i) A hole to the surface in the roof of a sea cave through which waves force air and water. (ii) A hole in the ground through which air blows in and out strongly, sometimes audibly; common in the Nullarbor Plain.

Bone Breccia A breccia containing many bone fragments.

Branchwork A dendritic system of underground streams or passages wherein branches join successively to form a major stream or passage.

Breakdown Fall of rock from the roof or wall of a cave.

Breccia Angular fragments of rock and/or fossils cemented together or with a matrix of finer sediment. cf. Bone breccia.

C

Calcite The commonest calcium carbonate ($CaCO_3$) mineral and the main constituent of limestone, with different crystal forms in the rhombohedral subsystem.

Canopy A compound speleothem consisting of a flowstone cover of a bedrock projection and of a fringe of stalactites or shawls on the outer edge.

Canyon (i) A deep valley with steep to vertical walls; in karst frequently formed by a river rising on impervious rocks outside the karst area. (ii) A deep, elongated cavity cut by running water in the roof or floor of a cave or forming a cave passage.

Carbide Calcium carbide, CaC_2, used with water to make acetylene in lamps.

Cave A natural cavity in rock large enough to be entered by people. It may be water-filled. If it becomes full of ice or sediment and is impenetrable, the term applies but will need qualification.

Cave Blister An almost perfect hemisphere of egg-shell calcite.

Cave Breathing (i) Movement of air in and out of a cave entrance at intervals. (ii) The associated air currents within the cave.

Cave Coral Very small speleothems consisting of short stalks with bulbous ends, usually occurring in numbers in patches.

Cave Earth Clay, silt, fine sand, and/or humus deposited in a cave.

Cave Ecology The study of the interaction between cave organisms and their environment, e.g. energy input from surface, climatic influences.

Cave Fill Transported materials such as silt, clay, sand, and gravel which cover the bedrock floor or partially or wholly block some part of a cave.

Cave Flower Syn. gypsum flower.

Cave Pearl A smooth, polished and rounded speleothem found in shallow hollows into which water drips. Internally it has concentric layers around a nucleus.

Cave Spring A natural flow of water from rock or sediment inside a cave.

Cave System A collection of caves interconnected by enterable passages or linked hydrologically or a cave with an extensive complex of chambers and passages.

Cavern A very large chamber within a cave.

Cavernicole An animal which normally lives in caves for the whole or part of its life cycle.

Caving The entering and exploration of caves.

Cenote A partly water-filled, wall-sided doline.

Chamber The largest order of cavity in a cave, with considerable width and length but not necessarily great height.

Chert A light grey to black or red rock, which fractures irregularly, composed of extremely fine crystalline silica and often occurring as nodules or layers in limestone.

Choke Rock debris or cave fill blocking part of a cave.

Column A speleothem from floor to ceiling, formed by the growth and joining of a stalactite and a stalagmite, or by the growth of either to meet bedrock.

Conduit An underground stream course completely filled with water and under hydrostatic pressure or a circular or elliptical passage inferred to have been such a stream course.

Coprolite Fossilised large excrement of animals, sometimes found in caves, especially those used as lairs.

Coprophage A scavenger which feeds on animal dung, including guano.

Corrasion The wearing away of bedrock or loose sediment by mechanical action of moving agents, especially water.

Corrosion Syn. solution.

Crawl (Way) A passage which must be negotiated on hands and knees.

Cross-Section A section of a cave passage or a chamber across its width.

Cryptozoa The assemblage of small terrestrial animals found living in darkness beneath stones, logs, bark, etc. Potential colonisers of caves.

Crystal Pool A cave pool generally with little or no overflow, containing well-formed crystals.

Current Marking Shallow asymmetrical hollows formed by solution by turbulent waterflow and distributed regularly over karst rock surfaces. cf. Scallops.

Curtain A speleothem in the form of a wavy or folded sheet, often translucent and resonant, hanging from the roof or wall of a cave.

D

Dark Adaptation A change in the retina of the eye sensitizing it to dim light (the eye 'becomes accustomed to the dark'). Loss of sensitivity on re-exposure to brighter light is 'light adaptation'.

Dark Zone The part of a cave which daylight does not reach.

Daylight Hole An opening to the surface in the roof of a cave.

Dead Cave A cave without streams or drips of water.

Decomposers Living things, chiefly bacteria, and fungi, that subsist by extracting energy from tissues of dead animals and plants.

Decoration Cave features as a result of secondary mineral precipitation, usually of calcite. Syn. Speleothem.

Detritivore Organisms which feed on organic detritus, such as the dead parts of plants or the dead bodies and waste products of animals.

Dig An excavation made (i) to discover or extend a cave or (ii) to uncover artefacts or animal bones.

Dip The angle at which beds are inclined from the horizontal. The true dip is the maximum angle of the bedding-planes at right angles to the strike. Lesser angles in other directions are apparent dips.

Dog-Tooth Spar A variety of calcite with acute-pointed crystals.

Doline A closed depression draining underground in karst, of simple but variable form, e.g. cylindrical, conical, bowl- or dish-shaped. From a few to many hundreds of metres in dimension.

Dolomite (i) A mineral consisting of the double carbonate of magnesium and calcium $CaMg(CO_3)_2$. (ii) A rock made chiefly of dolomite mineral.

Dome A large hemispheroidal hollow in the roof of a cave, formed by breakdown and/or salt weathering, generally in mechanically weak rocks, which prevents bedding and joints dominating the form.

Donga In the Nullarbor Plain, a shallow, closed depression, several metres deep and hundreds of metres across, with a flat clay-loam floor and very gentle slopes.

Driphole A hole formed by water dripping onto the cave floor.

Dripline A line on the ground at a cave entrance formed by drips from the rock above. Useful in cave survey to define the beginning of the cave.

Dripstone A deposit formed from drops falling from cave roofs or walls, usually of calcite.

Dry Cave A cave without a running stream. cf. Dead cave.

Dry Valley A valley without a surface stream channel.

Duck(-Under) A place where water is at or close to the cave roof for a short distance so that it can be passed only by submersion.

Dune Limestone Syn. Aeolian calcarenite.

Dye Gauging Determining stream discharge by inserting a known quantity of dye and measuring its concentration after mixing.

Dynamic Phreas A phreatic zone or part of a phreatic zone where water moves fast with turbulence under hydrostatic pressure.

E

Eccentric A speleothem of abnormal shape or attitude. cf. Helictite.

Epiphreatic Referring to water moving with some speed in the top of the phreatic zone or becoming part of the phreatic zone during floods.

Erosion The wearing away of bedrock or sediment at the surface or in caves by mechanical and chemical actions of all moving agents, such as rivers, wind, and glaciers.

Exsurgence A spring fed only by percolation water.

F

Fault A fracture separating two parts of a once continuous rock body with relative movement along the fault plane.

Fault Cave A cave developed along a fault or fault zone, either by movement of the fault or by preferential solution along it.

Fault Plane A plane along which movement of a fault has taken place.

Fissure An open crack in rock or soil.

Fissure Cave A narrow, vertical cave passage, often developed along a joint but not necessarily so. Usually, a result of solution but sometimes of tension.

Flattener A passage, which, though wide, is so low that movement is possible only in a prone position.

Floe Calcite Very thin flakes of calcite floating on the surface of a cave pool or previously formed in this way.

Flowstone A deposit formed from thin films or trickles of water over floors or walls, usually of calcite. cf. Travertine.

Fluorescein A reddish-yellow organic dye which gives a green fluorescence to water. It is detectable in very dilute solutions, so it is used in water tracing and dye gauging in the form of the salt sodium fluorescein.

Fluorometer An instrument for measuring the fluorescence of water; used in water tracing and dye gauging.

Fossil The remains or traces of animals or plants preserved in rocks or sediments.

G

Glacier Cave A cave formed within or beneath a glacier.

Gour Syn. Rimstone dam.

Grike A deep, narrow, vertical or steeply inclined rectilinear slot in a rock outcrop as a result of solution along a joint.

Grotto A room in a cave of moderate dimensions but richly decorated.

Groundwater Syn. Phreatic water.

Guano Large accumulations of dung, often partly mineralised, including rock fragments, animal skeletal material and products of reactions between excretions and rock. In caves, it is derived from bats and to a lesser extent from birds.

Gypsum The mineral hydrated calcium sulphate $CaSO_4.2H_2O$.

Gypsum Flower An elongated and curving deposit of gypsum on a cave surface.

H

Half-Blind Valley A blind valley which overflows its threshold when the streamsink cannot accept all the water at a time of flood.

Half-Tube A semi-cylindrical, elongate recess in a cave surface, often meandering or anastomosing.

Halite The sodium chloride mineral NaCl in the cubic crystalline system.

Hall A lofty chamber considerably longer than it is wide.

Helictite A speleothem, which at one or more stages of its growth changes its axis from the vertical to give a curving or angular form.

Heterotroph An organism that cannot manufacture food from inorganic raw materials, and therefore must feed on available organic compounds contained in the tissues of other organisms.

Histoplasmosis A lung disease which may be caught from the guano of some caves, caused by a fungus, Histoplasmosis capsulatum. Usually mild in effect, it can be fatal in rare cases.

Hydrology The scientific study of the nature, distribution, and behaviour of water.

Hydrostatic Pressure The pressure due to a column of water.

I

Ice Cave A cave with perennial ice in it.

Inflow Cave A cave into which a stream enters or is known to have entered formerly, but which cannot be followed downstream to the surface.

Interstitial Medium Spaces between grains of sand or fine gravel filled with water which may contain organisms.

Inverted Siphon A siphon of U-profile.

J

Joint A planar or gently curving crack separating two parts of once continuous rock without relative movement along its plane.

Joint-Plane Cave A cavity developed along a joint and elongated in cross-section.

K

Kankar (pronounced kunkar) A deposit, often nodular, of calcium carbonate formed in soils of semi-arid regions. Sometimes forms cave roofs.

Karren The minor forms of karst as a result of solution of rock on the surface or underground.

Karst Terrain with special landforms and drainage characteristics on account of greater solubility of certain rocks in natural waters than is common. Derived from the geographical name of part of Slovenia.

Karst Window A closed depression, not a polje, which has a stream flowing across its bottom.

Keyhole (Passage) A small passage or opening in a cave, which is round above and narrow below.

L

Lake In caving, a body of standing water in a cave. The term is used for what on the surface would be called a pond or pool.

Lava-Cave A cave in a lava flow; usually a tube or tunnel formed by flow of liquid lava through a solidified mass, or by roofing of an open channel of flowing lava. Small caves in lava also form as gas blisters.

Leucophor A colourless water tracer, which fluoresces blue.

Limestone A sedimentary rock consisting mainly of calcium carbonate $CaCO_3$.

Live Cave A cave containing a stream or active speleothems.

M

Maladie Verte Overgrowths of the green alga *Palmellococcus* on speleothems.

Marble Limestone recrystallised and hardened by pressure and heat.

Maze Syn. Network.

Meander An arcuate curve in a river course as a result of a stream eroding sideways.

Meander Niche A hemispherically roofed part of a cave formed by a stream meandering and cutting down at the same time.

Microclimate The climate (i.e. temperature, humidity, air movements, etc.) of a restricted area or space, e.g. of a cave or, on a lesser scale, of the space beneath stones in a cave.

Microgour Miniature rimstone dams with associated tiny pools of the order of 1 cm wide and deep on flowstone.

Moonmilk A soft, white plastic speleothem consisting of calcite, hydrocalcite, hydromagnesite, or huntite.

Mud Pendulite A pendulite with the knob coated in mud.

N

Natural Arch An arch of rock formed by weathering.

Natural Bridge A bridge of rock spanning a ravine or valley and formed by erosive agents.

Necrophage A scavenger feeding on animal carcases.

Network A complex pattern of repeatedly connecting passages in a cave.

Niche The unique life strategy of a species from an ecological viewpoint.

Nothephreatic Referring to water moving slowly in cavities in the phreatic zone.

O

Outflow Cave A cave from which a stream flows or formerly did so and which cannot be followed upstream to the surface.

P

Palaeokarst 'Fossil' karst – cave or karst features remnant from a previous period of karstification, characterised by the presence of ancient (buried) deposits, as lithified cave fills or breccias.

Passage A cavity which is much longer than it is wide or high and may join larger cavities.

Pendant Syn. Rock pendant.

Pendulite A kind of stalactite which has been partly submerged and the submerged part covered with dog-tooth spar to give the appearance of a drumstick.

Percolation Water Water moving mainly downwards through pores, cracks, and tight fissures in the vadose zone.

Permeability The property of rock or soil permitting water to pass through it. Primary permeability depends on interconnecting pores between the grains of the material. Secondary permeability depends on solutional widening of joints and bedding-planes and on other solution cavities in the rock.

Phreas Syn. Phreatic zone.

Phreatic Water Water below the level at which all voids in the rock are completely filled with water.

Phreatic Zone Zone where voids in the rock are completely filled with water.

Phytophage An animal that feeds on green plants.

Pillar A bedrock column from roof to floor left by the removal of surrounding rock.

Pipe A tubular cavity projecting as much as several metres down from the surface into karst rocks and often filled with earth, sand, gravel, breccia, etc.

Pitch A vertical or nearly vertical part of a cave for which ladders or ropes are normally used for descent or ascent.

Plan A plot of the shape and details of a cave projected vertically onto a horizontal plane at a reduced scale.

Plunge Pool A swirlhole, generally of large size, occurring at the foot of a waterfall or rapid, on the surface or underground.

Polje A large, closed depression draining underground, with a flat floor across which there may be an intermittent or perennial stream and which may be liable to flood and become a lake. The floor makes a sharp break with parts of surrounding slopes.

Polygonal Karst Karst completely pitted by closed depressions so that divides between them form a crudely polygonal network.

Pool Deposit (i) Any sediment which has accumulated in a pool in a cave. (ii) Crystalline deposits precipitated in a cave pool, usually of crystalline shape as well as structure.

Population Individuals of a species in a given locality which potentially form a single interbreeding group separated by physical barriers from other such populations (e.g. populations of the same species in two quite separate caves).

Porosity The property of rock or soil of having small voids between the constituent particles. The voids may not interconnect.

Pot(Hole) A vertical or nearly vertical shaft or chimney open to the surface.

Predator An animal which captures other animals for its food.

Projected Section The result of projecting a section composed of several parts with differing directions onto a single plane. Usually, the plane is vertical along the general trend of the cave. Only the vertical distance apart of points is correct, not the horizontal, so that slopes are distorted.

Pseudokarst Terrain with landforms which resemble those of karst, but which are not the product of karst processes.

R

Relict Karst Old cave forms produced by earlier geomorphic processes within the present cycle of karstification and open to modification by present-day processes such as deposition of speleothems, sediments, or skeletal deposits.

Relict Species Species belonging to an ancient group whose distribution is now restricted to a few locations and whose population is not increasing.

Resurgence A spring where a stream, which has a course on the surface higher up, reappears at the surface.

Rhodamine A red organic dye which gives a red fluorescence to water. It is detectable in very dilute solutions, so it is used in water tracing and dye gauging.

Rift A long, narrow, high, and straight cave passage controlled by planes of weakness in the rock. cf. Fissure.

Rimstone A deposit formed by precipitation from water flowing over the rim of a pool.

Rimstone Dam A ridge or rib of rimstone, often curved convexly downstream.

Rimstone Pool A pool held up by a rimstone dam.

Rising Syn. Spring.

Rock Pendant A smooth-surfaced projection from the roof of a cave as a result of solution. Usually found in groups.

Rock Shelter A cave with a more or less level floor reaching only a short way into a hillside or under a fallen block so that no part is beyond daylight.

Rockhole A shallow, small hole in rock outcrops, often rounded in form and holding water after rains. Well known on the Nullarbor Plain.

Rockmilk Syn. Moonmilk.

Rockpile A heap of blocks in a cave, roughly conical or part-conical in shape.

Roof Crust Thin speleothem on a cave precipitated from water films exuding from pores or cracks.

Room A wider part of a cave than a passage but not as large as a chamber.

S

Salt Weathering Detachment of particles of various sizes from a rock surface by the growth of crystals from salt solutions. Forms substantial features in desert caves.

Saprophage A scavenger feeding on decaying organic material.

Saturated (i) Referring to rock with water-filled voids. (ii) Referring to water which has dissolved as much limestone or other karst rock as it can under normal conditions.

Scallops Current markings that intersect to form points which are directed downstream.

Scavenger An animal that eats dead remains and wastes of other animals and plants (cf. Coprophage, Necrophage, Saprophage).

Sea Cave A cave in present-day or emerged sea cliffs, formed by wave attack or solution.

Section A plot of the shape and details of a cave in a particular intersecting plane called the section plane, which is usually vertical.

Sediment Material recently deposited by water, ice or wind, or precipitated from water.

Seepage Water Syn. Percolation water.

Selenite A crystalline form of gypsum.

Shaft A vertical cavity roughly equal in horizontal dimensions but much deeper than broad.

Shawl A simple triangular-shaped curtain.

Show Cave A cave that has been made accessible to the public for guided visits.

Siphon A water-filled passage of inverted U-profile which delivers a flow of water whenever the head of water upstream rises above the top of the inverted U.

Solution In karst studies, the change of bedrock from the solid state to the liquid state by combination with water. In physical solution the ions of the rock go directly into solution without transformation. In chemical solution acids take part, especially the weak acid formed by carbon dioxide (CO_2).

Solution Flute A solution hollow running down the maximum slope of the rock, of uniform fingertip width and depth, with sharp ribs between it and its neighbours.

Solution Pan A dish-shaped depression on flattish rock; its sides may overhang and carry solution flutes. Its bottom may have a cover of organic remains, silt, clay, or rock fragments.

Solution Runnel A solution hollow running down the maximum slope of the rock, larger than a solution flute and increasing in depth and width down its length. Thick ribs between neighbouring runnels may be sharp and carry solution flutes.

Species A group of actually or potentially interbreeding populations which is reproductively isolated from other such groups by their biology, not simply by physical barriers.

Speleogen A cave feature formed erosionally, or by weathering in cave enlargement, such as current markings (cf. Scallop) or rock pendants.

Speleology The exploration, description, and scientific study of caves and related phenomena.

Speleothem A secondary mineral deposit formed in caves, most commonly calcite.

Splash Cup A shallow cavity in the top of a stalagmite.

Spongework A complex of irregular, interconnecting cavities intricately perforating the rock. The cavities may range from a few centimetres to more than a metre across.

Spring A natural flow of water from rock or soil onto the land surface or into a body of surface water.

Squeeze An opening in a cave passable only with effort because of its small dimensions. cf. Flattener, Crawl (way).

Stalactite A speleothem hanging downwards from a roof or wall, of cylindrical or conical form, usually with a central hollow tube.

Stalagmite A speleothem projecting vertically upwards from a cave floor and formed by precipitation from drips.

Steephead A steep-sided valley in karst, generally short, ending abruptly upstream where a stream emerges or formerly did so.

Straw (Stalactite) A long, thin-walled tubular stalactite less than about 1 cm in diameter.

Streamsink A point at which a surface stream disappears underground.

Strike The direction of a horizontal line in a bedding-plane in rocks inclined from the horizontal. On level ground it is the direction of outcrop of inclined beds.

Stylolite Suture in rock formed where pressure solution has taken place, often leaving a thin lamina of insoluble material along it.

Subcutaneous Zone The uppermost layers of rock below the soil on a karst. This zone is distinguished from lower zones by a higher porosity and storage capacity for water as a result of the presence of many solutionally enlarged fissures.

Subjacent Karst Karst developed in soluble beds underlying other rock formations; the surface may or may not be affected by the karst development.

Sump A point in a cave passage when the water meets the roof.

Supersaturated Referring to water that has more limestone or other karst rock in solution than the maximum corresponding to normal conditions.

Survey In caving, the measurement of directions and distances between survey points and of cave details from them, and the plotting of cave plans and sections from these measurements either graphically or after computation of coordinates.

Swirlhole A hole in rock in a streambed eroded by eddying water, with or without sand or pebble tools.

Syngenetic Karst Karst developed in Aeolian calcarenite when the evolution of karst features has taken place at the same time as the lithification of dune sand.

T

Tafoni Roughly hemispherical hollows weathered in rock either at the surface or in caves.

Terra Rossa Reddish residual clay soil developed on limestone.

Threshold (i) That part of a cave near the entrance where surface climatic conditions rapidly grade into cave climatic conditions. Not necessarily identical with the twilight zone. (ii) Slope or cliff facing up a blind or half-blind valley below a present or former streamsink.

Through Cave A cave which may be followed from entrance to exit along a stream course or along a passage which formerly carried a stream.

Tower Karst Conekarst in which the residual hills have very steep to overhanging lower slopes. There may be alluvial plains between the towers and flat-floored depressions within them.

Tracer (i) A material introduced into surface or underground water where it disappears or into soil to determine drainage interconnections and travel time. (ii) A material introduced into cave air to determine cave interconnections.

Travertine Compact calcium carbonate deposit, often banded, precipitated from spring, river, or lake water. cf. Tufa.

Troglobite A cavernicole unable to live outside the cave environment.

Troglodyte A human cave dweller.

Troglophile A cavernicole which frequently completes its life cycle in caves, but is not confined to this habitat.

Trogloxene A cavernicole which spends only part of its life cycle in caves and returns periodically to the surface for food.

Tube A cave passage of smooth surface, and elliptical or nearly circular in cross-section.

Tufa Spongy or vesicular calcium carbonate deposited from spring, river, or lake waters. cf. Travertine.

Tunnel A nearly horizontal cave open at both ends, fairly straight and uniform in cross-section.

Twilight Zone The part of a cave to which daylight penetrates.

U

Uvala A complex closed depression with several lesser depressions within its rim.

V

Vadose Flow Water flowing in free-surface streams in caves.

Vadose Seepage Syn. Percolation water.

Vadose Water Water in the vadose zone.

Vadose Zone The zone where voids in the rock are partly filled with air and through which water descends under gravity.

Vauclusian Spring A spring rising up a deep, steeply inclined, water-filled passage into a small surface pool.

Vermiculation Pattern of thin, worm-shaped coatings of clay or silt on cave surfaces.

W

Water Table The surface between phreatic water, which completely fills voids in the rock, and ground air, which partially fills higher voids.

Water Tracing Determination of water connection between points of stream disappearance or of soil water seepage and points of reappearance on the surface or underground.

Watertrap A place where a cave roof dips under water but lifts above it farther on. cf. Duck(-under).

Well A deep rounded hole in a cave floor or on the surface in karst.

Further Reading

The following books may provide the curious with a greater depth of understanding of caves and karst. In particular, the excellent volume by Ford and Williams (2007), has become the single most referenced and comprehensive book on karst and caves in the English language.

Culver, D.C. and Pipan, T. (2009). *The Biology of Caves and Other Subterranean Habitats*. OUP Oxford.

Fairchild, I.J. and Baker, A. (2012). *Speleothem Science: From Process to Past Environments*. Blackwell Publishing Ltd.

Ford, D.C. and Williams, P.W. (2007). *Karst Hydrogeology and Geomorphology*. Chichester, England: John Wiley and Sons Ltd.

Gunn, J. (ed.) (2004). *The Encyclopedia of Caves and Karst Science*. New York: Fitzroy Dearborn.

Lace, M.J. and Mylroie, J.E. (eds.) (2013). *Coastal Karst Landforms*. Netherlands: Springer.

Migon, P. (ed.) (2010). *Geomorphological Landscapes of the World*. Springer.

Moldovan, O.T., Kováč, Ľ., and Halse, S. (eds.) (2018). *Cave Ecology*. Cham, Switzerland: Springer.

Palmer, A.N. (2007). *Cave Geology*. Dayton, OH: Cave Books.

Van Beynen, P. (ed.) (2011). *Karst Management*. Dordrecht: Springer.

White, W.B. and Culver, D.C. (2019). *Encyclopedia of Caves*, 3e. Academic Press.

Williams, P.W. (2008). *World Heritage Caves and Karst: A Thematic Study*. Gland: IUCN.

Williams, P.W. (2017). *New Zealand Landscape: Behind the Scene*. Amsterdam: Elsevier Inc.

Geographical Index

Caves: Processes, Development, and Management, Second Edition. David Shaw Gillieson.
© 2021 John Wiley & Sons Ltd. Published 2021 by John Wiley & Sons Ltd.

Subject Index

Caves: Processes, Development, and Management, Second Edition. David Shaw Gillieson.
© 2021 John Wiley & Sons Ltd. Published 2021 by John Wiley & Sons Ltd.